Energy Management Systems for Microgrids with Wind, PV and Battery Storage

Other related titles:

You may also like

- PBPO091 | Altaş | Fuzzy Logic Control in Energy Systems with Design Applications in MATLAB®/ Simulink® | 2017
- PBPO238 | Wang | AI for Status Monitoring of Utility Scale Batteries | 2022
- PBPO217 | Milano | Advances in Power System Modelling, Control and Stability Analysis, 2nd Edition | 2022

We also publish a wide range of books on the following topics:
Computing and Networks
Control, Robotics and Sensors
Electrical Regulations
Electromagnetics and Radar
Energy Engineering
Healthcare Technologies
History and Management of Technology
IET Codes and Guidance
Materials, Circuits and Devices
Model Forms
Nanomaterials and Nanotechnologies
Optics, Photonics and Lasers
Production, Design and Manufacturing
Security
Telecommunications
Transportation

All books are available in print via https://shop.theiet.org or as eBooks via our Digital Library https://digital-library.theiet.org.

IET ENERGY ENGINEERING SERIES 270

Energy Management Systems for Microgrids with Wind, PV and Battery Storage

Edited by
Badre Bossoufi

The Institution of Engineering and Technology

About the IET

This book is published by the Institution of Engineering and Technology (The IET).

We inspire, inform and influence the global engineering community to engineer a better world. As a diverse home across engineering and technology, we share knowledge that helps make better sense of the world, to accelerate innovation and solve the global challenges that matter.

The IET is a not-for-profit organisation. The surplus we make from our books is used to support activities and products for the engineering community and promote the positive role of science, engineering and technology in the world. This includes education resources and outreach, scholarships and awards, events and courses, publications, professional development and mentoring, and advocacy to governments.

To discover more about the IET please visit https://www.theiet.org/

About IET books

The IET publishes books across many engineering and technology disciplines. Our authors and editors offer fresh perspectives from universities and industry. Within our subject areas, we have several book series steered by editorial boards made up of leading subject experts.

We peer review each book at the proposal stage to ensure the quality and relevance of our publications.

Get involved

If you are interested in becoming an author, editor, series advisor, or peer reviewer please visit https://www.theiet.org/publishing/publishing-with-iet-books/ or contact author_support@theiet.org.

Discovering our electronic content

All of our books are available online via the IET's Digital Library. Our Digital Library is the home of technical documents, eBooks, conference publications, real-life case studies and journal articles. To find out more, please visit https://digital-library.theiet.org.

In collaboration with the United Nations and the International Publishers Association, the IET is a Signatory member of the SDG Publishers Compact. The Compact aims to accelerate progress to achieve the Sustainable Development Goals (SDGs) by 2030. Signatories aspire to develop sustainable practices and act as champions of the SDGs during the Decade of Action (2020–30), publishing books and journals that will help inform, develop, and inspire action in that direction.

In line with our sustainable goals, our UK printing partner has FSC accreditation, which is reducing our environmental impact to the planet. We use a print-on-demand model to further reduce our carbon footprint.

Published by The Institution of Engineering and Technology, London, United Kingdom

The Institution of Engineering and Technology (the "**Publisher**") is registered as a Charity in England & Wales (no. 211014) and Scotland (no. SC038698).

First published 2025

The Institution of Engineering and Technology
Futures Place,
Kings Way, Stevenage,
Herts, SG1 2UA, United Kingdom

www.theiet.org

British Library Cataloguing in Publication Data
A catalogue record for this product is available from the British Library

ISBN 978-1-83724-090-6 (hardback)
ISBN 978-1-83724-091-3 (PDF)

Typeset in India by MPS Limited

Cover image: Justin Paget/DigitalVision by Getty

I am honored to dedicate this book to family and my friends.

Contents

Acknowledgments

This book could not be that successful without the effort of authors and reviewers, especially Dr. Saad Motahhir. Therefore, I would like to express my sincere appreciation to all of you who generously supported this book.

Badre Bossoufi

About the editor

Badre Bossoufi (https://orcid.org/0009-0009-6506-8464) (Eng., Ph.D., IEEE Senior Member) received his Ph.D. in electrical engineering from the Faculty of Sciences at Sidi Mohamed Ben Abdellah University in Fez, and a joint Ph.D. from the Faculty of Electronics and Computer at the University of Piteşti, Romania, and the Montefiore Institute of Electrical Engineering in Liège, Belgium, in 2012. He was a professor of electrical engineering at the Faculty of Sciences Dhar El Mahraz at Sidi Mohamed Ben Abdellah University. His research interests include static converters, electrical motor drives, power electronics, smart grids, renewable energy, and artificial intelligence. He has published numerous papers in journals and conferences over the past few years, most of which relate to wind power control and microgrid systems. He has edited several books and served as a guest editor for various special issues and topical collections. He is a reviewer and is on the editorial boards of several journals. He has been associated with more than 20 international conferences as a program committee member, advisory board member, or review board member.

Chapter 1

AI and IoT for cost and emission reduction in microgrid energy management

Mohammed Amine Hoummadi[1], *Badre Bossoufi*[1], *Mohammed Karim*[1] *and Hana Zekraoui*[2]

Abstract

This chapter is dedicated to highly complex technologies and processes of microgrid energy management with an emphasis placed on artificial intelligence and Internet of Things in energy management system (EMS), energy trading and advanced batteries, and specific microgrid structures. In the first section of this chapter, microgrids are described together with their components, which once more emphasizes their potential toward improvement of resilience and reliability of energy. Consequently, the presentation of the microgrid roles goes on emphasizing their crucial enabler nature of distributed renewable energy generation, off-grid backup power, and local energy independence through examples. The chapter then progresses in the discussion of how metaheuristic and the basic approaches to microgrid energy management are carried through battery management systems (BMS), EMS, as well as making the best of the microgrid components. That is why the further enhancement of the microgrid design with the help of metaheuristic optimization has been suggested in this study. In this work, the cost saving of 29%, as well as reduction of CO_2 emission, and improvement of the microgrid performance parameters have been verified on the basis of a case study of 100 houses in Fez, Morocco. Lastly, the chapter also recalls the purpose of this as a source for researchers and scientists so that more cooperation for improving the microgrid could be established, along with the knowledge of the optimal microgrid mix considering EMS and BMS for achieving the lowest energy cost and highest CO_2 emission mitigation.

Keywords: Microgrid; control system; renewable energy; energy management methods

[1]LIMAS Laboratory, Faculty of Sciences Dhar El Mahraz, Sidi Mohamed Ben Abdellah University, Morocco
[2]LGEM Laboratory, Higher School of Technology, Mohamed First University, Oujda, Morocco

1.1 Introduction

Today, microgrids are considered keystone solutions in distributed energy systems that provide improvements in reliability and resiliency. This chapter overviews the largely complex realm of microgrid energy management while placing a strong focus on emergent disruptive technologies, including artificial intelligence (AI) and Internet of Things (IoT). The first part of this chapter offers a clear and detailed definition of microgrids with pinpointing their meaning as localized energy systems incorporating distributed energy resources (DER). These resources include a changing list of components that consists of different generation sources, storage, control systems, and loads. Some of the most important characteristics and functions that can be associated with the use of microgrids are as follows: the ability to include various types of renewable energy sources, support the delivery of backup electricity during grid interruption events, support the grid and its flexibility and stability, and, finally, providing a certain measure of autonomy to local communities [1,2].

On this basis, this chapter continues the analysis of today's energy systems by focusing specifically on the importance of microgrids. Relating the experiences and incorporating cases from real-world applications, the story builds to show how microgrids themselves are living up to their promises as responsive solutions to the transforming expectations of the energy end-users. The elaboration of this chapter proceeds further to present diverse approaches used for enhancing energy control in microgrids. Of these, two groups are defined for the approaches that are used to tackle various aspects of microgrid energy management, namely metaheuristic methods such as genetic algorithms (GA) and particle swarm optimization (PSO), and classic optimization techniques for example linear programming and nonlinear programming (NLP) [3]. The above-mentioned theoretical background provides the background for the subsequent sections where the use of these methods in critical microgrid parameters, namely BMS, energy management system (EMS), suitable combinations of microgrid elements, and optimal energy trading strategies are clearly explained in detail [4,5].

As the chapter progresses, it takes a deep dive into the practical realm by presenting a compelling case study: Microgrid Design for 100 Houses and Load Calculations for FEZ, Morocco. This real-world problem provides an application arena by which different metaheuristic optimization approaches can be demonstrated and their usefulness explained. Both the quantitative and qualitative findings of cost, CO_2 emissions, and performance demonstrate the practical benefits of these methodologies in microgrid designs and optimizations. Lastly, the chapter draws general conclusions about the study for the researchers and scientists, specifically the role and importance of this work for subsequent studies. The plea for cooperation and knowledge sharing highlights the need for improving the current state of the art in microgrid technology thus enabling more sustainable resilient and cost-efficient management of energy supplies. Besides, the introduction presents an outlook to the subsequent section that is devoted to the identification of the optimal pairings for microgrids, emphasizing the contributions of EMS and BMS, as well as their cumulative influence toward the decrease of energy consumption price, and minimization of CO_2 emission in microgrid configurations [6,7].

In other words, the inclusion of sophisticated optimization techniques, such as metaheuristic algorithms, AI-driven tools, and IoT-based solutions, greatly enhances the performance and efficiency of the microgrid EMS. It is due to these innovations that complexities introduced by the integration of renewable energy sources, dynamic energy demands, and storage systems are more easily handled. The Fez microgrid case study has marked an application that showcases, in practice, the capability of these methods to reduce operational costs, minimize carbon emissions, and improve energy reliability. This really requires further research and collaboration in refining these technologies for various configurations of microgrids across different parts of the world. This work innovation adds value by pushing the edge of the state-of-the-art in microgrid optimization for resiliency, sustainability, and cost-effectiveness within future energy solution options [8,9].

The first part briefly introduced microgrids as the localized areas of energy systems integrated into the main using DERs like solar panels and wind turbines, including energy storage. They can operate in a connected mode to the main power grid or in an "islanded" mode. Microgrids are very important in their contribution toward enhancing energy resilience, reliability, and sustainability. They ensure the effective utilization of renewable resources while offering backup power during grid outages. The next section reviewed various methods of energy management of microgrids, focused on the balance between energy supply and demand, the reduction of operation costs, and the optimization of the use of available resources. Energy management techniques range from simple rule-based strategies up to more complex optimization ones, those able to optimize microgrid performance under time-varying conditions. Finally, we explored how metaheuristic algorithms work on the design of microgrids. Inspired by nature, metaheuristics such as evolution and swarm intelligence yield innovative solutions to nonlinear optimization problems commonly faced in the configuration of microgrids. Methods like genetic algorithms, PSO, and simulated annealing efficiently explore large solution spaces in their effort to find near-optimal configurations in terms of designing, operating, and allocating resources toward microgrids. These approaches will enable microgrids to reach cost-effective, adaptable configurations that will be able to meet effectively the challenges of variable energy demand and integration of renewable energy.

1.2 Microgrid definition and role

1.2.1 Microgrid definition

Microgrid is a modern and sophisticated power system to produce, transmit, and distribute electric power individually in a small bounded area or a number of houses which makes microgrid a new concept in comparison to the conventional large-scale macrogrid system. Microgrids are different from the large and complex utility systems that encompass power plants and deep and commanding transmission and distribution centers; instead, they are built on DERs and have the abilities to function in parallel to, or in isolation from, the main utility. Some of the examples of these DERs include a number of technologies that are used to generate energy renewable energy technologies like solar PV wind power or small hydropower; non-renewable energy technologies like diesel generators, natural gas turbines, and others. By adopting such an approach to the generation of energy, this microgrid is ready for a shift of environmental and energy demands besides supporting the provision of energy self–supply in the structure of the communities. By harnessing a balanced mix of renewable and conventional energy sources, microgrids empower communities to diversify their energy portfolios, reduce dependence on centralized infrastructure, and chart a resilient path toward a sustainable and self-reliant energy future, thus fundamentally reshaping the landscape of energy distribution and consumption on both local and global scales as shown in Figure 1.1. Thus, the choice of the various energy mixes is enabled by microgrids, and the community's dependence on the centralized structure is discouraged; for development of the sustainable and secure power systems as depicted in Figure 1.1.

In addition to the diversity of generation sources, the storage systems are even more essential components for proper and secure operation of microgrids [10]. These storage technologies, including batteries, fly wheels, and others, also have the additional vital role of putting more energy storage for the extra energy produced during off-peak hours or when the generation or demand level is high. At some point of time, this stored energy is nicely utilized whenever there is a need to provide the energy to meet the demand or generation, which forms the supply-demand balance in a microgrid. Besides the use in controlling energy use on the microgrid level, energy storage systems (ESS) are vital in improving the stability of the power grid/information services such as frequency regulation and voltage. Further, sectors that employ storage systems in microgrid offer an effective backup system each time there is a disruption in the supply of electricity to crucial entities

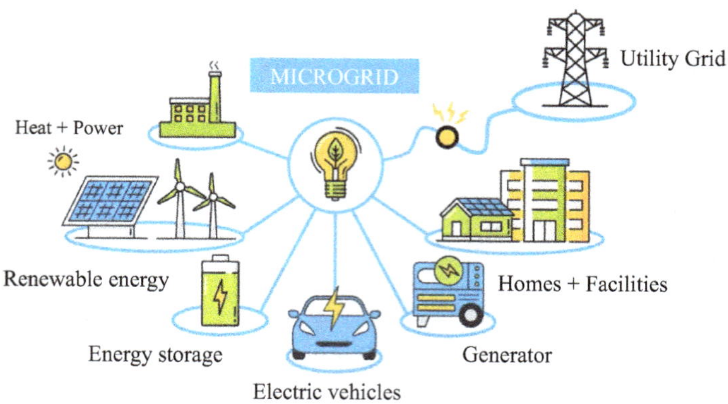

Figure 1.1 Microgrid components

and structures. It is especially so for facilities that cannot afford interruption of power as this is life-affecting; this includes among the following; hospitals, emergency centers, Phone Company's electrical equipment, and others. Therefore, ESS are presented as indispensable solutions, which enable the enhancement of microgrids' characteristics such as power, stability, and reliability – the basic components that shape the future power system in the contemporary world. Controlling systems are the essence of a microgrid in a conductor position to efficiently manage and coordinate the DERs in respect of power generation and consumption. Sophisticated monitoring tools, the possibility of automatic control of the transmission and distribution grid with the help of optimization procedures make it possible to control the grid in real time, indicating the conditions in which the grid can be optimized depending on fluctuations in the production of renewable energy sources or in load demand. These control systems also ensure that the microgrid can interface with the main grid so that resources of the microgrid can be synchronized and fed to the main grid while maintaining its stability and security.

In addition, a microgrid operates by arranging connections among electricity consumers, including those in the residential, commercial, and industrial sectors, each with varying power demands and usage timings. These loads mimic the other energy usage difficulties added in the microgrid ambiance whereby the utilization familiar range from the domestic use of electrical energy to industrial use. As per the demand response programs and strategic load management, the microgrid is ready on the consumer site and it manages energy and grid control. Therefore, by encouraging consumers to use different demand responses, especially during peak demand or periods of renewable energy generation, microgrids have a significant impact lowering the general energy use and pressure on the existing grid structures, thereby strengthening the energy system resilience and capacity. At the same time, the structure of microgrids is localized, which gives all the necessary conditions for the implementation of measures that improve energy efficiency and the use of intelligent smart-grid solutions for a new effective one. Such an approach would allow microgrids only a partial reinforcement of the management of energy within a locality as other proper strategies can be developed for a community because of its features. Microgrid advancement comprises artificial lighting, energy efficient utensils, intelligent building controls that helps not only boosting the performance and reliability of microgrid for the consumer and local utility grid but also offers many economic benefits to everyone more directly or indirectly. Thus, the demand-side management with energy conservation and the optimal grid design make microgrids pioneers of change of an efficient, sustainable, and equitable energy supply system to support change within the local/global environment [11,12].

All in all, microgrids play an irreplaceable role in efforts to improve energy availability, reliability, and efficiency through decentralizing energy generation and consumption, integrating renewable energy resources into the electrical networks, and supporting communities' autonomy to determine their energy future. Standing in the middle of many changes that occur in the global energy context, microgrids are ready to become game changers on the worldwide scale of energy systems evolution. Given the fact that microgrids are inherently flexible, adaptive, and robust, they are poised to become the solutions of choice in meeting the complex demand in contemporary power systems. Thus, microgrids that incorporate a community-level approach to managing and controlling energy improve the strength and overall stability of the grid as well as make the energy systems of the world more sustainable and just for the future generations. Consequently, microgrids are seen as a revolution in energy management, allowing communities to have an exploited control of how the energy systems are developed, moving toward a more decentralized and sustainable energy framework [13,14].

1.2.2 Model of microgrid

1.2.2.1 Generation sources

Renewable energy sources like solar PV panels and wind turbines are crucial components of microgrids. The power generated by these sources can be modeled using the following equations:

For solar PV panels:

$$\mathrm{PPV} = A \times G \times \eta_{\mathrm{PV}} \tag{1.1}$$

where

PPV = power output of the PV panels (W)

A = area of PV panels (m^2)

G = solar irradiance (W/m^2)

η_{PV} = efficiency of PV panels

For wind turbines:

$$P_{\mathrm{wind}} = 1/2 \times \rho \times A \times v\text{^}3 \times C_p \tag{1.2}$$

where

P_{wind} = power output of the wind turbine (W)

ρ = air density (kg/m^3)

A = swept area of the turbine blades (m^2)

v = wind speed (m/s)

C_p = power coefficient of the wind turbine

In the case of renewable energy systems, both PV and wind turbine technologies are imposed on non-stable and random factors such as solar radiation and wind velocity. For PV systems, nominal power is therefore realized under nominal conditions, commonly referred to as standard test conditions involving an array of solar irradiance. Nevertheless, variations from nominal rates of irradiance affect the PV electricity production in the same proportion. In particular, the decrease of the solar irradiance below nominal decreases the PV power rate correspondingly. However, wind turbine power availability depends on the wind velocity whereby power is a cubic function of wind velocity. Hence, variations with reference to nominal wind speeds considerably affect the power output. For instance, when wind speed is half of the wind turbine, nominal, it is just one eighth of the nominal power making the relationship between wind speed and power not linear. As a result, PV systems are equally sensitive to alterations in environmental inputs, while wind turbines also show similar fluctuation indicating a need for efficiency in input prediction to enhance renewable energy productivity [14,15].

Also, there are conventional sources like diesel generator or biomass gasifier that aid the generation of power in microgrids. Thus, their useful power output can be described by a rather simple formula with reference to the fuel consumption and conversion efficiency [15,16].

1.2.2.2 Energy storage systems

ESS, typically consisting of batteries, play a crucial role in maintaining grid stability and resilience. The charging and discharging of batteries can be modeled using the following equations:

Battery charging:

$$P_{\text{charge}} = V_{\text{bat}} \times I_{\text{charge}} \tag{1.3}$$

where

P_{charge} = charging power (W)
V_{bat} = battery voltage (V)
I_{charge} = charging current (A)

Battery discharging:

$$P_{\text{discharge}} = V_{\text{bat}} \times I_{\text{discharge}} \tag{1.4}$$

where

$P_{\text{discharge}}$ = discharging power (W)
$I_{\text{discharge}}$ = discharging current (A)

The output power of a battery is dependent on its state of charge (SOC), which is the amount of energy stored in the battery relative to its maximum capacity. The mathematical model for the output power of a battery in the function of SOC [15,17] can be expressed as

$$P = V * I * f(\text{SOC}) \tag{1.5}$$

Consequently, the function f(SOC) tertains to the type of battery chemistry and its discharge character. For instance, lead acid battery has a fairly low voltage discharge profile that is fairly flat; this means that the output power of the battery is rather constant at various SOC levels. Lithium-ion batteries have a discharge curve with a form of exponent, indicating that their output power provides less SOC value as compared to the same of lithium ferro-phosphate (LFP) chemistry [17]. Thus, manufacturers often perform numerous tests in order to evaluate f(SOC) of a certain battery chemistry by measuring the discharge characteristics of batteries at varying SOC levels. They then apply this data in deriving an empirical equation that most closely approximates the observed behavior [15,16]. The battery SOC however is the measure of battery power that is available in terms of a percentage. It is thus possible to calculate the SOC in ways such as by voltage, current, temperature, and impedance. The simplest way to estimate SOC is to use the open circuit voltage (OCV) method and it is commonly used by most of the researchers as well. In this method, the SOC is estimated by using the OCV of the battery.

$$\mathrm{SOC} = (\mathrm{OCV} - \mathrm{OCVmin})/(\mathrm{OCVmax} - \mathrm{OCVmin}) * 100\% \tag{1.6}$$

where SOC is the battery SOC in percentage, OCV is the open-circuit voltage of the battery, OCV_min is the minimum OCV value corresponding to 0% SOC, and OCV_max is the maximum OCV value. The values of OCV_min and OCV_max depend on various factors such as battery chemistry, temperature, and age [18,19].

1.2.3 Control systems

Control systems in microgrids utilize various algorithms and control strategies to optimize energy flow and maintain grid stability. One common approach is the use of power flow equations to ensure power balance within the microgrid:

$$\sum P_{\mathrm{generation}} - \sum P_{\mathrm{consumption}} = \sum P_{\mathrm{storage}} \tag{1.7}$$

where

$P_{\mathrm{generation}}$ = power generated by all sources (W)
$P_{\mathrm{consumption}}$ = power consumed by all loads (W)
P_{storage} = power stored or discharged by ESS (W)

The articles present the discourses of the energy control systems of many microgrids which are involved in managing and maintaining the energy supply with the aid of various algorithms and control techniques. One of the most used strategies is based on the active/reactive power flow used to enforce the power balance in a microgrid. Therefore, the target power allocation is given by $\sum P$generation – $\sum P$consumption = $\sum P$storage where the symbols point at their respective description. This equation can evolve the process by which energies in the microgrid can be utilized locally and at what time the big grid could make up the microgrid. Moreover, in this regard, EMS is considered significant as it takes the responsibility for controlling the supply of energy resources. EMS also speaks to the use of renewable energy in as much as they are cheaper and have negligible impact on the environment. But if the renewable power generation is inadequate to meet the need, EMS has a storage – batteries – between supply and load. The grid management of EMS is only done when the renewable sources and storage cannot supply a location's demand and the backup entails diesel or biomass gasifiers. This hierarchical mode ensures that the renewable energy sources are fully utilized to the maximum and hence the equipment is more or less independent of conventional backup systems [17,19,20].

1.2.3.1 Loads

Loads within a microgrid represent the energy demand from various consumers. The total load can be determined by summing up the power demand of individual loads:

$$P_{load} = \sum P_{individual\ loads}$$

Table 1.1 shows a real example of a house power

Table 1.1 The power of one house

	Load type	Number	Power (Watts)	Summer h/day	(May–Sep) Watt-h/day	Winter h/day	(Oct–April) Watt-h/day
Domestic load							
1	Compact fluorescent lamps (CFL)	4	20	10	880	7	620
2	CFL	1	13	8	96	11	132
3	Ceiling fan	1	120	20	2,440	0	0
4	Kitchen fan	1	28	6	174	0	0
5	Cooler	2	122	10.5	2,500	0	0
6	TV	1	97	9	890	6	591
7	PC	1	298	8	2,392	9	2,690
8	Exhaust fan	1	18	5	80	3	48
9	Table fan	1	18	8	124	0	0
10	Room heater	1	94	0	0	12	1,104
11	Room heater	1	147	0	0	7	1,036
12	Electric blanket	1	126	0	0	3	375
13	Vacuum cleaner	1	212	2	430	1	215
14	Bulb	1	102	1	102	2	204
	Total (one house)				10,114		6,980

Furthermore, techniques like smart grid and OR method in demand-side management are equally applicable to manage and reduce energy load in case to be utilized to manage the energy load of microgrid and in case of excessive energy production to transfer the dump load to avoid pressure on the microgrid. The dump load formula [21], a mathematical imperative, can be applied to find the amount of power required to be fed to a dump load to protect the electrical system against over-voltage or over-current conditions in a given electrical system. Dump loads are frequently used where regulation is needed in renewable systems such as those applied in wind turbine circuits and solar panels among others. Therefore, the dump load equation is

$$P = V^2/R \tag{1.8}$$

The next equation shows that the power that should be dissipated by the dump load is directly proportional to voltage and inversely proportional to the impedance. Therefore, to ensure that the dump load dissipates the required power, voltage will have to be raised, or resistance has to be lowered.

It should be noted that in practice there is a need to choose a dump load with an appropriate level of resistance [22], based on the electrical network. If the resistance is too low, the current that will be passing through the dump load is high and may cause the dump load to heat up or fail. If the resistance is high, then the amount of power dissipated is low and so the electrical system may still be at risk.

Therefore, these equations present the mathematical representation of each component of a microgrid to enable comprehension of the behavior of every one of them. The above-mentioned equations can be implemented in the simulation or in other optimization techniques where the engineers or the investigators can anticipate the behavior, enhance performance, and design efficient and dependable microgrid systems for specific applications and certain climate conditions [5,19,23].

1.2.4 Role of microgrids

If we talk about modern discussions about future energy, microgrids are one of the key technologies that mark the dawn of the new age in the generation, distribution, and consumption of electricity. The general definition about microgrids can be stated in the way that it is an electric network formed from distributed electricity production, storage, and supply system that is capable of operating autonomously or interconnected with the other grid system. It is not just the case that those are some of the advanced pieces of technology for which those solutions have their value, but the holistic transformation they foster in various aspects associated with energy and energy security [24].

1.2.4.1 Enabling renewable energy integration

The outlooks for the microgrids therefore provide a relevant predicate to the incorporation of renewable energy into the grid, thus addressing the challenges facing today's power source. Living in the 21st century any country that is aspiring to move forward into the future and at the same time, be environmentally conscious, then they need to harness solar, wind, and hydroelectric power in the most prudent manner as is possible. They perform even better in this regard because microgrids provide a highly flexible grid structure on which the new type of renewables can readily be incorporated into energy systems. For example in arid regions with high levels of solar heat, solar microgrids use reversed photovoltaic panels to harness the solar energy, and the excess power is stored in batteries for use during times that the sun is weak. Likewise, wind microgrids self-generate power during windy conditions which are climatic and contribute toward the source of supply security of renewable power. These examples vividly illustrate how microgrids play a vital role in facilitating the transition to clean energy by maximizing the efficient utilization of renewable resources, thereby reducing carbon emissions and mitigating environmental impact as shown in Figure 1.2. These examples can all clearly describe what kind of role microgrids is serving as the transition toward cleaner energy by offering a way to dispatch renewables most optimally than other options, and thus minimize the carbon footprint to the environment as illustrated in Figure 1.2.

Further, microgrids have been proven to have efficient proactive responses to the challenge of intermittency of resources, which is a characteristic of most RE

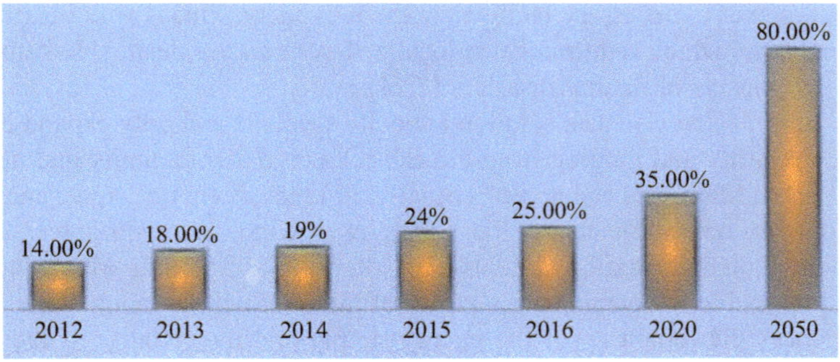

Figure 1.2 The evolution of renewable energy adoption in the world

technologies as shown in Figure 1.2. The more, the majority of renewable energy technologies such as solar and wind power have many inconveniences in terms of the fluctuations of the girds. Nevertheless, owing to highly developed microgrid control systems and the application of analytical tools to predict the generation of renewable energy sources in the future, microgrids possess the capability of accurately controlling energy storage and distribution procedures. Regarding the supply chain, microgrids employ intelligent resource management to provide for demand adequately at the same time as not expending fresh renewable power resources wastefully. Besides, integrated smart metering, IoT sensors, and machine learning incorporated in the microgrids enable these to make modifications continually, making energy networks more robust. In addition to enhancing the flexibility of rising renewable power integration, this capability enhances the reliability of the grid and helps advance the construction of infrastructure for additional sustainable power systems in the future [23–25].

1.2.4.2 Enhancing resilience through backup power

The possibility of having microgrid during grid failure or in case of emergency continues to be one of the biggest strengths in today's world where everything is interconnected. Although the traditional centralized grids are efficient at regulating normal conditions, such grids are highly vulnerable to faults that stem from natural disasters, cyber-terrorism, or else equipment failure. Such weaknesses lead to effects such as large-scale failure of electricity supply and the absence of basic social amenities a society requires. Microgrids emerge as a unique distributed architecture during calamity to solve this problem. Because of sharing energy sources and optimizers in the regulation of load, microgrids offer the required electricity at a time when the main grid has contracted. Some of the few facilities that may not be affected by power fluctuations due to power generated by microgrid are hospitals, emergency centers, telecommunication centers among others since they can continue performing their functions without interruption. For instance, when the storm surge of Hurricane Sandy attacked New Jersey in 2012, the Princeton University microgrid functioned independently and delivered the demanded power to necessary facilities in the university. This is well illustrated by a practical case, which reinforces that localized systems are desirable as insurance against the impacts of natural disasters [7,24].

Moreover, microgrid as a solution are quite versatile and only expand in terms of energy security and ranging from the cabin level to a community and up to the regional level. Microgrid can accept virtually all kinds of energy inputs and can be designed to use renewable energy like solar power, wind, and hydropower, and can include distributed generation systems like batteries and fuel cells which altogether make it impossible to contain single points of failure. Such distributed architecture also enhances the reliability of the energy supply and the stability of the system addressing errors and quick recovery from catastrophes. But smart grid technologies and analytics that predict conditions enable microgrid to quickly shift to the latter and offer the utmost efficient use of energy and resources. By providing microgrid for energizing communities, which will enable people to become proactive in creating their energy infrastructure that is not vulnerable to volatility and disasters that define today's world [25,26].

1.2.4.3 Optimizing grid stability and flexibility

Microgrid is an element of the modern energy system without which it is impossible to solve the problem of increasing stability and flexibility of the main grid; at the same time, it is an element that performs several functions on their solution. Microgrid can therefore be described to be dissimilar to a conventional centralized grid system since it is possible to separate control of the generation, distribution, and use of power in a given area. Nevertheless, this approach of energy management is possible because of efficiency control from computer-controlled systems and smart grid technology to optimize the utilization of the microgrid and fulfill its intended aim of controlling the energy flow, voltage, and frequency levels. Therefore, the microgrid manage and observe the parameters mentioned above which minimize the transmission losses and make sure that there is no real or potential instability on the grid for the efficient supply of electricity by the microgrid. One clear example of this capability is described by the University of California, San Diego's microgrid, in which highly advanced forecasting techniques are used to predict energy requirements and subsequently manage the usage of electricity proficiently. Through such preventive measures, the microgrid at UCSD has not only earned lots of savings but has also gone far in fulfilling its environmental mandate, which makes microgrid technologies an effective weapon for enhancing the capability and reliability of existing grid [27].

Further, they allow for the augmentation of the modernization of devices such as EVs and flexible demand response programs hence making the microgrid more reliable for the grid. As a result of battery charging and discharging of EVs and through demand-side management strategies, microgrids are to a large extent capable of controlling supply and demand balance during peak load periods. This approach does not only assist in sharing the load on the grid but it also takes a proactive approach in using more sustainable and interconnected energy systems. For instance, microgrid can schedule electrical power demand during the current high energy usage times to cause system changes in the charging of electric vehicles that instead harmonize with the utilization or receipt of renewable energy supply. Additionally, for the same consumers, through the demand response programs, microgrids can be utilized with efficiency to increase the flexibility and reliability of the grid to make the energy system more dynamic. However, as further development is carried out to enhance the microgrid and incorporate new solutions into them, microgrids are fundamental elements within future energy systems and their future work of offering higher flexibility, and efficiency as well as collectively contributing to improving the advancement of sustainability aspects of the energy market for the dynamic energy needs and challenges [27,28].

1.2.4.4 Empowering communities with energy independence

Microgrids can be considered as an example of the alteration in the energy industry because the local communities get a precious thing – the ability to control energy. In fact, because of the described structure, microgrids transform the traditional process of electric power generation and distribution; thus, the residents, businesses, and institutions are able to at least partially "gain some control" over their energy future. It also enables the communities to liberate themselves from cooperation with large-scale centralized grids and fosters the connection of clean and renewable energy installations. A clear evidence of these positive impacts of decentralization is the local ownership of microgrid, and this is evident from the case study of Loiza, Puerto Rico. In this case, the residents have enlightened up with the vision of renewable energy using microgrid as the tool to free themselves from the challenges resulting from fossils. Thus, despite the conservation of the non-renewable resources, the integrated microgrid of Loiza has offered a positive economic revolution and a ray of hope into the lives of the people and has revolutionized the viewpoints of the communities regarding energy-producing systems [29,30].

Moreover, a microgrid is a transformative enabling mechanism that encourages innovation in participation at the grassroots level for community-based engagement and innovation, thus engaging people in the creative process of planning and management of energy systems. From the involvement-based planning approaches to the methodologies for microgrid planning and the cooperative ownership model, the communities unleash the transformative power of having requisite local energy needs met and at the same time catalyze economic and social advancement. This inclusion does not only contribute to support of community but also will enhance development of the culture of ownership of energy resources in the community systems. This is where through coordinated actions such as in the creation of the community energy coops and the local energy trading hubs microgrid knit thick and strong webs of energy democracy, into which the communities work to actualize sustainable and resilient power systems. As microgrid initiatives continue to gain momentum, both in their quantity and their variety, it is possible to decentralize a part of the energy system and allow for various communities to start redesigning how they and various forms of technologies and applications engage with electricity and proactively begin to build a better energy system [29,31].

1.2.4.5 Case studies and examples

These case studies and examples from different parts of the world provide a clear picture of the efficiency and versatility of microgrids in various situations suggesting that microgrids have the potential to transform energy systems across the world. In several villages in the territory of India, where concerns about the electricity supply have been general for many years, microgrids become life essential, providing electric power to the rural people to the first time. These microgrids are not merely lighting homes but also important services such as schools, health centers, and petite businesses making them central to socioeconomic improvement and human welfare. Furthermore, microgrids operating in India act as an enabler for achieving rural electrification; this means that people can free themselves from the vise of energy poverty and use electricity to unlock new opportunities for further development [28].

However, microgrids are adopted in European cities to control and reduce the energy and carbon utilization according to smart and sustainable city theory. Proper microgrids in Amsterdam and Copenhagen show how renewable energy, energy storage, and demand-side management solutions are progressively getting integrated into cities' energy systems. Worth noting is that these measures do not only come with a number of advantages in the enhancement of the reliability and flexibility of the grid but also aid in the attainment of a new and improved urban energy system for the reductions in carbon emission and protection against climate change impacts in the future. However, microgrids in urban environment support dynamic energy management trends such as load shifting and peak shaving to reduce demand on the grid during critical periods, which ensures the availability of affordable and reliable energy to consumers and businesses [29,30].

Another work that provides the positive description of the microgrid is the MG Systems Integration Demonstration project carried out in Japan. This effort renders it feasible to integrate renewable energy systems alongside ESS and electric vehicles as a microgrid. The embedded control of microgrid loses equilibrium of supply and demand and timely utilization and stability of energy offered by the conventional grid technologies and the standard control methods but the project and its planning support how the control system of microgrid is deterministic in achieving the proper balance of supply and demand, utilization of energy, and stability of the grid. Also, electric vehicles integration as part of the microgrid environment offers the V2G prospectively, which refers to the capability of the electric vehicle to interact in both directions with the grid. As a result, the MG Systems Integration Demonstration project has been established as a model for microgrids for different sustainable energy systems in the world since it has established the immense role of microgrids in the formulation of the new sustainable energy framework [21,31].

In total, microgrids are a symptom of emergent trends and future potential in a constantly transforming power sector. In this way, microgrids are becoming the key to the future of energy with the potential for renewable integration, resilience enhancement, growth of the stability of the grid, and coordinated execution of community-based projects. Microgrids seem to be the way to a new chapter in the sustainable, accessible, and secure energy system by analyzing climate change, limited resources, and uncertainty in international relations [22].

1.3 Methods for microgrid energy management

1.3.1 Overview of energy management methods

In the context of intense and progressive competition between consumers and producers of energy along with the consumer's increasing control over energy use, microgrids have emerged at the heart of the prospects of the change in the generation and distribution of energy. Nonetheless, the dynamics of the microgrid system call for optimization algorithms for the energy resources, load demand, and operational capacities of the microgrid. These challenges and objectives are as follows: All the optimization techniques used in microgrid energy management have been considered in the broader context of analysis as shown in Figure 1.3, while encompassing all the optimization methods, this chapter focuses on the aspects mentioned below.

1.3.1.1 Metaheuristic methods

Metaheuristic methods is a type of optimization algorithms that are well known for their ability to search through vast areas of solutions space and achieve reasonably good solutions. As for these methods, GA and PSO are believed to be two major techniques to be employed in the microgrid EMS.

- Genetic algorithm: From the natural selection and evolution theory, GA successively optimizes a set of solutions using selection, crossover, and mutation. Microgrid under the smart grid framework assists GA in solving optimization issues such as energy control and management, load flow control, and demand

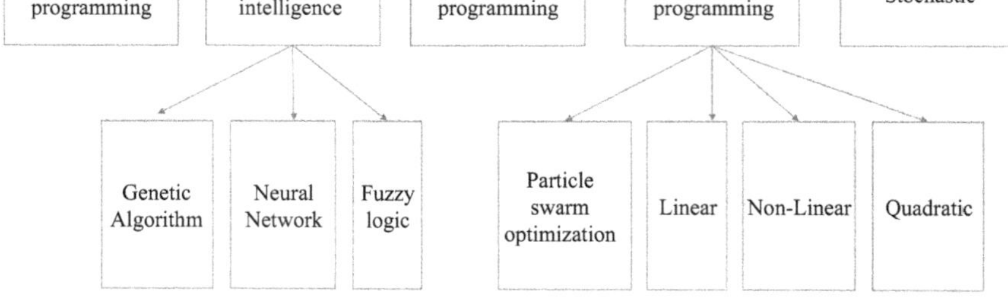

Figure 1.3 Diverse optimization techniques used in microgrid management

response. Since the candidate solutions in GA are represented in the form of chromosomes containing energy generation, energy storage, and energy consumption-related information, GA naturally searches toward optimal or sub-optimal solution space. The conventional goal-setting approach is advantageous as it can accommodate opposing objectives for instance the lowest cost and the highest incorporation of renewable energy [22].

- PSO: Like the behavior of birds flying in groups or schools of fish in a cyclical motion, the particles are in continuous movement as they search for the best solution in PSO. Thus, in microgrid energy management, PSO determines battery charging/discharge profile, power control, and energy trading control. Through an update of a particle's location using its own experience and experience of the swarm and the best particles, PSO in an optimal and self-tuning manner searches for the best solution in the solution space and achieves the best solutions of an operation, regardless of the change of environment or system conditions [17,22].

In addition to GA and PSO, several other metaheuristic methods have shown promise in addressing microgrid energy management challenges: from GA and PSO, there are several other meta-heuristic approaches that were used in addressing the micro-grid energy management problems some of which include the following:

- Ant colony optimization (ACO): Depending on the most well-known arrangement of ants, ACO may supplement the routing and scheduling problems affecting microgrid through the creation of pheromones to enhance the search process.
- Simulated annealing: Actually analogously to what annealing is in metallurgy, the simulated annealing (SA) algorithms can search the solution spaces where worse solutions can be allowed in the beginning before they become better over time.
- Tabu search: Tabu search employs memory structures in order that it does not revisit the previously searched solutions and hence it can move to other regions in the solution space without being stuck by the local optimum traps. GA and PSO can be used in conjunction with the various metaheuristic methods used above which will enhance the decision-making ability of the microgrid operators and make better energy management strategies [19,20,32].

1.3.1.2 Classic optimization techniques

Other optimization methodologies that are used for microgrid energy management optimization other than linear programming (LP) and NLP include the following traditional optimization methods that improve the solution for various classes of problems and characteristics of the system. Below are further developments on LP and NLP followed by descriptions of additional classic optimization methods: Here are some more refinements of LP and NLP along with the description of a few more optimization techniques:

- Linear programming: Therefore, it can be observed that LP is applicable for microgrid control when generation cost and power flow restrictions are linear.

Other forms or LP include mixed-integer linear programming (MILP) that contain integer decision variables especially for conditions where the decision taken are binary for instance, ON/OFF status of the generators or switches. Regarding contingency of forecast loads and generation, there exist possibilities of utilizing robust optimization or stochastic programming to develop the strategies optimal under the contingency of uncertainty [8].

- NLP: As a matter of fact, in applied analysis, many functions that reflect cost and constraints, integrated in microgrid, are nonlinear. NLP can model the efficiency maps of generators, which are functions; SOC of the battery, which is nonlinear, and voltage stability margins which are also nonlinear. However, the difficulties in NLP could be then computational and may be exacerbated by the fact that the systems are large in scale, or nonlinearity leads to non-convexity. Nevertheless, having in mind the recent advances in the field of algorithms and computing, NLP is now a feasible solution to many of the optimization challenges in the microgrid management [9,11].
- Dynamic programming (DP): DP is a robust optimization technique best suited in problem solving in a number of phases of the decision-making process. This form of solving operates on less complex problems and as for the solutions to these less complex problems it employs the use of recursion. Regarding the control strategies of microgrid, DP has been used to take the decision in human steps on energy storage and generation dispatch meanwhile taking into account the temporal aspect [15].
- Integer programming and mixed-integer programming (MIP): Mathematical programming with integer variables (MPIV), integer programming (IP), and MIP. If these decisions are discrete, then IP is required where discrete decisions are involved, for instance, commitment of generators or routing of electricity around the network. MIP combines linear programming with integrality and is most often utilized in scheduling and planning issues that take place in microgrid.
- Stochastic programming: Thus, the essential objectives of optimization in conditions of risk are met by stochastic programming. This goes on and allows the decision-maker to stack several outcomes and select the most desirable performance. Such a technique is quite useful in microgrid optimization due to fluctuating and uncertain patterns in energy demand, and renewable energy sources [20,21].
- Robust optimization: The aim is to develop such improved solutions that satisfy feasible conditions in all possible cases. Out of these approaches used in controlling loads, the control approach is highly suitable in microgrid energy management since in most cases energy is generated from renewable sources in microgrid and hence the supply is normally erratic.
- Network flow optimization: Network flow models on the other hand are mainly aimed at mostly improving the flow of commodities within a network. In the context of microgrid energy management, this can be interpreted as the optimal power flow model – the computational tools specifying how the electricity from several sources could be supplied to several demands under the laws of physical and constraints of the system [33].

All methods above indicate that each of the used techniques has its advantages and it should be used where suitable. The criterion for selection of a particular method would be based on the type of microgrid system, the objective formulated for the analysis, the level of details and variations incorporated in the model, and the extent of variability and randomness in the components of the microgrid systems [31,32].

1.3.1.3 AI methods

AI today is gradually being considered to be a new powerful tool that will be used in microgrid management enabling the overall efficiency of the energy sector, as well as improving the integration of renewable energy sources with the overall energy production process. Here's a deeper insight into the AI methodologies applicable to microgrid optimization: Here is a bias analysis of the AI methodologies that may be applied to microgrid optimization:

- Reinforcement learning (RL): Reinforcement learning (RL) is special as it can get hold of a policy that provides the maximum action every time the environment is in a certain state. Regarding an RL agent, an entity tries out a number of actions and is rewarded or penalized based on the received outcome. Such learning is referred to as the trial-and-error learning mechanism; with the aid of Q-learning or policy gradients, the system is able to determine plans that would yield the best outcome in the future. In microgrid contexts, this means changes in its operational strategies that the system applies for the management of microgrid, and the learning that such system points for instance, to avoid certain costs or distinct demand peaks over a period [29].
- Deep learning and deep neural networks (DNNs): Compared to traditional machine learning, deep learning (DL) is suitable in representing functions and searching for patterns in high-dimensional data with the assistance of more complicated structures called neural networks. Hence, the structure equipping DNNs with multiple interconnected nodes makes it possible to calculate the key values required for the microgrid decision-making process, including load demand, energy from variable renewable sources, and price forecasting. In this respect, when the DNNs are trained by historical and real-time data, it makes a significant improvement in the level of the predictive analytics in the microgrid system and assists the operators for optimal scheduling and dispatch of energy resources.
- Swarm intelligence algorithms: Swarm intelligence attempts to mimic the highly organized peer-to-peer behavior of thousands of individual agents, which are functioning in self-organized manner similar to ants, birds, etc. ACO and PSO algorithms are examples of the algorithms, which are based on active cooperation between numerous agents in their problem-solving. The latter is possible, for example, with the help of algorithms that define the path of electric flows or the schedule of DERs in a microgrid to some degree of optimality and flexibility that resembles natural ecosystems [21].
- Fuzzy logic systems: Fuzzy logic therefore provides a chance of reaching decisions where information is quite unclear or least certain as may be

observed in energy production. They can therefore offer strong control principles strange with the uncertainty that defines the operations of microgrid.

- Support vector machines: In cases where systems are categorized under the state or the future state, then this is under the classification or regression problem, respectively; therefore, the control of microgrid systems through support vector machines can be of help.
- Evolutionary computation: There is use of such as GAs and the evolution strategy (ES) among others, just like with the RL and SI in handling of difficult and challenging objectives of optimization. It generates solutions; the technique is used to get at almost the best solutions of problems that could not be solved using normal algorithms; solutions are developed through generations.

Energy management and AI are key enablers for the evolution of smart grid systems when used together. Sophisticated EMS go beyond the standard functionality in areas such as demand response, peak shaving, predictive maintenance, and real-time control of DERs leveraging on two main advantages that AI has over all other methods: its capacity to manage and learn from massive input data. AI is instrumental involved in effectively converting microgrid from passive "lifelines" into very active systems that can not only ensure reliable energy supply and continually feasible large-scale economic operations such as energy trading and supply of ancillary services to the mainstream grid [27,33].

1.3.1.4 Application of optimization methods

These optimization methods find versatile applications in various facets of microgrid energy management. All these optimization methods come in handy.

- Optimal resource allocation: Planning of energy resources distribution including generation units, storage, and demand response resources in such a manner that will reduce operational expenses or carbon emissions for maximum renewable energy.
- Energy scheduling and dispatch: From a demand-side management perspective, the issue relates to the timing and dispatch of energy products and services over a time horizon for demand calls and for dealing with or controlling congestion together with the stability and reliability of the power system.
- Energy trading and market participation: Offering incremental improvements in energy exchange decisions on the microgrid level, with other microgrid, with neighboring grids and/or energy markets, for the generic aim of maximizing income or efficiency.
- Grid integration of renewable energy: The possible integration of renewable energy sources including solar photovoltaic, wind turbines, and hydro generators in the microgrid architecture without influencing the stability, reliability issues, and the use of green power resources [5,33,34].

1.3.1.5 Comparison of some common energy strategies and principles

Table 1.2 shows a comparative analysis of the common energy optimization techniques featuring AI methods, and classical and metaheuristic methods used in

Table 1.2 Compare optimization mathematical models

Model	Advantages	Disadvantages
MILP	It uses simple steps to overcome complex problems. Its ability to provide numerous optimal solutions gives it an advantage over the MILP formulation.	Reliability and economic stochastic analysis. Limited functionality for applications whose goal functions are continuous or non-differentiable.
MINLP	Either all or one of the parameters in the mathematical function is nonlinear.	Many iterations with a large computing load.
Dynamic programming	Delved into minor issues to gradually resolve a major one.	Due to the several recursive techniques and challenging implementation.
Genetic algorithms	Population-based evolutionary algorithms that use crossover, mutation, and selection to find the optimal solution. An adequate rate of convergence. extensively used in a variety of sectors.	Population-based evolutionary algorithms that use crossover, mutation, and selection to find the optimal solution.
Particle swarm optimization	Outstanding results in difficulties related to scattering and optimization.	An adequate rate of convergence. extensively used in a variety of sectors.
Artificial bee colony	A robust population-based technique that is simple to implement. a sufficiently quick convergence.	Harder development due to greater computational complexity.
Artificial Fish Swarm	Accuracy, speedy convergence, reduced parameters, and adaptability.	Keeps the advantages of GA but adds further disadvantages (crossover and mutation).
Bacterial foraging algorithm	The non-linearity and scale of the problem have less of an impact. When methods of analysis come together to produce the optimal answer.	Broad and intricate search space.

energy optimization that helps in drawing useful information about their specific application and usefulness in facing energy management challenges. Two classical optimization techniques are linear programming and NLP that offer a standardized mathematical framework to solve all the optimization challenges that are well-defined in terms of objectives, zones, and constraints. These methods are especially suitable to be used when the structure of a problem is known and parameters can be quantified easily. Other metaheuristic methods such as genetic algorithm and PSO which are flexible and robust methods are favorable in searching the large solution space and obtaining the solutions to the problems. These methods are most suitable when the optimization problem is nonlinear or non-convex or at times even a combinatorial problem where traditional methods easy linear programming techniques and optimization algorithms do not offer the best of solutions. In addition, the AI-based methods like RL, DL use algorithms and machine learning to manage more power and effectively handle the data of its performance and deliver the best-suited algorithms for the distribution and management of system resources. These AI methods present adaptive and data-driven methods to deal with uncertainties and dynamics in the energy systems to cause higher efficiency, resilience, and sustainability. Therefore, it can be stated that it may not be wrong to include the classical optimization methods as a part of metaheuristic methods and AI methods to address and solve the energy optimization issues of today's complex systems as they could be complementary to solve the complex issues [13,33].

Table 1.2 represents different types of optimization strategies with pros and cons where it is clear that schedule, resource management, and market participation for microgrid energy management are essential for microgrid application. As a result, metaheuristic methods, as well as other conventional optimization methodologies, help microgrid operators deliver the aforementioned levels of performance that lead to increased microgrid reliability and readiness to shift more consumers toward green energy. These methodologies can be considered as invaluable tools in the armory of microgrid management and opening up possibility of more efficient, reliable, and sustainable energy systems across the world [35,36].

1.3.2 Application of methods to microgrid parameters

Optimization of microgrid can be said to depend on the minute adjustments of parameters in different components for the best outcome. Examining further for BMS, EMS, selecting microgrid components, combination of components along with energy trading presents the multifaceted approach of incorporating these parameters along with optimization algorithms for consistent improved performance [36].

1.3.2.1 Battery management systems

The lifespan and performance of microgrid wholly depend on depending on the efficiency of a BMS system. The former's important parameters include charge/discharge frequency; types of charging and discharging as well as temperature control; and all of these aspect highly affect battery efficiency. Some of these parameters can be fine-tuned by microgrid and it is allowed to reach this optimum with the aid of the optimization methods of GA and RL. However, by considering the trends of power consumption, the climate, and the cost of energy over a time period in GA, nearly the most effective charge/discharge schedules are possible; in real-time learning, with climate prediction, the most efficient BMS stratagem and real battery performance can be learnt. However, it is also useful to enhance the battery duration and also to utilize the energy storage efficiently which is very crucial for microgrid stability and reliability [37,38].

Figure 1.4 shows some charging methods and the common type of batteries.

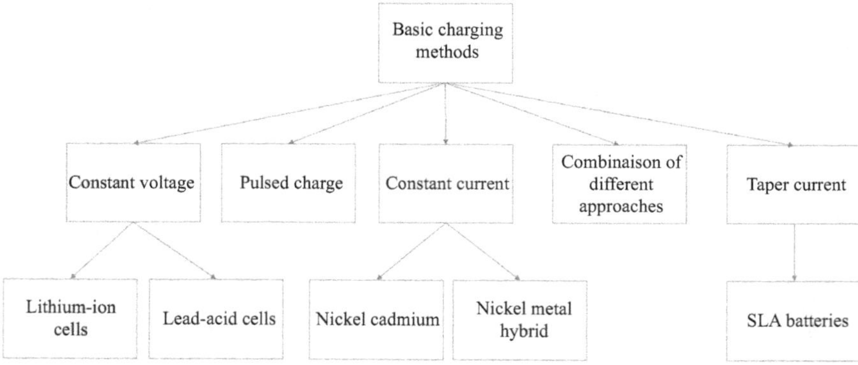

Figure 1.4 Some basic charging methods and their appropriate types of batteries

1.3.2.2 Energy management systems

Demand response mechanisms integrated within EMS allow microgrid to dynamically adjust energy consumption in response to supply-demand dynamics, grid conditions, and pricing signals. Optimization techniques such as LP and metaheuristic methods optimize demand response strategies, considering factors like energy tariffs, system constraints, and user preferences [39]. By incorporating demand response parameters into EMS optimization models, microgrids can effectively balance energy supply and demand, curtail peak loads, and optimize grid operation in real time. This integration ensures adaptability to fluctuating conditions, enhancing the overall grid performance and efficiency. Figure 1.5 shows the interaction of EMS with a microgrid.

1.3.2.3 Optimal combination of microgrid components

The level of coupling between the microgrid's constitutive components, such as the generation resources, the storage systems, and the load-controlling devices, is relevant to the outcome. Techniques including simulated annealing and PSO undertake an extensive study on various parameters in an endeavor of identifying optimum setting of the components. Costs, reliability and performance among others are therefore carefully weighed in order to provide predetermined goals such as the highest level of renewable energy share for instance, or the least carbon-friendly electricity mix for instance. This has been done to ensure that the optimal microgrid will operate efficiently, and at the same time will ensure that optimal solutions available in the models of optimization take into account other operational issues that may be attached to the microgrid, and questions of feasibility. That kind of approach enhances the stability, sustainability, and efficiency of the grid and properly anticipates the functionality of microgrids [25,39].

In the deployment of microgrid, the main idea has been to analyze factors in a bid to find out which microgrid components would help in the design of a cheap

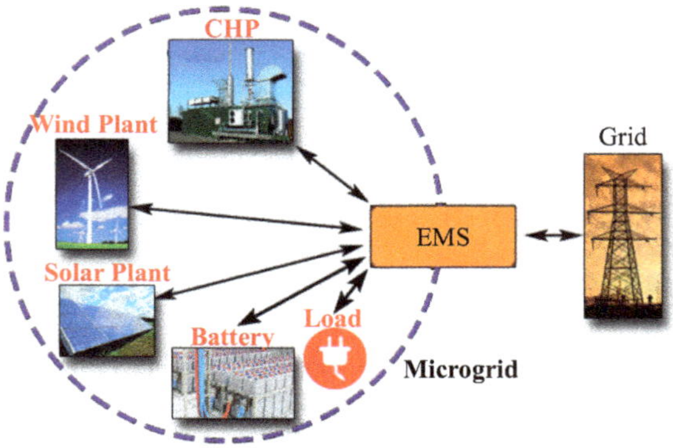

Figure 1.5 EMS interaction with a microgrid

and efficient microgrid system. In the middle of this process, dimensioning takes place which includes the number of solar panels and wind turbines as well as batteries and other aspects in need of energy based on the demand for energy and at the same time the costs involved. Major inputs include the cost of procurement of the components, the life span of the components, the costs of maintenance and part replacement, installation costs, the location of the power plant, and the availability of materials at the particular location.

Energy demand requirements play the role of dimensioning and it is very important to evaluate the power consumption in the past and tendencies for the future. Component prices are also large where an economic decision is made on which component has to be used as the most economical one without having to be affected efficiently. While considering the lifecycle of a component, it is possible to make long-term planning and cost reduction because a definite component may take more time in comparison to the other, so it may be cheaper in the long run [40,41].

This also takes into account of other expenses such as the general maintenance, repair, and replacement costs of the vehicles. When the necessary maintenance has been established for all of the constituents, this will help the designers in identifying the most suitable time to incorporate the schedule and cost value of the maintenance. Furthermore, the cost for installation including labor cost as well as the cost of installation facilities is well estimated with regard to site characteristics including the geographic nature of the land whether flat or sloping. This is due to the fact that where a microgrid is located plays a big role in its design because the chances and potential of renewable resources vary. Resource audits of renewable resources in conjunction with climatic studies of the location enable the designer to properly scale and locate the components so that they harness as much of the REN energy as possible and use as little of the energy as possible.

Thus, it enables integration of the critical microgrid inputs into the equations; following detailed analysis, efficient microgrid systems capable of satisfying the energy demand can be developed affordably with high reliability and performance. All of this secures the delivery of reasonably priced, efficient, and sustainable energy services tailored to the needs and circumstances of clients that help to design a proper way toward a cleaner energy future [40,42].

1.3.2.4 Integration of parameters with optimization techniques

One has to be able to incorporate highly detailed activity constraints including the dimension of demand response, component coupling, and intricate management of battery storage with the numerous high-level optimization algorithms to provide a tremendous upgrade in the efficiency, resilience, and prowess of microgrid. It provides an integrated approach, which enables microgrid operators to make corrections with reasonable levels of accuracy with regard to goals, requirements, and conditions that may transpire volatile with time and at the same time it creates a real-time optimization of the microgrid resources [17].

These complex parameters are incorporated into the complex weaves of optimization models and thereby the microgrid operators eliminate the hurdles posed by static structural frameworks. They possess the capability of handling the new conditions; they are capable to work under the conditions offered by the varying demand from the grid or its complexity. For instance, integration of the demand response parameters into the state-of-the-art RL algorithms helps the microgrid in handling the energy consumption challenges. This active interaction allows the simultaneous application of the principles of energy necessary for user preferences in combination with dynamics in the grid. Therefore, not only does it enhance the rate of the level of energy-efficient improvement process but also provides stability to the whole system [16].

However, the fine-tuning of the battery management parameters through such methods as, for example, the genetic or PSO takes the process to a new level of possibilities. It enhances the effectiveness of battery use by the microgrid system; it also enhances the battery life and the effectiveness of the microgrid system. Synchronization of these optimization methods ensures that microgrid operators get the maximum performance of the microgrid since efficiency, reliability, and sustainable ways of managing power systems are achieved [19,32].

In essence, this comprehensive and sophisticated integration of parameters and optimization techniques empowers microgrid operators to navigate the intricate landscape of energy management with unparalleled finesse. The result is a robust, adaptive, and resilient energy ecosystem, poised to meet the challenges of the ever-evolving energy landscape with unwavering precision and efficacy. Through this approach, microgrid operators can not only meet but surpass the expectations of efficiency, reliability, and sustainability, heralding a new era in the domain of energy infrastructure [43].

1.3.2.5 Energy trading strategies within the microgrid

Energy trading strategies are strategic in the microgrid when it comes to making specific particular grids to be as efficient and as economical as possible. The integration of energy trading parameters and microgrid with internal and external microgrid gives a new dimension of flexibility and stability here. Besides, the integration of IoT enhances the capabilities of the microgrid to address these problems and unveil new areas of integration. On the internal side, microgrid can enable trades between one and another, demand and supply, consumers, producers, and storage systems as per the needs of microgrid members in short intervals of time. Exact methods include elementary game theory, market equilibrium models, and others, which, for example, can be used, in order to analyze trading patterns, for determining the most effective bidding strategy in order to maximize the revenue or minimize the cost. For instance, blockchain-enabled smart contracts which act as reliable energy contracts in the microgrid system enhance the percentage reliability. Some of the challenges, which are transaction latency and scalability among others, are some of the other challenges that have to be addressed to make full use of it. Externally microgrid can sell energy to the neighboring grids or in energy markets for further profit or even buy electricity as and when needed. Approaches like metaheuristic algorithms and RL can be applied to enhancing the selection of energy exchange based on factors such as the price, the state of the grid, and other regulatory frameworks. Integration with energy market platforms and real-time decision-making on spot price arbitrage helps microgrid earn more revenues. Some of these complications include regulatory constraints and market volatility which need to be well managed through risk management strategies and fluctuations trading models [22,33,43].

1.3.2.6 Integration of IoT for challenges and upgrades

The application of IoT solutions in microgrid offers new solutions to such challenges with a focus on enhancing efficiency. By using IoT sensors, one is able to get data instantly of the consumption and generation of energy, and the condition of the grid to facilitate better decision-making and also to prevent breakdowns. It is easy to improve the IoT information regarding energy trading using artificial neural networks for a better trading strategy to look for loopholes in energy trading or even to predict the frequency of energy usage in specific time intervals. Besides this, the IoT smart meters and grid sensors assist in increasing the performance of the demand response programs and grid monitoring that contribute toward increased stability of the grid reliability. Among the things that require the solutions, there are data protection, security, and compatibility and these can be addressed when IoT governance frameworks and standards are developed. IoT-connected devices and analytics when applied in the microgrid systems make control and operation of microgrid; hence, efficiency, reliability, and sustainability are improved [44].

Therefore, getting acquainted with energy trading solutions to be included to internal and external microgrid is an excellent opportunity to improve the functionality of the grid as well as gain the maximum profit. Chasing away the current challenges and unveiling the new possibilities within the microgrid, IoT technologies boost its strengths and promises. By means of such improvements, the operators of the microgrid which were provided in this chapter can reveal the energy market conditions, adapt to them continuously, and make sure that the microgrid possesses potential that aims at positive energy transitions [40,42,45].

1.3.3 Some real-world examples and contributions of the optimization methods

1.3.3.1 Some real examples of the use of optimization methods

With regard to optimization methods, the literature reveals cases that depict a significant shift in microgrid that have diverse functions in distinct regions of the world. In a rural setting of the sub-Saharan region where people have no regular supplies of electricity, a reliable way of solving energy problems has been done in this chapter by applying GA as optimization techniques. That meant that, in order to minimize the running costs while meeting the energy demand, the microgrid designers had to optimize microgrid topology with respect to factors such as the amount of solar irradiation, the energy demand profile, and the available resources in the location of the microgrid in terms of installing solar panels and batteries and incorporating diesel generators. This approach also greatly reduced the extent of the expensive diesel fuel and contributed to the increase of availability of resources for the people, thus assisting in developing the economy and civil society [17,44]. However, in the New York City, the efficiency of the optimization algorithms used offers transient solutions within microgrid to energy trading difficulty. Applied to structures under microgrid, linear programming, and meta-heuristic algorithms for example, the structures can mutually engage in transactions like buying and selling commodities in the market depending on prices and conditions in the grid system. This helps in maintaining the load factor energies used which are conserved in this manner, whereas this option ensures increased reliability and stability of the grid while considering the ever-increasing population that is found in the urban facilities [46]. Furthermore, where the environment is remote and at times extreme as in Alaska, microgrid can be a life support to some communities; therefore, optimization methodologies are important in increasing the reliability of energy delivery. Similarly, RL will help microgrid leverage battery management with factors such as weather conditions, seasons, or energy demand with the operators of the microgrid. The above flexibility ensures microgrid to cope up with fluctuations in extreme weather conditions; ensure constant electricity, security concern facilities and structures, and in general, support in making the community more stable [3,46].

These examples underscore the versatility and efficacy of optimization methods in addressing multifaceted challenges inherent to microgrid deployment, offering tailored solutions that optimize resource utilization, enhance economic viability, and promote sustainability across diverse environments. By embracing optimization methodologies, microgrid stakeholders can unlock new avenues for innovation, resilience, and progress, ultimately driving the transition toward a more sustainable and inclusive energy future for communities worldwide [47].

1.3.3.2 Some contributions based on previous funding

Previous funding initiatives have played a pivotal role in catalyzing advancements in microgrid technology and fostering innovative solutions to energy challenges in the light of common optimization strategies shown in Figure 1.6.

Thus, with investment and grants, the researchers, entrepreneurs, and other stakeholders have created incentives to find out new ideas, invent new technologies, and initiate the innovative projects. Such advances have supported major developments in the configuration, optimization, and implementation of microgrids as per identified developments across the domains. For example, funding support has enabled research institutions to come up with research that focuses on the outcomes of the incorporation of renewables, various ESS and buffering systems, as well as the capability and readiness of the grid to cope with any change. Likewise, the funding sources have supported both the startups and the small businesses for the introduction of new microgrid solutions, testing projects, and broader commercialization to revolutionize the microgrid market. Similarly, the funding for community-based interventions means that the needy and poor segment of the community has been empowered through microgrid funding to have access to reliable and cheap electricity for socioeconomic uplift. Overall, the contributions stemming from previous funding have laid a strong foundation for advancing microgrid technology, fostering collaboration, and shaping the future of sustainable energy systems as shown in Table 1.3. On balance, the provided outcomes of previous funding have established a solid base for microgrid progress, cooperation,

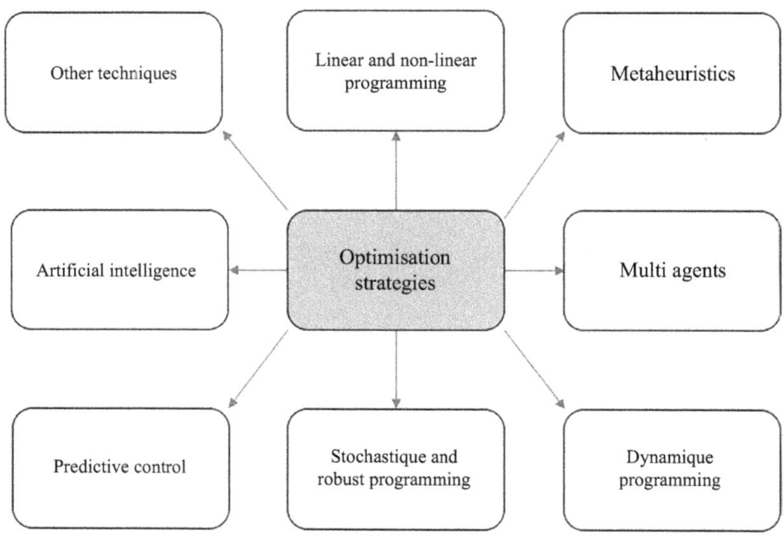

Figure 1.6 The most common optimization strategies used in microgrid

Table 1.3 An examination optimization of microgrid methods

Reference	Optimization strategy	Contributions	Constraints	Drawbacks	Multi/single goal
[22]	Mixed integer and non-linear	A robust optimal EMS MPC-based approach is used to determine the optimal power scheduling given each generator, which incorporates dump and deferrable loads.	Power ratio Battery Generator Renewable sources Loads	It does not account for demand or power outages.	Multi-goals
[32]	Mixed integer and linearLinear programming	A method of energy management that integrates power sharing, ON/OFF, and continuous run modes – three possible techniques.	Battery Generation dispatch	The costs of battery degradation are not taken into account.	Multi-goals
[33]	Nonlinear programming	Reduced overall running costs while maintaining the standalone MG's safe functioning	AC power DC power Converter power Load Distributed generators power	Battery storage that is systematic is not considered. When producing biomass in a distributed manner, the cost of emissions is not considered.	Mono-goal
[33]	Linear programming	Combining linear programming with AI-based methods to address multi-objective optimization	High complexity of computation. Battery degradation is not taken into account.	Dispersed generation's limitations in power balance	Multi-goals
[34]	Particle swarm algorithm (PSO)	Combination of two perfect energy storage devices. Shorter computing time than GA	The power of the generators Energy conversion to the grid and storage unit charging and discharging equilibrium between supply and demand	The emission costs of the conventional generator are not considered.	Multi-goals

(Continues)

[35]	PSO with Gaussian mutation	PSO variant new algorithm.	Active power Voltage Current	Power losses and Emissions of distributed generation are not considered.	Mono-goal
[36]	Artificial bee colony	To lower the running costs of a microgrid, a two-layer control approach is used.	Power balance resources that are easily accessible. Non-transferable assets keeping parts in storage	Formulation is challenging. A dispatchable microturbine's emission cost is not determined	Mono-goal
[37]	Fuzzy logic (Grey Wolf Optimization)	Optimization of the battery Evolutionary algorithm	Power balance energy storage size and of the generation plan.	Cost of battery deterioration is not the generators' power battery charge	Mono-goal
[38,39]	EA and the PSO algorithm	Application of an energy hub model for optimization of a multicarrier MG.	Power balance voltage		Multi-goals
[40]	Artificial fish swarm optimization	A MG's energy management schedule, which takes into account storage for the entire day and dynamic pricing, is optimized	Power parity traditional methods of generating power standard power generators	Battery degradation cost is not counted	Mono-goal
[41]	Particle swarm algorithm	Three different objectives are considered: Reliability, operation cost, and environmental impact.	Unknown	Degradation cost of battery is not taken into consideration.	Multi-goals
[45]	Bacterial foraging algorithm	Optimized the exchanging power with the battery setpoints, the grid, and the generator setpoints. quick convergence.	Power balance Generation limits of distributed generators Storage limits	Losses of power not considered	Multi-goals
[42]	Mixed-integer nonlinear programming (MINLP)	Information. Different battery types compared.	Charge flow Sending out of generators Programming of generator on/off Battery charge and discharge	The prediction of battery life is disregarded	Multi-goals

(Continues)

Table 1.3 (Continued)

Reference	Optimization strategy	Contributions	Constraints	Drawbacks	Multi/single goal
[48]	Dynamic rules	MG management system uses different limits for the SOC of the batteries bank.	Battery Power balance	Battery cost and degradation are not considered.	Mono-goal
[43]	Dynamic programming	Different limitations are used by the MG management system for the batteries bank's SOC. Energy management strategy for PV. Batteries to stabilize and permit PV to run at a stable and constant output power	Charge discharge of batteries	Battery degradation and lifetime prediction are not considered	Multi-objective
[44,46]	Multi-agents	Fantastic for instantaneous Power imbalance is effectively compensated for with energy storage management. Multiple-layer control system with coordinated control A multi-agent system, a distributed intelligence-based optimization problem, and a battery energy storage system	Battery discharge and charge Power scheduling and load scheduling Control system with many layers and coordinated control	Battery ageing is not counted complex control scheme Dual stage optimization. First stage	Multi-goals
[49]	Stochastic	Establishes the generators' hourly unit commitment, while the second stage involves economically dispatching the generators and batteries.	Power balance Dispatchable Distributed generation Renewable power generation load Charge/discharge of batteries	No consideration is given to the battery ageing model or the DG's emissions costs.	Multi-goals

(Continues)

[50]	Strong programming	Hybrid wind-battery-diesel system load management is optimized	Battery bank with wind turbine power source for the diesel generator	Shifting of controllable loads may not be ideal	Mono-goal
[51]	Mixed-integer quadratic programming	Demand-side management and unit commitment for generators are both taken into account by an integrated stochastic energy management model.	Power balance Generation Demand Reserve capacity	The deterministic model requires more processing time. Costs associated with emissions from traditional generators and DG are not considered	Mono-goal
[52]	Model predictive control	For noncritical loads, automated load shedding is advised when power imbalances could jeopardize the stability of the MG.	Power distributed generators	The battery's charging and discharging rates are not taken into account. Similarly, communication lags	Multi-goal
[53]	Model predictive control	A comprehensive mathematical characterization of the optimal energy management system (EMS) for independent microgrids that considers realistic power flow and unit commitment constraints	Power balance Reserve Unit commitment Energy storage Grid	Higher complexity and computing. The cost of emissions from traditional sources is not computed.	Mono-goal
[54]	Model predictive control	This work's primary contribution is a daily optimizer that accounts for capacity losses while determining how quickly lead-acid batteries deteriorate.	Not specified	Model of lithium battery is considered.	Multi-goal
[55]	Genetic algorithm	A unique cost function includes the startup costs of distributed resources as well as the costs of selling and purchasing power.	Balance of power Emissions Battery power Generator start	Battery status of charge and distributed sources are not taken into account.	Mono-goal

(Continues)

Table 1.3 (Continued)

Reference	Optimization strategy	Contributions	Constraints	Drawbacks	Multi/single goal
				In their energy generation, MGs and customers' uncertainties are not taken into consideration.	
[56]	Game theory	Distributed energy management schedule in various MGs.	Energy transfer to the grid and the MG's generation capacity	Computational complexity is not counted.	Multi-goal
[57]	Artificial intelligence (Fuzzy logic)	Simple enhanced grid power profile implementation, quality	Discharge/charge of batteries	Only the battery charger/grid-connected inverter is under control. Battery degradation is not considered.	Multi-goal
[57]	Game theory	Decrease cost of fuel and trading power.	Power balance DG traditional The maximum generator power that can be transferred between the main grid and the MG	There is no discussion of the conventional generators' emission costs.	Multi-goal
[58]	Markov decision process	Linear model to evaluate the MG lifetime cost.	Gas turbine capacity Gas turbine emissions	Limited number of possible combinations of sizes	Mono-goal
[58]	Rule-based	Analysis of the anticipated costs associated with the hybrid system, taking battery deterioration into account. A levelized cost of electricity (LCOE) research is carried out subsequent to the development of a hybrid operating regime. Reliability of energy storage deterioration expenses	• Power balance • SOC battery	Do not consider temperature in capacity modelingConditions for dynamic state-of-charge cycling are not taken into consideration	Multi-goal

and subsequent developments of future sustainable energy systems are shown in Table 1.3.

1.4 Metaheuristic results for microgrid design

1.4.1 Microgrid best combination

To construct an accurate model for the price per kilowatt-hour (kWh) of a micro-grid, a meticulous consideration of multiple factors is essential. These factors encompass the initial investment cost, operational and maintenance costs, potential fuel costs (if applicable), and the influence of any available subsidies or incentives. Furthermore, the geographical location plays a crucial role in determining the pricing structure. Additionally, various tariff structures, such as time-of-use pricing or demand-based pricing, can introduce variability into the overall cost for consumers.

The mathematical model devised for optimization is expressed as follows:

$$\left(N_{\text{wind}} \cdot C_{\text{wind}} + N_{\text{panel}} \cdot C_{\text{panel}} + N_{\text{batt}} \cdot C_{\text{batt}} + C_{\text{convert}} + C_{\text{biomass}}\right) / \left(\langle P_{\text{moy}} \rangle\right)$$

with

$\langle P_{moy} \rangle = 47$ kW. 365.24.20

$= 8{,}2$ GWh

$\langle P_{moy} \rangle = N_{wind} \langle P_{wind} \rangle + N_{panel} \langle P_{panel} \rangle$

$C_{panel} = 1{,}200+1{,}200+4*20 = 2{,}480$ USD/kW

$C_{wind} = 2300+1500+20*4 = 3{,}880$ USD/kW

$C_{batt} = 234*4+1.67+20 = 970$ USD/unit

$C_{convert} = 150*20+127*150 = 22{,}000$ USD

$C_{biomass} = 3{,}880*50 = 194{,}000$ USD

and $N_{wind} < 19$; $N_{panel} < 125$; $N_b < 14$

Constraints are set on the number of wind turbines ($N_{wind} < 19$), solar panels ($N_{panel} < 125$), and batteries ($N_{batt} < 14$). These constraints are derived from dimensioning, where the power of one wind turbine is considered 10 kW, and the nominal power for one photovoltaic panel is set at 1 kW under standard conditions.

The following components are too many to list but the equality includes the following components to form the basis of optimization for microgrid design: This mathematical model gives not only the real concept of microgrid design but also a base where all the desired costs, energy generation methods, as well as the limitations can be incorporated. With the aid of this model, one is capable to identify certain design that is able to offer an optimum cost/sustainability solution in establishing efficient and cost-efficient microgrids.

Renewable energy resources connected to a micro-grid system are perhaps one of the most effective approaches that can be employed if the objective is to realize substantial savings on electricity bills. It has however been seen that the cost per kilowatt hour can be lowered significantly and comes down from the first standardized rate of 0, 115 USD/kWh down to which it got an impressively low range of 0.037 to 0.055 USD/kWh. This means shifting the cost range to lower which is done by excluding the use of non-renewable energy sources. However, rather frequently, the potential cost savings become apparent only when using more complex strategies that involve the use of AI. Thanks to the AI this will assist in determining which stand of electrical systems are effective for achieving goals and, due to the application of other methods of managing the electricity flow within the microgrid, the minimum value of 0. This way the LCOE of 037 USD/kWh is feasible and can be achieved in the future. This essentially means that the amount of billing was reduced by 67% from the normal billing rate of 0.115 USD/kWh for Fez.

That is why the given "0.037 USD/kWh" value has been verified with the help of several AI techniques. Further, explaining the reason for the lesser cost has been supported by the HOMER tool which is internationally recognized and is an efficient microgrid system analysis tool. The examples of implementing the combination of AI techniques of advanced tools clearly illustrate the benefits and prospects of the microgrid that use renewable energy and the crucial importance of AI in tuning and optimizing the microgrid setting and bringing significant economic value and positive changes to the energy future [43,44,58].

1.4.2 Energy management system contribution to microgrids

EMS is the central component responsible for maintaining balance and diverting excess energy to the dump load, ensuring that all generated energy is utilized in dynamic mode. This system is based on the principle that the purer the energy, the more effectively it can be utilized in the system. It is for this reason that the below priorities are important for the generation of solar and wind energy to be free from CO_2 emissions and be free during the generation of power. These sources are used where the energy is used in real time for increasing the use of green energy and helping the environment.

In the second position, the EMS has duties concerned with the kind of stored energy present in batteries. This requires that there be mechanisms of managing any excess energy that may be generated from the renewable energy form and this can safely be done using batteries. This also helps in managing the distribution of stored energy within the system whenever there is a scarcity of renewable energy or high demand, hence the correct utilization of energy and at the same time minimizing the use of conventional sources of energy.

When it comes to the structure of priorities, the given object – biomass gasifier – is represented as the third most significant contingency plan just like the second most important concern. This backup system is initiated in conditions where renewable sources and stored battery energy are unable to meet fluctuating energy needs. In a gasifier, biomass energy is created and is then transformed into usable or utilizable energy, a nice supplement source but not really a disruptive change to the system for clean energy.

In other words, if the EMS is to be maintained, it ensures the preference of renewable energy sources in meeting energy demand and a real-time schedule of energy consumption and utilization, an aspect enhanced by storage or backup systems when deemed necessary. In addition, it optimally employs clean energy and increases the chances that the energy management will be dependable and secure – which will help to achieve the set goals of sustainability and efficient utilization of resources [33,36,57].

The organization chart of EMS is shown in Figure 1.7.

For the results, a comparative analysis of three scenarios unfolds: a basic microgrid, a microgrid augmented with an EMS, and an advanced microgrid that incorporates EMS, AI, and the IoT. The progression from a basic microgrid to an advanced one underscores the potential for more efficient and economically viable EMS.

In the absence of an EMS, a basic microgrid directs excess energy not utilized by loads or stored in batteries to a dump load, resulting in wasteful energy loss and associated financial costs. The introduction of an EMS optimizes energy distribution, reducing losses compared to the basic microgrid. However, the integration of AI and IoT in the most advanced microgrid enables dynamic adaptation to changing conditions, real-time decision-making, and sophisticated energy predictions. This leads to minimal energy losses and substantial cost savings.

For a microgrid serving 100 houses with an average load of 47 kW, the following adaptive power (ADP) values are observed:

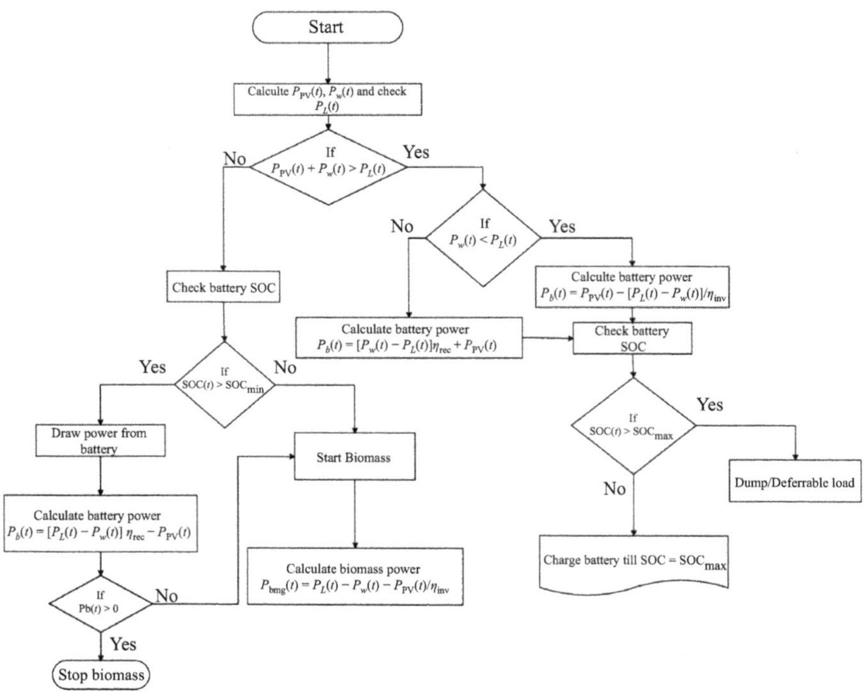

Figure 1.7 Organization chart of EMS used in microgrids

- ADP = 27.5 kW for the basic microgrid without EMS
- ADP = 18.3 kW for a microgrid with EMS
- ADP = 1.8 kW for a microgrid with EMS, AI, and IoT

The results demonstrate that the addition of EMS alone reduces energy losses by 35%. However, the incorporation of AI and IoT achieves a remarkable 93% reduction in energy losses. In a typical scenario without EMS, where 27.5 kW is lost for 69 kW of stored and used energy, losses account for 28% of the total energy produced. The introduction of EMS alone reduces this figure to 20%, resulting in an 8% reduction in the microgrid's total cost. Furthermore, the addition of AI and IoT to the EMS further reduces losses to 2.5% of the total energy produced, leading to an impressive 25% reduction in the microgrid's overall costs.

This comprehensive approach not only emphasizes the pivotal role of EMS but also highlights the transformative impact of AI and IoT in achieving optimal energy management, substantial cost savings, and a more sustainable and resilient energy ecosystem for the microgrid serving residential areas [3,46,56].

1.4.3 Battery management system's contribution to microgrid

The differences existing between the basic microgrid and the microgrid with complex technologies to control energy are described and compared through three unrelated cases in the understanding of microgrid energy. The first case represents a basic microgrid that does not incorporate an EMS. Here, the surplus energy cannot be controlled or effectively used and is simply tied into a dump load where a great deal of unnecessary losses and associated expenses are incurred. The latter, an improvement to an EMS as seen in the second emerged situation, is a major progress. As for the comparison with the basic microgrid architecture, the use of EMS leads to improved energy management so that the surplus energy can be utilized in a better way, thereby decreasing energy losses. This means that it has been able to find out what the current energy demand and supply is and then plot a course different from that followed by what was said to be a simple method.

However, the paradigm shifts significantly more in the third case which occurs in an advanced future EMS-based microgrid incorporating AI and IoT. AI's integration and IoT produce dynamic adaptability, real-time decision-making, and computational energy prediction. They help in handling the changes that occur in the microgrid by reducing the energy losses to the minimum, hence reducing the costs. The rationale for this level is well captured in the fact that the central system is complex enough to assess and predict the energy needs; second, the ability of the new advanced system to harness renewable resources when the weather is conducive and turn to conventional resources during adverse weather conditions. Clearly, the compatibility of AI and IoT pushes the microgrid's performance further up, thus making it a beacon of sustainability, resilience, and economic feasibility.

The level of detail pertaining to its application in a microgrid of 100 houses with a photovoltaic capacity of 10 kW has provided numeric values of ADP to explain energy loss. In the proposed microgrid without having the EMS, the ADP is

27.5 kW, and this is a clear sign that the energy in the building is being wasted most of the times. This is reduced to 18 the moment an organization has put in place an EMS. The newly dumped power of 3 kW is more efficient, with energy losses reduced to 35%, as noted here. The reason for this enhancement is found in the fact that the EMS is able to monitor energy flow according to intelligent aspects and to manage new and renewable power sources in a more optimal manner. However, the highest level of innovation is achieved when both AI and IoT are integrated into the highly complex microgrid in which ADP reduces to 1. To compare the data, the energy losses in the network after the transformation were 8 kW with a reduction of 93% compared to the initial setting. The reason for such a quantum leap must exist in the capabilities of the advanced system, which makes the real-time decision based on its analytical data and defines the flow of energy as precise and as multifaceted as possible. It clearly signifies that with this new preferred approach, energy requirements can be effectively met while at the same time defining the new wave of technologies and how they modify the prospects of microgrids for maximum sustainability as well as economic feasibility [21,47,58].

1.4.4 Discussion of results

This graph illustrates the improvement in the use of the EMS within the proposed microgrid with 100 houses at a photovoltaic capacity of 10 kW. In the absence of the EMS, the ADP stood at as high as 27.5 kW, depicting huge energy losses. When this was subjected to the EMS, this value was reduced to 18.3 kW in order to reduce losses by 35%. Such is explained by the fact that EMS can monitor and intelligently control energy flux, hence optimizing the consumption of renewable sources. In fact, if there is more integration of AI and IoT in the EMS system, even better results would have been achieved when it is able to reduce ADP to 1 kW, therefore reducing energy losses by 93%. This huge gain in efficiency underlines the possibility of the system making decisions in real time, with analytics involvement, and with precision, to manage the energy distribution and flow.

Besides, enhanced EMS will enable the microgrid to not only manage the pattern of energy consumption to be optimally determined but also, when necessary, allow any surplus energy of the microgrid to be directed with much better efficiency. Surplus energy may be sold back into the grid to make sure economic benefits accrue or be put to other uses like crypto currency mining. It helps the system achieve maximum financial returns with optimized energy utilization by utilizing surplus energy for such purposes. This capability for both energy trading and the use of excess energy for decentralized activities shows the flexibility and potential that advanced EMS systems have in modern microgrids to open ways toward sustainability and profitability.

1.5 Conclusion and future perspectives

Apparently, this chapter has provided comprehensive demonstrations of different approaches and strategies that can be used in energy management in microgrids particularly by integrating IoT and AI systems. The main premise that entailed a brief on the capacity of microgrid, microgrid ability, and some framework components that form the understanding of the microgrid model, its reliability, its sustainability, and the robustness provided by the microgrid for enhanced energy resilience. Continuing the conversation, they elaborated on how microgrid is essential for the integration of renewable energy systems and as a backup power during outages together with decentralizing the power systems for the benefit of certain communities as examples have shown. The next presentation on the microgrid energy management methods, metaheuristic, and classic optimization demonstrated various possible uses of such methods as BMS, EMS, and optimal sets of selected microgrid components. The relevance of these methodologies was illustrated with the help of a real-life example of the Fez city in Morocco which will help to demonstrate the potential for reasonable cost savings, reduced carbon footprint, and increased performance parameters. The results also wave with the researcher's impression that high technological innovation is helpful in developing and fine-tuning the microgrid.

Concerning the future outlooks regarding the research and development of microgrid, the prospects seem to be rather promising. This reports even more enhancements that are expected to be achieved with the integration of AI with the IoT in managing microgrids with even other advanced characteristics like real-time decision-making, and even predictive analysis as well as adaptive energy systems. Such optimization techniques that may be updated perhaps after a certain period based on advanced studies in that area may also enhance the prospects of making microgrid implementation more feasible and flexible. Furthermore, the opportunities of integration of such hopeful new technologies as blockchain in a specific microgrid in a way to provide more secure and transparent energy trading have the potential to significantly change the manner of how energy trading happens. Furthermore, collaborative efforts among researchers, industry stakeholders, and policymakers will be crucial to establishing standardized frameworks for microgrid deployment and ensuring interoperability across diverse systems. This collaborative approach will facilitate the seamless integration of microgrid into existing energy infrastructures, contributing to broader sustainability goals and enhancing the resilience of energy ecosystems.

The advancements that are realized in this chapter are to a greater extent a step forward toward constructing better, cheaper and more effective microgrid EMS. Further research and innovation in this field will be able to pursue challenges that are new now today, push for attaining energy independence, and open the way to more sustainability and robustness in systems.

Acknowledgment

The Ministry of Higher Education, Scientific Research, and Innovation (MESRSI) provided funding for this research as part of the LEAP-RE program for the MIDINA project.

References

[1] Hoummadi MA, Bouderbala M, Alami Aroussi H, Bossoufi B, El Ouanjli N, and Karim M. Survey of sustainable energy sources for microgrid energy management: a review. *Energies* 2023;16:3077.

[2] Adam V. *All Volvo cars to be electric or from hybrid 2019* [EB/OL]. https://www.theguardian.com/business/2017/jul/05/volvo-cars-electrichybrid-2019; [Accessed September 15, 2024].

[3] Yang G, Zhou Q, and Hou M. Analysis of China's electric vehicles and its prospect. *Int Pet Econ* 2017;25(4):59–65.

[4] Zheng Z, Zhang Z, and Tang T. Recent development of lithium ion batteries. *ChemWorld* 2004;45(5):270–3.

[5] Hannan MA, Lipu MS, Hussain A, and Mohamedb A. A review of the lithium ion battery state of charge estimation and management system in electric vehicle applications: challenges and recommendations. *Renew Sustain Energy Rev* 2017;78:834–54.

[6] Shen WX, Tu vo T, and Ajay K. Charging algorithms of lithium-ion batteries: an overview. In: *7th IEEE Conference on Industrial Electronics and Applications*. ICIEA; 2012.

[7] Sun Q. Electric vehicles: the breakthrough of "core" is the key – talking about three bottlenecks in the development of the Li-ion battery. *Adv Mater Ind* 2010;12:62–3.

[8] Zou C, Hu X, Wei Z, and Tang X. Electrothermal dynamics-conscious lithium ion battery cell-level charging management via state-monitored predictive control. *Energy* 2017;141:250–9.

[9] Liu K, Li K, Yang Z, Zhang C, and Deng J. An advanced lithium ion battery optimal charging strategy based on a coupled thermoelectric model. *Electrochim Acta* 2017;225:330–44.

[10] Shahriari M, and Farrokhi M. Online state-of-health estimation of VRLA batteries using state of charge. *IEEE Trans Ind Electron* 2013;60(1):191–202.

[11] Lin HT, Liang TJ, and Chen SM. Estimation of battery state of health using probabilistic neural network. *IEEE Trans Ind Inf* 2013;9(2):679–85.

[12] Xiong R, Li L, and Tian J. Towards a smarter battery management system: a critical review on battery state of health monitoring methods. *J Power Sources* 2018;405:18–29.

[13] Song JY, Wang Y, and Wan C. Review of gel-type polymer electrolytes for lithium ion batteries. *J Power Sources* 1999;77(2):183–97.

[14] Lu L, Han X, Li J, Hua J, and Ouyang M. A review on the key issues for lithium ion battery management in electric vehicles. *J Power Sources* 2013; 226:272–88.

[15] Sun F, and Xiong R. A novel dual-scale cell state-of-charge estimation approach for series-connected battery pack used in electric vehicles. *J Power Sources* 2015;274:582–94.

[16] Tarascon JM, and Michel A. Issues and challenges facing rechargeable lithium batteries. *Materials for Sustainable Energy: A Collection of Peer-Reviewed Research and Review Articles From*. London: Nature Publishing Group; 2011. p. 171–9.

[17] Meena N, Baharwani V, Sharma D, *et al.* Charging and discharging characteristics of Lead acid and Li-ion batteries. *Power Energy Syst Conf: Sustain Energy* 2014. pp. 1–3.

[18] Feng X, Ouyang M, Liu X, Lu L, Xia Y, and He X. Thermal runaway mechanism of lithium ion battery for electric vehicles: a review. *Energy Storage Mater* 2017;10:246–67.

[19] Mathew M, Kong Q, Mcgrory J, and Fowler M. Simulation of lithium ion battery replacement in a battery pack for application in electric vehicles. *J Power Sources* 2017;349:94–104.

[20] Dung LR, and Yen JH. ILP-based algorithm for Lithium ion battery charging profile. In: *IEEE International Symposium on Industrial Electronics*; 2010.

[21] Zou C, Hu X, Wei Z, Wik T, and Egardt B. Electrochemical estimation and control for lithium ion battery health-aware fast charging. *IEEE Trans Ind Electron* 2018;65(8):6635–45.

[22] Chu Z, Feng X, Lu L, Li J, Han X, and Ouyang M. Non-destructive fast charging algorithm of lithium ion batteries based on the control-oriented electrochemical model. *Appl Energy* 2017;204:1240e50.

[23] Anseán D, González M, Viera JC, García VM, Blanco C, and Valledor M. Fast charging technique for high power lithium iron phosphate batteries: a cycle life analysis. *J Power Sources* 2013;239:9–15.

[24] Bhatt M, Hurley WG, and Wolfle WH. A new approach to intermittent charging of valve-regulated lead-acid batteries in standby applications. *IEEE Trans Ind Electron* 2005;52(5):1337–42.

[25] Li J, Murphy E, Winnick J, and Kohl PA. The effects of pulse charging on cycling characteristics of commercial lithium ion batteries. *J Power Sources* 2001;102(1):302–9.

[26] Vo T, Chen X, Shen WX, and Kapoor A. New charging strategy for lithium ion batteries based on the integration of Taguchi method and state of charge estimation. *J Power Sources* 2015;273:413–22.

[27] Liu YH, Hsieh CH, and Luo Y. Search for an optimal five-step charging pattern for Li-ion batteries using consecutive orthogonal arrays. *IEEE Trans Energy Convers* 2011;26(2):654–61.

[28] Luo Y, Liu Y, and Wang S. Search for an optimal multistage charging pattern for lithium ion batteries using the Taguchi approach. In: *TENCON 2009 – 2009 IEEE Region 10 Conference*. IEEE; 2009.

[29] Legrand N, Knosp B, Desprez P, Lapicque F, and Ra L. Physical characterization of the charging process of a Li-ion battery and prediction of Li plating by electrochemical modelling. *J Power Sources* 2014;245:208–16.

[30] Huang SJ, Huang BG, and Pai FS. Fast charge strategy based on the characterization and evaluation of LiFePO batteries. *IEEE Trans Power Electron* 2013;28(4):1555–62.

[31] Guo Z, Liaw BY, Qiu X, Gao L, and Zhang C. Optimal charging method for lithium ion batteries using a universal voltage method accommodating aging. *J Power Sources* 2015;274:957–64.

[32] Wen F, Jiang J, and Zhang W. Charging method for Li-ion battery pack in electric vehicles. *Automot Eng* 2008;30(9):792–5.

[33] Leitman S. *Modern Electric, Hybrid Electric, and Fuel Cell Vehicles*. Boca Raton, FL: CRC Press, 2018.

[34] Chen L, Wu S, Shieh D, and Chen T. Sinusoidal-ripple-current charging strategy and optimal charging frequency study for Li-ion batteries. *IEEE Trans Ind Electron* 2013;60(1):88–97.

[35] Partovi A. System and method for charging or powering devices, such as robots, electric vehicles, or other mobile devices or equipment. U.S. Patent: US9722447B2, 2017.

[36] Hare WM, Assenmacher MA, Hainer NJ, and Twarog CJ. Dynamic power limit adjustment in a battery charging process. U.S. Patent: US10065526B2, 2018.

[37] Hussein Ala A, Abbas A, and Samantha S., *et al.* An online frequency tracking algorithm using terminal voltage spectroscopy for battery optimal charging. *IEEE Trans Sustain Energy* 2016;7(1):32–40.

[38] Chen X, Shen WX, Cao Z, and Kapoor A. Adaptive gain sliding mode observer for state of charge estimation based on combined battery equivalent circuit model. *Comput Chem Eng* 2014;64:114–23.

[39] Chen X, Shen WX, Cao Z, and Kapoor A. A novel approach for state of charge estimation based on adaptive switching gain sliding mode observer in electric vehicles. *J Power Sources* 2014;246(3):667–78.

[40] Hua AC, and Syue BZ. Charge and discharge characteristics of lead-acid battery and LiFePO4 battery. In: *The 2010 International Power Electronics Conference (ECCE ASIA)*. 2010. pp. 1478–1483.

[41] Konopka JG. Constant current source and battery charger. U.S. Patent: US4862013A, 1989.

[42] Li Z, Xiong R, Mu H, He H, and Wang C. A novel parameter and state-of-charge determining method of the lithium ion battery for electric vehicles. *Appl Energy* 2017;207:363–71.

[43] Xiong R, Zhang Y, Wang J, He H, Peng S, and Pecht M. Lithium ion battery health prognosis based on a real battery management system used in electric vehicles. *IEEE Trans Veh Technol* 2019;68(5):4110–21.

[44] *Engineers edge, trickle charging – battery – battery application & technology* [EB/OL]. https://www.engineersedge.com/battery/trickle_charging.htm; [Accessed September 15, 2024].

[45] Xiong R, Tian J, Shen WX, and Sun F. A novel fractional order model for state of charge estimation in lithium ion batteries. *IEEE Trans Veh Technol* 2019;68(5):4130–9.

[46] Hussein A, and Batarseh I. A review of charging algorithms for nickel and lithium battery chargers. *IEEE Trans Veh Technol* 2011;60(3):830–8.

[47] Dong Q, Kang HX, and Yan YB. Effect of charge cutoff voltage on electrochemical performance of lithium ion battery. *Chem Ind Eng Prog* 2008;27 (5):770–2.

[48] Yao L, and Wang Z. A research on the charging methods of lithium ion traction battery. *Automot Eng* 2015;37(1):72–7.

[49] Hoque M, Hannan M, and Mohamed A. Optimal CC-CV charging of the lithium ion battery for charge equalization controller. In: *International Conference on Advances in Electrical, Electronic and Systems Engineering*; 2016.

[50] Ayoub E, and Karami N. Review on the charging techniques of a Li-Ion battery. In: *Third International Conference on Technological Advances in Electrical, Electronics and Computer Engineering (TAEECE)*; 2015. p. 50–5.

[51] Liu C, Wang S, Liu Y, and Tsai M. An optimum fast charging pattern search for Li ion batteries using particle swarm optimization. In: *The 6th International Conference on Soft Computing and Intelligent Systems, and the 13th International Symposium on Advanced Intelligence Systems*; 2012. p. 727–32.

[52] Sikha G, Ramadass P, Haran B, White R, and Popov B. Comparison of the capacity fade of Sony US 18650 cells charged with different methods. *J Power Sources* 2003;122(1):67–76.

[53] Chen L, Roy CH, and Liu C. A design of a grey-predicted Li-ion battery charge system. *IEEE Trans Ind Electron* 2008;55(10):3692–701.

[54] Xu J. Design of constant current-constant voltage mode regulation in Li-ion battery charger. *Semicond Technol* 2011;36(4):291–5.

[55] Liu Y, Teng J, and Lin Y. Search for an optimal rapid charging pattern for lithium ion batteries using ant colony system algorithm. *IEEE Trans Ind Electron* 2005;52(5):1328–36.

[56] Wang S, and Liu Y. A PSO-based fuzzy-controlled searching for the optimal charge pattern of Li-ion batteries. *IEEE Trans Ind Electron* 2015;62 (5):2983–93.

[57] Ikeya T, Sawada N, Murakami J, *et al.* Multi-step constant-current charging method for an electric vehicle nickel/metal hydride battery with high-energy efficiency and long cycle life. *J Power Sources* 2002;105(1):6–12.

[58] Wang H, Li G, Jiang Z, and Shen J. Research on fast charge method for lithium ion battery. *Chin J Power Sources* 2012;36(11):1616–19.

Chapter 2

Improvement of AC network performance using shunt and hybrid FACTS devices

Mankour Mohamed[1], *Ghomri Leila*[2], *Miloudi Mohamed*[1] *and Sahnoun M'hamed*[3]

Abstract

This chapter provides an investigation into AC power grid behavior as impacted by shunt flexible AC transmission systems (FACTS) devices. The primary objective is to demonstrate the necessity of nodal voltage control and reactive power compensation. The study examines three different shunt FACTS devices: the thyristor valve-based static var compensators (SVC); the insulated gate bipolar transistor/gate turn-off thyristor (IGBT/GTO) valve-based static synchronous compensator (STATCOM); and the unified power flow controller (UPFC). To assess the response of the FACTS devices to voltage profile and reactive power, three test cases are conducted, incorporating computations into the load flow both with and without FACTS devices. Overall, the findings of this study provide valuable insights into the Shunt FACTS devices' impact in improving the performance of AC power grids. The correction of voltage violations can be achieved through the use of transformer ratios or by adjusting the capacitor banks' reactive power output. Additionally, the placement of FACTS devices in a certain area can be utilized to demonstrate and examine these devices' impact in comparison to static equipment like capacitor banks and synchronous compensators.

In order to test these methodologies, two AC networks were analyzed using the Newton–Raphson method and MATLAB®/Simulink® environment. The first network was evaluated on the IEEE-05 bus, while the second network was assessed on the IEEE-57 bus. The results indicate the efficacy of the approach with a rapid number of iterations. Furthermore, this study explores the impact of shunt and hybrid FACTS devices on the AC system behavior, particularly when such devices are installed in an optimal location.

[1]Department of Electrical Engineering and Automation, GIDD Laboratory, University of Relizane, Algeria
[2]Department of Electrical Engineering, University of Abdelhamid Ibn Badis, Mostaghanem, Algeria
[3]Department of Electrical Engineering, University of Ghardaia, Algeria

Keywords: Voltage control; reactive power compensation; shunt FACTS devices (SVC; STATCOM; UPFC)

2.1 Introduction

The demand for electrical energy has been rising steadily and consistently, necessitating the distribution of this energy to various points of consumption in order to satisfy consumer needs. Due to the thermal limits of existing lines and transit capacity, expanding and adding new lines to the electrical network were essential. However, conventional control devices were not capable of providing reliable energy control in highly interconnected networks. As reviewed by Gandoman *et al.* (2018) and Biswas and Nayak (2018), this posed a challenge to ensuring optimal energy distribution. Therefore, novel solutions were needed to address this issue and ensure the efficient and reliable distribution of electrical energy.

The progress of power electronics has had a significant impact on enhancing the administration of AC power networks, specifically concerning their maximum functional limit, through the regulation of their parameters. This is accomplished by regulating the AC power grid's voltage, active power, and reactive power, while keeping these parameters at predetermined values. The use of flexible AC transmission systems (FACTS), which are control devices based on power electronics, has made this possible (Gomis-Bellmunt *et al.*, 2019; Mathur and Varma, 2002). In the AC systems, the role of FACTS devices is crucial in managing power transits and ensuring safe operations during various conditions (Reddy, 2018).

In 1988, EPRI, an American company, initiated a research project aimed at examining the FACTS systems for optimal regulation of power transmission in power lines, thereby augmenting their transfer capabilities and introducing a fresh prospect of power electronics in control (Hingorani, 2007).

The application of traditional control using electromechanical control systems has resulted in a range of operational problems faced by large AC electrical networks. This results from the slower reaction time of various control systems when compared to the newer FACTS systems that rely on power electronic components and static switches. Unlike traditional systems, FACTS devices are capable of providing a short response time in all states including static, transient, and dynamic conditions. Based on the makeup of their power electronic components, the two kinds of FACTS devices that are present in real power systems are categorized, which can either be based on thyristor or IGBT/GTO semiconductors, as found in the works of Biswas and Nayak (2018), Grainger (1999), and Yazdani and Iravani (2010).

The examination of power dissemination within an electrical network comprises multiple generators, transmission lines, and loads holds significant importance in the investigation, preparation, and operation of an electrical network. The estimation of power flow encompasses the determination of all the power transitions and voltages within the network for a specific load. Each node of the system is associated with four variables: active and reactive powers, module, and voltage phase. In a bus, only two variables are discernible, while the remaining two are ascertained during computation (Sauer *et al.*, 2017).

In order to address this issue, it is imperative to ascertain the prerequisites for the stable functioning of a power system. These prerequisites include the development of an appropriate mathematical framework, identification of a set of variables and constraints within the system, and the computational resolution of the system (Mathur and Varma, 2002).

The shunt FACTS devices are the main topic of this research, namely SVC, STATCOM, and unified power flow controller (UPFC), and examine their configuration and reaction toward AC system behavior. A comparison of these devices is presented, with particular emphasis on their ability to decrease voltage flicker in high-voltage stations, as described by Samet and Jarrahi (2015) and García-Cerrada *et al.* (2000). Additionally, Demirovic (2016) demonstrates a comparison between the two devices in voltage boosting versus wind farms. This work aims to highlight the role of shunt FACTS devices in improving AC system performance by keeping reactive power and voltage within reasonable levels across large power networks. Each scenario study uses a specific shunt FACTS device to showcase its influence on the system behavior.

This chapter investigates the power flow, voltage profile, active and reactive power, as well as losses for two distinct networks, namely IEEE-05 bus and IEEE-57 bus, with and without the integration of shunt FACTS devices. The calculations are carried out using the Newton–Raphson method due to its strength and rapid convergence. MATLAB and MATLAB/Simulink environment are utilized to model the different systems.

2.2 Mathematical representation of the power system

The mathematical expressions that establish the relationship between the nodal voltages and the currents that are introduced into a network comprising n buses are as follows:

$$I = Y.V \; I_i = \sum_{j=1}^{n} Y_{ij}.V_i \quad i = 1, \ldots, n \tag{2.1}$$

$$S_i = P_i + Q_i = V_i.I_i^* \tag{2.2}$$

$$S_i^* = P_i - Q_i = V_i.\sum_{j=1}^{n} Y_{ij}.V_i \tag{2.3}$$

2.2.1 Newton–Raphson method

The approach utilized in this study is founded upon the Taylor series expansion as outlined in the work of Acha *et al.* (2004). Using this method, consecutive first-order approximations are obtained.

$$f(x) \approx f\left(x^{k}\right) + f'\left(x^{k}\right).\left(x^{k}+^{1} - x^{k}\right) = 0 \tag{2.4}$$

where $f' = \frac{\partial f}{\partial x}$ is the Jacobin of the $f(x))$ from an initial value x^{0}, corrections Δx^{K} arise from the linear system's solution:

$$-f'\left(x^{k}\right).\Delta x^{k} = f\left(x^{k}\right) \tag{2.5}$$

and the new value of x^{k+1} is:

$$x^{k+1} = x^{k} + \Delta x^{k} \tag{2.6}$$

The voltage's angle and magnitude in the network have been adjusted according to the following equations:

$$\Delta P_i = P_i^{spe} - P_i^{cal} = V_i \sum_{j=1}^{n} V_j(G_{ij} \cos\ \theta_{ij} + B_{ij} \sin\ \theta_{ij}) \tag{2.7}$$

$$\Delta Q_i = Q_i^{spe} - Q_i^{cal} = V_i \sum_{j=1}^{n} V_j(G_{ij} \sin\ \theta_{ij} - B_{ij}\cos\ \theta_{ij}) \tag{2.8}$$

By utilizing the aforementioned notation and partitioning the Jacobian into smaller sub-matrices, (2.5) can be effectively transformed into a matrix system, specifically one that is expressed in (2.9) and (2.10), which is particularly useful for executing load flow calculations:

$$\begin{bmatrix} \Delta P \\ \Delta Q \end{bmatrix}^k = \begin{bmatrix} H\ N \\ M\ L \end{bmatrix}^k \cdot \begin{bmatrix} \Delta\theta \\ \Delta V \end{bmatrix}^k \tag{2.9}$$

The variable ΔV can be divided by V:

$$\begin{bmatrix} \Delta P \\ \Delta Q \end{bmatrix}^k = \begin{bmatrix} H\ N \\ M\ L \end{bmatrix}^k \cdot \begin{bmatrix} \Delta\theta \\ \frac{\Delta V}{V} \end{bmatrix}^k$$
$$H_{ij} = \frac{dP_i}{d\theta_j}, M_{ij} = \frac{dQ_i}{d\theta_j}, N_{ij} = \frac{dP_i}{dV_j}.V_j, L_{ij} = \frac{dQ_i}{dV_j}.V_j \tag{2.10}$$

And (2.6) is transformed into a matrix as follows:

$$\begin{bmatrix} \theta \\ V \end{bmatrix}^{k+1} = \begin{bmatrix} \theta \\ V \end{bmatrix}^k \cdot \begin{bmatrix} \Delta\theta \\ \Delta V \end{bmatrix}^k \tag{2.11}$$

The Jacobian matrix may be found using the following equation. For $i = j$,

$$H_{ii} = -Q_i - B_{ii}.V_i^2 \tag{2.12}$$

$$Mii = Pi - Gii.Vi2 \tag{2.13}$$

$$Nii = -Pi + Gii.Vi2 \tag{2.14}$$

$$Lii = -Qi + Bii.Vi2 \tag{2.15}$$

when $i \neq j$,

$$H_{ij} = V_i.V_j\left(G_{ij}sin\theta_{ij} - B_{ij}cos\theta_{ij}\right) \tag{2.16}$$

$$N_{ij} = V_i.V_j\left(G_{ij}cos\theta_{ij} + B_{ij}sin\theta_{ij}\right) \tag{2.17}$$

$$L_{ij} = H_{ij} \tag{2.18}$$

$$M_{ij} = N_{ij} \tag{2.19}$$

Passive compensating devices such as shunt capacitors, shunt reactors, and series capacitors are commonly represented as admittance elements or fixed values in the context of power flow and stability assessments. Synchronous condensers, however, are represented as synchronous generators that lack steady-state active power output. These modeling strategies have been outlined in prior literature (Alassouli, 2020; Mathur and Varma, 2002).

2.3 FACTS devices

2.3.1 General concept and definition

The IEEE provides a definition for FACTS, which pertains to the AC transmission system and involves the use of power electronic devices or a static controller to enhance the control, stability, and line transfer capacity of the AC system. These beneficial attributes stem from the FACTS devices' capacity to govern the parameters denoted in (2.20) and (2.21) and depicted in Figure 2.1.

$$P = \frac{E_1 E_2}{X}\sin\left(\delta_1 - \delta_2\right) \tag{2.20}$$

$$Q = E_1 \frac{\left(E_1 - E_2\sin\left(\delta_1 - \delta_2\right)\right)}{X} \tag{2.21}$$

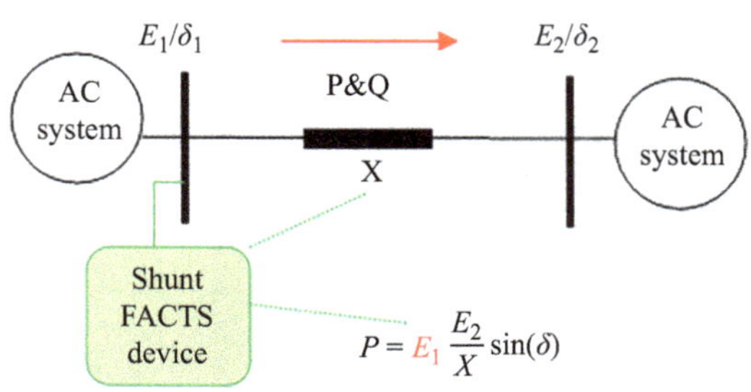

Figure 2.1 Shunt FACTS device regulation

When contemplating the management of voltage and compensating for reactive power, it is necessary to address the shunt device, which is elaborated upon in Figure 2.1. Significantly, the regulation of bus voltage has an indirect and dynamic influence on the active power flux.

The organization of a device is determined by its connection within the AC network. Typically, FACTS devices can be divided into three major groupings of compensation: series, shunt, and hybrid (series-shunt) (Sood, 2006).

An alternative classification scheme for FACTS devices is possible based on their specific type or constituent components. This categorization divides the devices into two distinct categories, the first of which comprises traditional high voltage direct current (HVDC) devices consisting of passive elements and semiconductor thyristors such as static var compensators (SVCs), thyristor-controlled reactors (TCRs), thyristor switched reactors, thyristor control breaking resistors, thyristor controlled series capacitors, thyristor switched series capacitors, thyristor controlled series reactors, thyristor switched series reactors, thyristor controlled phase angle regulators, and static phase shifters (Gomis-Bellmunt *et al.*, 2019; Noroozian *et al.*, 1997).

The second classification involves FACTS devices which predominantly comprise IGBT/GTO semiconductors. These devices commonly include the STATCOM, static synchronous series compensator (SSSC), interline power flow controller, and UPFC (Acha *et al.*, 2004; Biswas and Nayak, 2018; Yazdani and Iravani, 2010).

This chapter delves into an analysis of two distinct varieties of shunt FACTS devices, namely those equipped with thyristors and those with IGBT/GTO components, with the aim of controlling the AC power network's reactive power and voltage profile.

2.3.2 Modeling of the FACTS devices

2.3.2.1 Static var compensator

SVCs are devices that are connected in parallel with the AC power system to either generate or absorb reactive power, thereby regulating specific parameters of the system (Dash *et al.*, 2020; Mankour *et al.*, 2021). An SVC comprises two main components, namely TCR and thyristor-switched capacitor (refer to Figure 2.2).

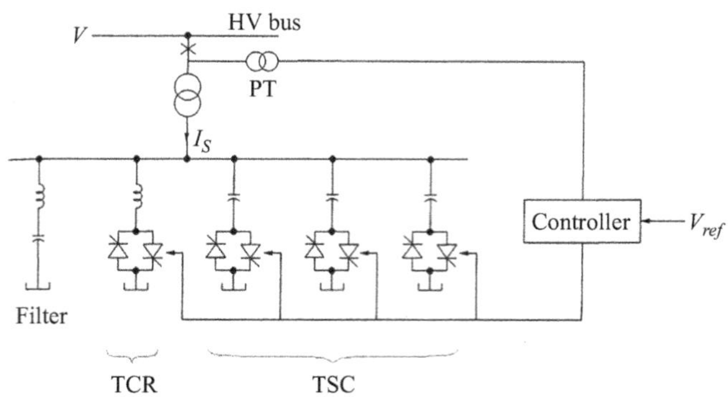

Figure 2.2 Schematic of typical SVC

In addition, a hybrid compensator includes capacitor banks and a harmonic filter that is used to filter out TCR-generated harmonics.

The SVC is constructed by utilizing thyristor valves that are inherently commutated in the frequency of the AC system. In actual installations, two types of SVC exist: industry SVC, which is almost always associated with unbalanced loads, and SVCs for transmission lines in AC systems. SVCs enable constant and swift regulation of reactive power and voltage, and also offer an efficient and economical solution to address voltage flicker problems (Mankour *et al.*, 2021). The graphical representation of the steady-state *V/I* and *V/Q* features of the SVC can be observed in Figure 2.3. The linear control range is established by the reactor maximum susceptance (B_{Lmax}). The comprehensive capacitive susceptance (B_c) is determined by the combination of filter capacitance and capacitor banks (Acha *et al.*, 2004).

In this particular study, as evidenced by the visual aid provided in Figure 2.4, the SVC device comprises a controlled variable that pertains to the overall susceptance ($b_S V\ C$) and assumes a temporal regulator as depicted in Figure 2.5. Within this framework, the differential equation takes the following form:

$$\dot{b}_{SVC} = \frac{K_r\left(V_{ref} - V\right) - b_{SVC}}{T_r} \tag{2.22}$$

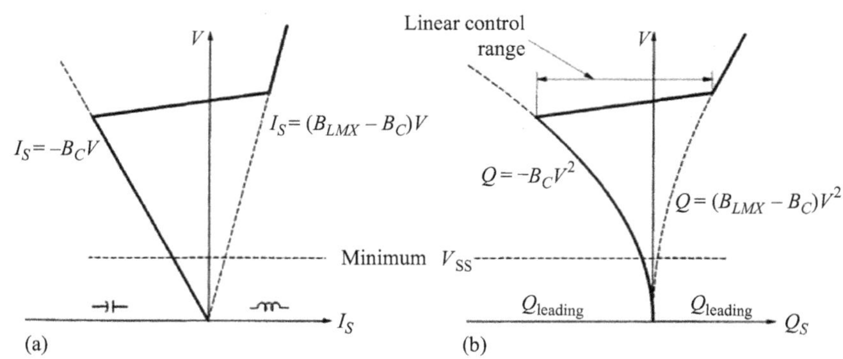

Figure 2.3 SVC steady-state characteristic. (a) Voltage–current characteristic and (b) voltage–reactive power characteristic

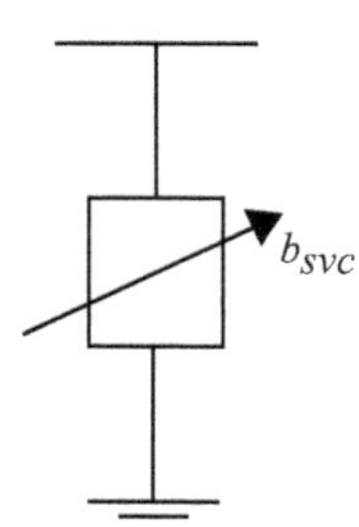

Figure 2.4 SVC equivalent circuit

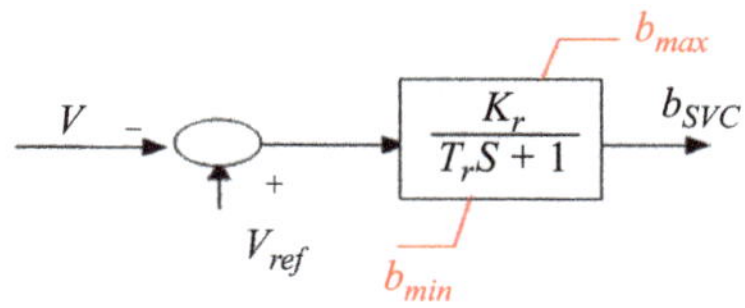

Figure 2.5 SVC regulator (type 1)

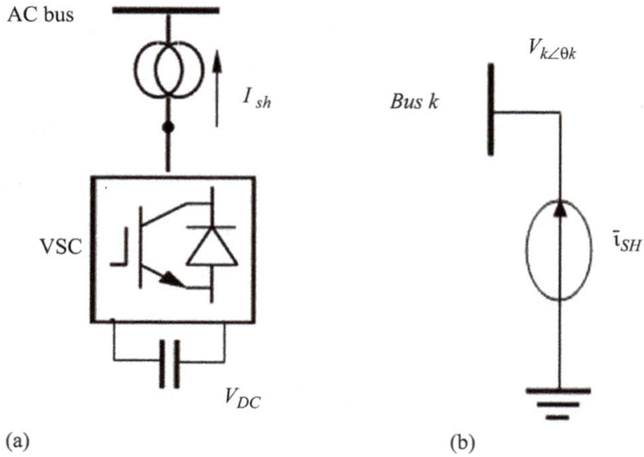

Figure 2.6 STATCOM (a) model, (b) equivalent circuit

The variables present in this system include K_r, which is the regulator gain, T_r, the regulator time constant V_{ref}, the voltage reference, and b_{max} and b_{min}, which represent the maximum and minimum susceptance, respectively.

The present model is finalized through the incorporation of the algebraic equation that represents the reactive power injected at the specific bus where the SVC is linked.

$$Q = -bsvcV^2 \tag{2.23}$$

2.3.2.2 STATCOM (STATic COMpensator)

The IGBT or GTO semiconductor valves link the shunt device, known as the STATCOM. It is recognized as the voltage source converter (VSC) VSC-FACTS device, as per the works of Chun *et al.* (1998) and Yazdani and Iravani (2010). The STATCOM device bears a functional resemblance to the SVC equipment; yet its internal design is somewhat distinct. Specifically, the STATCOM comprises three primary components, as indicated in Figure 2.6, namely (i) the DC network, (ii) the VSC, and (iii) the controllers.

The STATCOM has been suggested as a simplified current injection model by Yazdani and Iravani (2010) and Haque (2004). Keeping the STATCOM current in

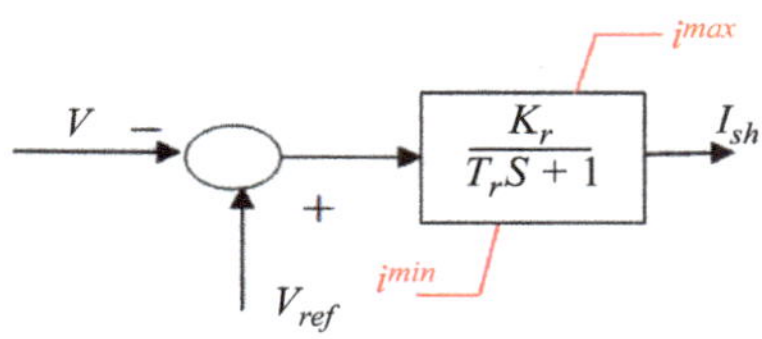

Figure 2.7 STATCOM voltage regulator

quadrature with the bus voltage guarantees that the only power transferred between the AC power grid and the STATCOM device is reactive power (Acha *et al.*, 2004). The equivalent circuit of the model can be observed in Figure 2.7.

The depiction of the voltage regulator can be observed in Figure 2.7, while the equations governing the differential current and reactive power of the STATCOM are, respectively, presented as follows:

$$\dot{i}_{SH} = \frac{K_r(V_{ref} - V) - i_{SH}}{T_r} \tag{2.24}$$

$$Q = -i_{SH}V \tag{2.25}$$

The variables involved in the equation are as follows: I_{SH} indicates the current that is injected, V_{ref} denotes the voltage reference, K_r represents the regulator gain, T_r signifies the regulator time constant, and I_{min} and I_{max} correspond to the minimum and maximum currents, respectively.

2.3.2.3 Unified power flow controller

The UPFC, also called the universal regulator phase shifter or universal load dimmer, is considered as one of the most robust FACTS components. It is composed of two compensators, namely the series and parallel compensators, as well as a phase shifter. Essentially, the UPFC has the capacity to provide the functionality of three compensators. This has been highlighted by Mathur and Varma (2002).

The UPFC model is achieved through the integration of STATCOM and SSSC devices. This model comprises a series voltage source (v_s) and a shunt current source (parallel) i_{SH}, as depicted in Figure 2.8. The parallel and serial sources are defined as follows: the voltage angle $\delta(cr)$ that stems from the serial converter plays a vital role in determining the power distribution control mode and voltage in the different nodes (Gandoman *et al.*, 2018; Gupta *et al.*, 2017; Mathur and Varma, 2002). The illustration of the UPFC operation's fundamental notion is depicted in Figure 2.9, and it can be succinctly conveyed as follows:

1. if δ_{cr} in phase with θ_k: UPFC regulates voltage at bus k, UPFC behaves as an voltage regulator (Figure 2.9(a)).
2. if δ_{cr} is in quadrature with θ_k: UPFC controls the active power, UPFC behaves as a phase shifting transformer (Figure 2.9(b))

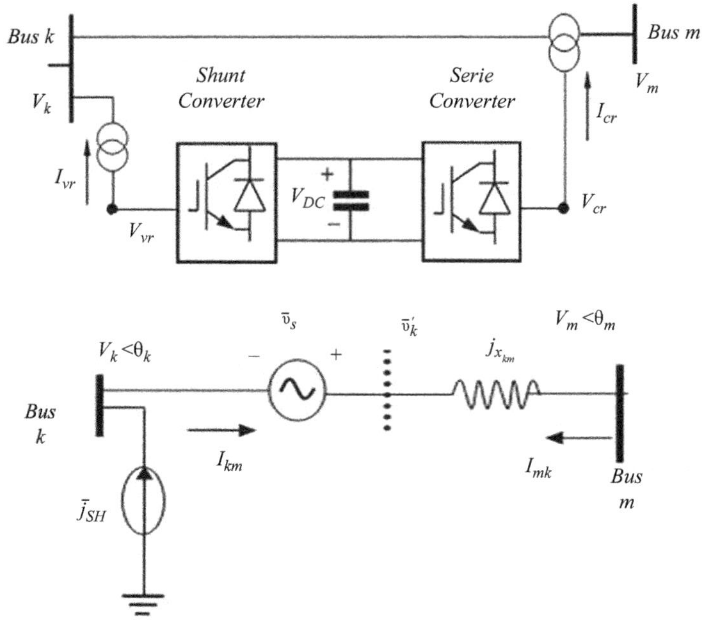

Figure 2.8 (a) Basic schema of the UPFC and (b) UPFC circuit equivalent

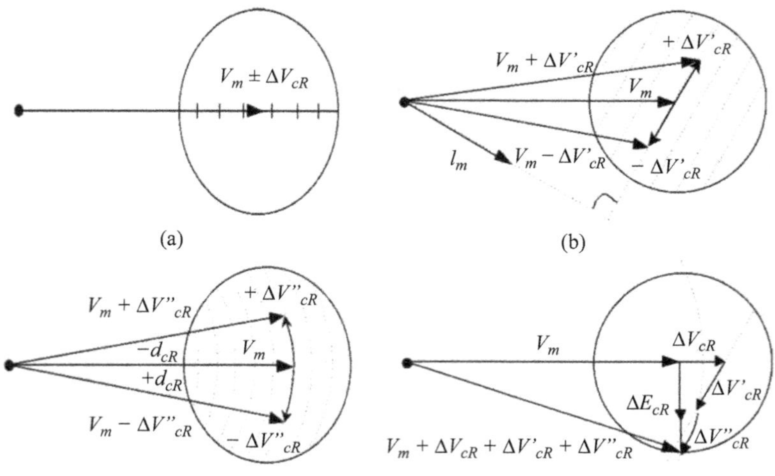

Figure 2.9 General concept of UPFC operations

3. if δ_{cr} is in quadrature with line current angle: UPFC controls the active power, UPFC behaves as an impedance regulator (series compensator) (Figure 2.9(c)).
4. if δ_{cr} is another value: The UPFC integrates the functions of the three previous cases: voltage regulator, series compensator, and phase-shifting transformer. (Figure 2.9(d)).

The UPFC governs the principal power distributions which are defined by the voltage module that is inserted in a sequential manner.

The vector diagram circuit that is equivalent to the series voltage source is illustrated in Figure 2.8(a) and (b), correspondingly. The power equations that depict the power injection model of the UPFC are articulated as follows:

$$P_{km} = b_r V_k V_m sin(\gamma + \theta_k - \theta_m) \tag{2.26}$$

$$Qkm = brVk2cos\gamma - iqVk \tag{2.27}$$

$$P_{km} = -b_r V_k V_m sin(\gamma + \theta_k - \theta_m) \tag{2.28}$$

$$P_{km} = -b_r V_k V_m cos(\gamma + \theta_k - \theta_m) \tag{2.29}$$

The UPFC comprises two three-phase voltage inverters, wherein one is linked in parallel to the grid via a transformer and the other is attached in series through a second transformer. The inverters are interlinked by a DC bus that is denoted by a capacitor (V_{dc}).

The UPFC has the ability to manage active power and reactive power in a simultaneous and independent manner, as reported by a previous study (Gupta *et al.*, 2017). In addition, this device can regulate the three fundamental parameters linked to power transmission, namely:

- The line voltage.
- The line impedance.
- The phase shift of the voltages at the ends of the line.

2.4 Implementation and interpretation of results

The utilization of the Newton–Raphson methodology is initially executed on the IEEE-05 bus network and subsequently on the IEEE-57 bus network. The conception, modeling, and simulation of both networks are proficiently conducted using the MATLAB and MATLAB/Simulink environment.

Table 2.1 provides a comprehensive overview of the two systems. The study presents distinct outcomes, scrutinizes the power distribution, and evaluates voltage control within the network both with and without the integration of FACTS devices such as SVC, STATCOM, and UPFC. All the shunt FACTS devices examined in this research aim to preserve the voltage at 1 pu for the connected buses while disregarding the reactive power produced or absorbed at these locations.

- **AC GRID 01**: The IEEE-05 bus network has been documented by Acha *et al.* (2004) and Gandoman *et al.* (2018). Active and reactive power generation after three different testing scenarios, with and without FACTS device integration, are presented in Table 2.2. The initial test case, which is the base-load flow

Table 2.1 Networks description

	IEEE-05 bus	IEEE-57 bus
Load bus	4	48
Compensation bus	2 bus (1,2)	4 bus (1, 3, 8, and 12)
Compensation bus	–	3 (2, 6, and 9)
Transformer	–	18
Lines	7	60
Condenser batteries	–	3 batteries (15, 25, and 53)

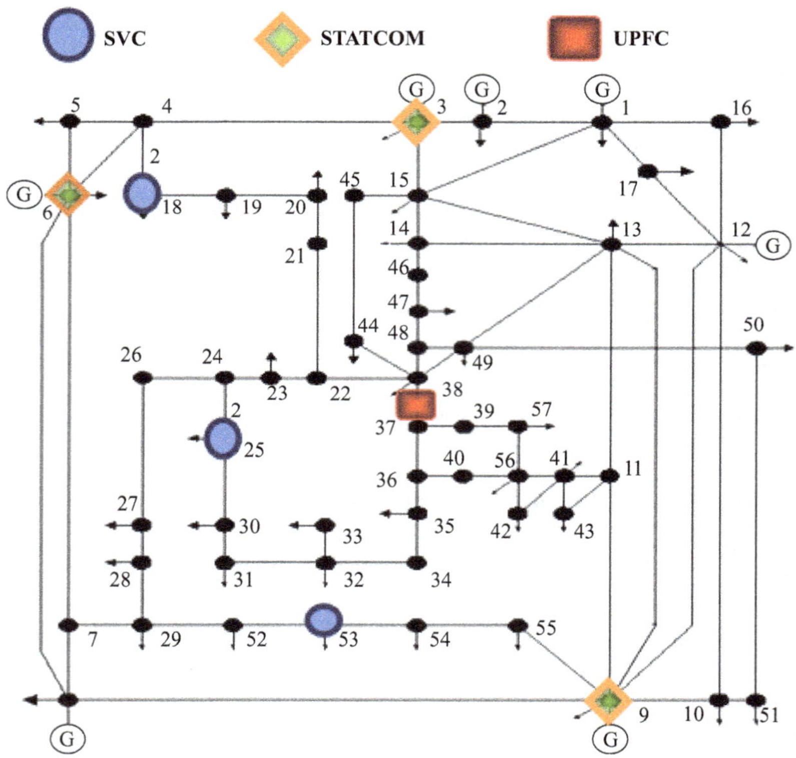

Figure 2.10 IEEE 57-bus network

case, has produced results that are complicated with the outcomes reported by Acha *et al.* (2004). The simulation has been conducted in three different scenarios as mentioned below:

1. Scenario 1: First findings (load flow basis case).
2. Scenario 2: Adjusting the voltage by installing a bank of capacitors in bus number 3.

Table 2.2 Active, reactive power generation, and active power losses

Case number			Bus number		P_{Losses} (MW)
			1	2	
Scenario 1		P_G (MW)	130.12	40	6.122
		Q_G (MVAR)	90	−61.59	
Scenario 2		P_G (MW)	131.06	40	6.056
		Q_G (MVAR)	85.33	−77.11	
Scenario 3	(a)	P_G (MW)	131.06	40	6.056
Q_G (MVAR)		85.34	−77.07		
	(b)	P_G (MW)	131.06	40	6.056
		Q_G (MVAR)	85.34	−77.07	
	(c)	P_G (MW)	131	40	6.056
		Q_G (MVAR)	85.3	−75.35	

3. Scenario 3: Outcomes following the addition of FACTS devices.
   (a) substituting SVC for the capacitor bank.
   (b) substituting STATCOM for the capacitor bank.
   (c) substituting UPFC for the capacitor bank.

- **AC Grid 02**: The IEEE-57 bus network (depicted in Figure 2.10) is subjected to the load flow method for the purpose of exploring, investigating, and analyzing the performance of the AC network (specifically the voltage profile and reactive power compensation) when shunt or hybrid FACTS devices are incorporated. The limits of the system (including voltages, transformer ratios, reactive powers, and capacitors) are presented in Table 2.3. The simulation outcomes are conducted in three scenarios as follows:
  1. Scenario 1: Initial results (base case-load flow) (Mankour and Sami, 2019)
  2. Scenario 2: Results after voltage corrections (Mankour *et al.*, 2021).
  3. Scenario 3: Results after integration of FACTS devices.
     (a) Replacing the capacitors (18, 25, and 53) by SVCs.
     (b) Replacing the compensators (3, 6, and 9) by STATCOMs.
     (c) Including an UPFC into the system between 37 and 38 buses.

Scenario Nbr 01: In this study, we showcase the outcomes of the base scenario which pertains to the load flow. A comprehensive analysis of the initial condition and the voltage profile is presented in Figure 2.11.

In this instance, as depicted in Figure 2.11, it is observed that buses 25, 30, 31, 32, 34, 35, 36, 37, 39, 40, 42, 56, and 57 have surpassed their prescribed minimum threshold of 0.9 pu. The most egregious violation is exhibited by bus 31.

Table 2.3 Limits of voltage, reactive power, and capacitors

	Min	Max
$V_1 \ldots\ldots\ldots V_{57}$	0.9	1.1
$a_1 \ldots\ldots\ldots a_{18}$	0.9	1.1
$Q2$	−17.00	50.00
$Q3$	−10.00	60.00
$Q6$	−8.00	25.00
$Q8$	−140.00	200.00
$Q9$	−15.00	90.00
$Q12$	−50.00	155.00
$C18$	0	20
$C25$	0	25
$C53$	0	10

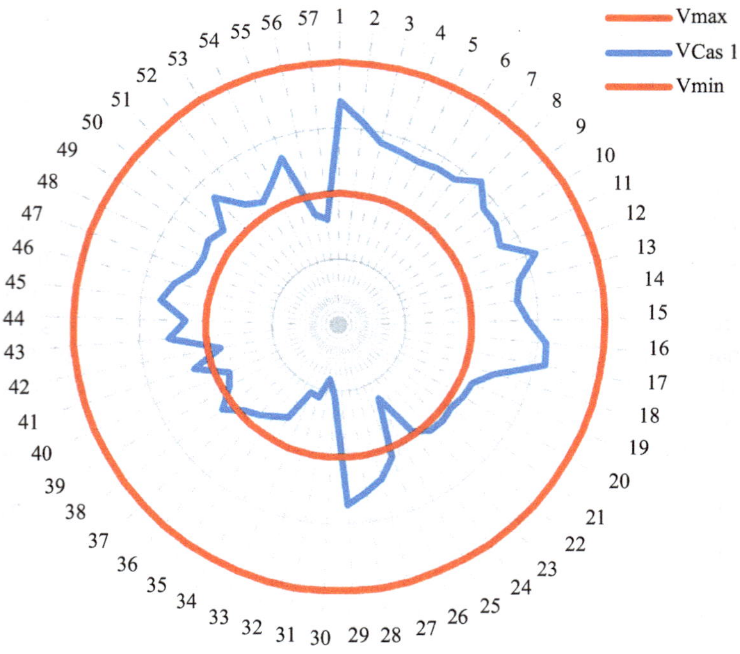

Figure 2.11 Voltage profile of Scenario Nbr 1

Scenario Nbr 2: The powers generated and voltage profile are presented after voltage correction in Figure 2.12.

In an attempt to regulate the voltage profile within the network, we have endeavored to manipulate the reactive power output of both capacitor banks and transformers. Specifically, we initiated corrective measures by injecting reactive power into the system via a 1 MVAR step to the capacitor located at bus 25, ultimately resulting in the production of 20.5 MVAR of reactive power (Table 2.4).

The subsequent course of action entailed the modification of the transformers situated between the bus numbers 11-41, 11-3, and 56-57. The transformation ratio underwent a decrease of 0.04, with the alteration being carried out in steps of 0.01. The outcomes of this endeavor are depicted in Figure 2.12 and Table 2.5.

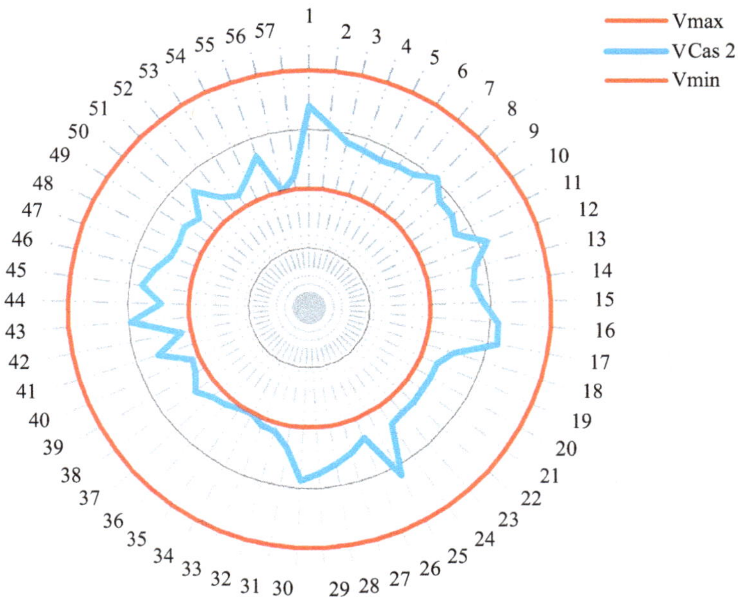

Figure 2.12 Voltage profile of Scenario Nbr 2

Table 2.4 Scenario Nbr 1 power results

	Active power [MW]	**Reactive power [MVAR]**
Generation	1280.09	395.66
Load	1250.8	319.12
Losses	29.29	76.5

Table 2.5 Scenario Nbr 2 power results

	Active power [MW]	**Reactive power [MVAR]**
Generation	1279.63	375.29
Load	1250.8	301.85
Losses	28.827	73.447

Scenario 3: Integration of FACTS devices

1. **Integration of SVCs**: Here, we demonstrate how SVCs may be used in place of the capacitors (18, 25, and 53). Figure 2.13 the voltage profile is displayed in graphical form, with accompanying results for active and reactive powers generated as well as losses, as shown in Table 2.6.

Upon correction of nodal voltage violations, we observe changes in the values of both active and reactive powers in comparison to the second scenario. Additionally, we note a substantial increase in reactive power generation of approximately 36 MVAR, which is offset by the implementation of SVCs in place of capacitor banks. Specifically, at buses 15, 25, and 53, the reactive powers are, respectively, generated by the SVCs which amount to 19.7, 18.52, and 38.54 MVAR.

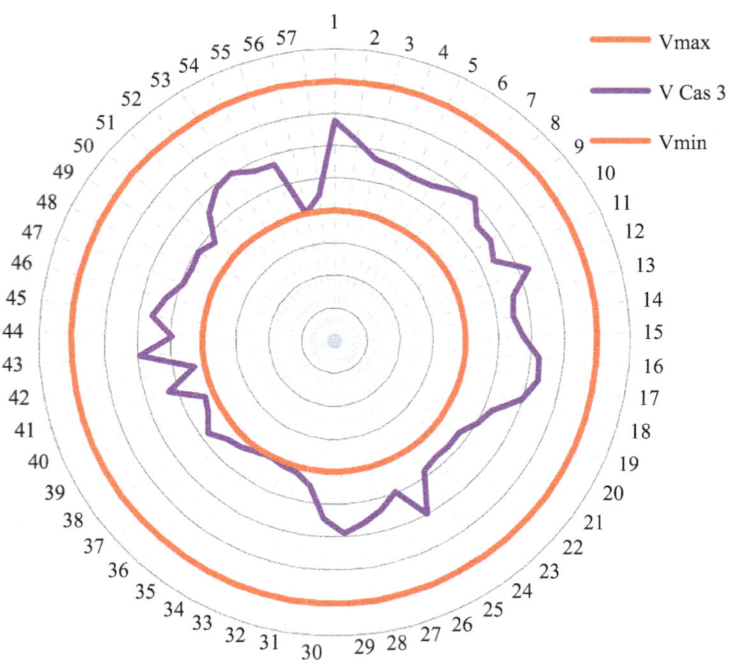

Figure 2.13 Buses voltages of scenario Nbr 3-1

Table 2.6 Scenario Nbr 3-1 Power results

	Active power [MW]	Reactive power [MVAR]
Generation	1280.32	410.91
Load	1250.8	337.4
Losses	29.51	73.51

2. **Integration of STATCOMs**: During this experience, STATCOMS are replaced by compensators in the respective buses (3, 6, and 9). The implementation of STATCOMs in place of synchronous compensators (2, 6, and 9) has ensured the maintenance of bus voltage at 1 pu in the presence of these devices, as illustrated in Figure 2.14. This step has yielded results that are in line with the previous step, given that the principles underlying the characteristics of the SVC and STATCOM devices are similar (Table 2.7).

In addition, this replacement aims to assess how synchronous compensators and capacitor banks affect the voltage profile of the AC system in comparison to the parallel connected FACTS devices.

3. **Integration of a UPFC**: In the present scenario, we offer the integration of a sole UPFC apparatus connecting the two junctions numbered 37 and 38 (Table 2.8).

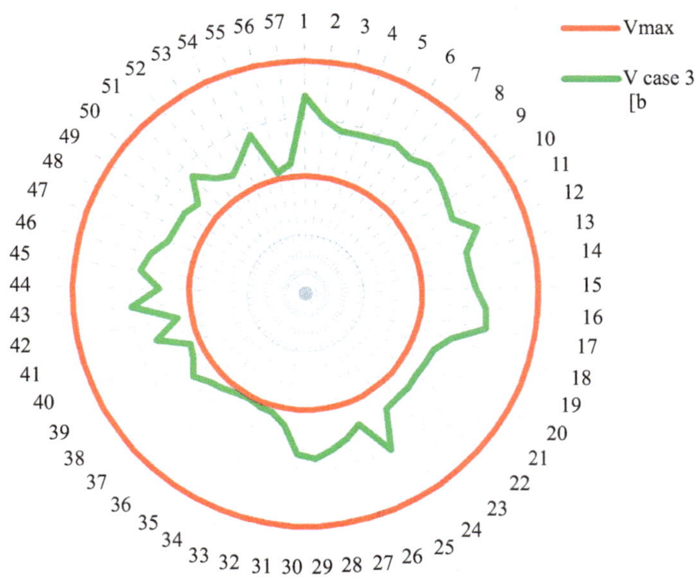

Figure 2.14 Buses voltages of scenario Nbr 3-2

Table 2.7 Scenario Nbr 3-2 power results

Active power [MW]	Reactive power [MVAR]	
Generation	1280.05	374.82
Load	1250.8	301.09
Losses	1250.8	301.09

Table 2.8 Scenario Nbr 3-3 power results

	Active power [MW]	**Reactive power [MVAR]**
Generation	1278.89	367.41
Load	1250.8	297.82
Losses	28.15	69.59

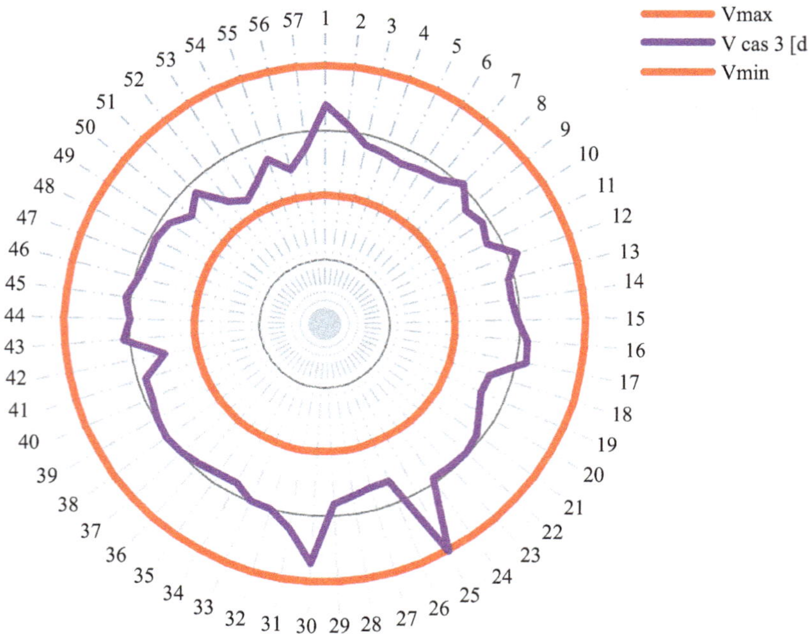

Figure 2.15 Voltage profile of Scenario Nbr 3-3

To improve the voltage profile of the entire power network while maintaining acceptable levels of reactive power, it is crucial to strategically install the UPFC at an optimal location in the network (Figure 2.15). This is given that the UPFC has a major effect on all system parameters, including active and reactive power, voltage, and current, as well as losses. Given the high cost of this equipment, it is imperative to conduct a thorough analysis of the implementation of the device for a real AC power grid.

Several simulations were conducted for various locations of the UPFC in the AC network, with only six positions depicted in Figure 2.16. The voltage profile of the system was found to be optimal when the UPFC was situated in branches 37–38 and 38–44, as evidenced by the results obtained. Additionally, these positions yielded the lowest power losses relative to the other branches, as indicated in Table 2.9. Therefore, it can be concluded that the UPFC is optimally placed within the network.

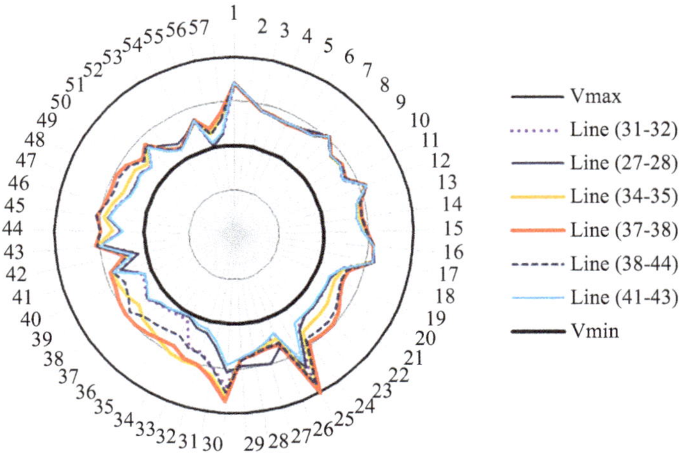

Figure 2.16 Voltage profile for different location of UPFC

Table 2.9 Power of the system according to UPFC location

	Location of the UPFC						
	21–22	**27–28**	**31–32**	**34–35**	**37–38**	**38–44**	**41–43**
Reactive power generated by UPFC	42.18	25.13	5.39	31.85	69.76	61.29	4.06
Active losses [MW]	28.60	28.67	28.37	28.97	28.15	28.02	28.83
Reactive losses [MVAR]	71.44	72.66	72.87	72.41	**69.59**	69.85	72.25

Note: The bold value indicates the best minimum reactive power losses.

The line 37–38 experiences a 3% decline in active power and a 60% decrease in reactive power. This reduction is due to the UPFC device and the system's response to the modified active power, and is also influenced by the device's shunt characteristic. The UPFC device is capable of producing its own reactive power, resulting in a 1% and 20 MVAR decrease in reactive power generated at buses 3 and 9, respectively, with only minimal changes in power observed at other buses.

The observed variations in reactive powers among different scenarios can be attributed to the response of the FACTS devices, namely SVC, STATCOM, and UPFC, to the reactive power. Additionally, a marginal decline in active losses is detected subsequent to the rectification of the voltage profile, which is evident in Scenario 2, as depicted in Figure 2.17. It is worth noting that the implementation of SVC and STATCOM devices (Scenarios 3-a and 3-b) resulted in a 1 MW surge in power consumption, owing to the individual components of the devices such as static switches, thyristors, IGBT/GTO, and controllers (Table 2.10).

Through the examination of various scenarios involving FACTS devices in electrical systems, it has been determined that shunt FACTS devices have the

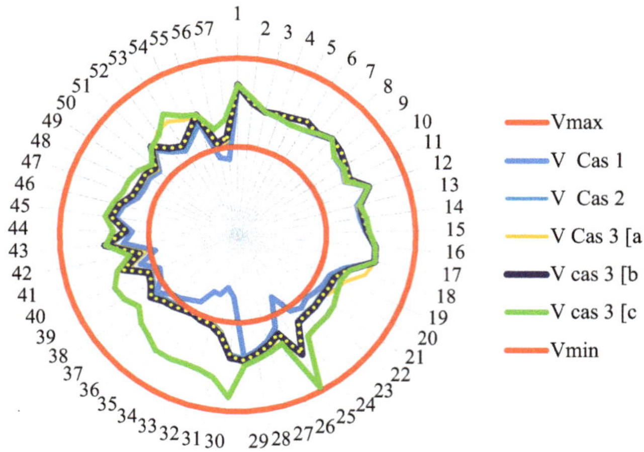

Figure 2.17 Buses voltages of the three scenarios 1, 2, and 3

Table 2.10 Total active power losses

Scenarios	**Scenario 1**	**Scenario 2**	**Scenario 3**		
			(a)	**(b)**	**(c)**
$P_{Gen,tot}$ [MW]	1280.09	1279.6	1280.3	1280.0	1278.89
P_{Losses} [MW]	29.29	28.827	29.51	29.245	28.15

ability to effectively manage both voltage profiles and reactive power, while maintaining a high level of flexibility and stability in the system. This is particularly important, given the ongoing increase in electrical load. Additionally, while the installation of UPFC devices can be complex, it is still a valuable tool for controlling voltage and reactive power in electrical networks, as demonstrated both in simulations and real-world installations. This chapter focuses specifically on the control of voltage and reactive power using UPFC devices, due to the intricate nature of their operation. The results obtained from the data presented in Table 2.11 indicate that the reactive power compensation plays a significant role in both the steady-state and dynamic performance of the AC power system.

To assess the impact of various static equipment such as capacitor banks and synchronous compensators on the behavior of the AC system, these devices are often substituted for shunt FACTS devices. However, it is important to note that such substitutions are made without taking into account the cost of the equipment, additional factors, particularly dynamic stability in the event of a failure or voltage collapse (Figure 2.18).

Table 2.11 Active and reactive powers generated

	Scenario 1		Scenario 2		Scenario 3					
					(a)		(b)		(c)	
Bus Nbr	P_G **[MW]**	Q_G **[MVAR]**	P_G **[MW]**	Q_G **[Mvar]**	P_G **[MW]**	Q_G **[Mvar]**	P_G **[MW]**	Q_G **[Mvar]**	P_G **[MW]**	Q_G **[Mvar]**
1	480.09	142.9	479.62	140.12	480.31	139.63	480.05	173.83	478.9	129.45
2	0.0	5.72	0.0	5.72	0.0	5.74	0.0	−40.75	0.0	5.68
3	40	15.74	40	10.43	40	4.15	40	0.683	40	−9.34
6	0.0	8.83	0.0	7.15	0.0	−2.62	0.0	45.99	0.0	4.12
8	450	68.55	450	66.40	450	55.78	450	3.43	450	63.38
9	0.0	12.77	0.0	9.64	0.0	−3.69	0.0	80.46	0.0	−9.24
12	310	141.12	310	135.81	310	135.13	310	111.19	310	113.59

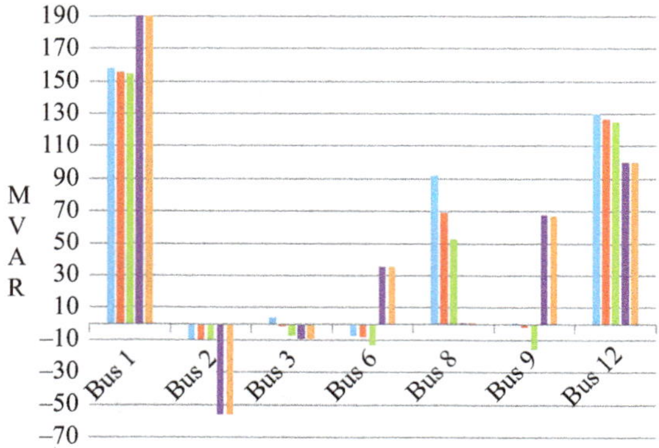

Figure 2.18 Produced reactive powers in (bus 1, 2, 3, 6, 8, 9, and 12): Scenarios 1, 2, and 3

2.5 Conclusion

This chapter undertook three scenario studies on two distinct AC networks, namely the IEEE-05 and IEEE-57 benchmark, to execute load flow utilizing the s method within the MATLAB framework. Additionally, the MATLAB/Simulink was utilized to model the AC networks.

The initial scenario assessed the base-scenario load flow, while the second involved correcting the buses voltages by modulating the transformer ratios and reactive power produced by capacitor banks. The third scenario incorporated the substitution of capacitor banks with SVCs and synchronous compensators with STATCOMs. In this case, an UPFC device was introduced into the network, with its location being determined through the simulation of various scenarios to determine the optimal voltage profile while minimizing both reactive power generation and power loss. The findings indicate that the FACTS equipment has the capacity to enhance the stability of AC power grids by regulating the bus voltage at a predetermined level, optimizing the reactive power produced/conveyed within the power system, and reducing power losses in the system, particularly when the placement of these devices is optimal.

In actual AC power networks, FACTS devices exhibit a dynamic response to changes in system behavior such as active loads, AC faults, flickers, instability, voltage collapse, power electronics, harmonic production, and other similar circumstances. The implementation of FACTS devices in AC networks is aimed at providing resistance and flexibility to ensure system stability under such conditions. Our future research will concentrate on two specific areas: (i) an examination of how AC behaves in dynamic situations, especially when various fault types are present; and (ii) the inclusion of the expenses for both FACTS equipment and generator fuel. This will enable the integration of these AC networks into the optimal power flow calculation.

References

Acha, E., Fuerte-Esquivel, C. R., Ambriz-Perez, H., and Angeles-Camacho, C. (2004). *FACTS: Modelling and Simulation in Power Networks*. New York: Wiley.

Alassouli, H. M. (2021). *Control of Flexible Alternating Current Transmission System (FACTS) for Power Stability Enhancement and Power Quality Improvement*. https://www.amazon.in/Alternating-Transmission-Stability-Enhancement-Improvement/dp/1716043751.

Biswas, S. and Nayak, P. K. (2018). State-of-the-art on the protection of FACTS compensated high-voltage transmission lines: a review. *High Voltage*, 3(1):21–30.

Chun, L., Qirong, J., Xiaorong, X., and Zhonghong, W. (1998). Rule-based control for statcom to increase power system stability. In *POWERCON'98. 1998 International Conference on Power System Technology. Proceedings (Cat. No. 98EX151)*, volume 1, pages 372–376. Piscataway, NJ: IEEE.

Dash, S., Subhashini, K. R., and Satapathy, J. (2020). Efficient utilization of power system network through optimal location of FACTS devices using a proposed hybrid meta-heuristic ant lion-moth flame-salp swarm optimization algorithm. *International Transactions on Electrical Energy Systems*, 30(7):e12402.

Demirovic, N. (2016). Impact of statcom and SVC to voltage control in systems with wind farms using induction generators (IG). *Mediterranean Conference on Power Generation, Transmission, Distribution and Energy Conversion (MedPower 2016)*, Belgrade, pp. 1–6.

Gandoman, F. H., Ahmadi, A., Sharaf, A. M., *et al.* (2018). Review of FACTS technologies and applications for power quality in smart grids with renewable energy systems. *Renewable and Sustainable Energy Reviews*, 82:502–514.

García-Cerrada, A., Garcia-Gonzalez, P., Collantes, R., Gómez, T., and Anzola, J. (2000). Comparison of thyristor-controlled reactors and voltagesource inverters for compensation of flicker caused by arc furnaces. *IEEE Transactions on Power Delivery*, 15(4):1225–1231.

Gomis-Bellmunt, O., Sau-Bassols, J., Prieto-Araujo, E., and Cheah-Mane, M. (2019). Flexible converters for meshed HVDC grids: from flexible AC transmission systems (FACTS) to flexible DC grids. *IEEE Transactions on Power Delivery*, 35(1):2–15.

Grainger, J. J. (1999). *Power System Analysis*. New York: McGraw-Hill.

Gupta, S., Singh, M., and Sharma, H. D. (2017). Installation of UPFC for enhancing overall performance of electrical power system. *Russian Electrical Engineering*, 88(5):314–320.

Haque, M. (2004). Improvement of first swing stability limit by utilizing full benefit of shunt FACTS devices. *IEEE Transactions on Power Systems*, 19(4):1894–1902.

Hingorani, N. G. (2007). FACTS technology-state of the art, current challenges and the future prospects. In *IEEE Power Engineering Society General Meeting*, volume 2, page 11.

Mankour, M., Merzoug, Y., and Sami, B. (2022). New strategy to enhance AC grid performance with SVC device in an optimal location. *Journal of Ambient Intelligence and Humanized Computing*, 13(12):5873–5884.

Mankour, M. and Sami, B. S. (2019). Voltage and reactive power enhancement in a large AC power grid using shunt FACTS devices. In *International Conference on Smart Innovation, Ergonomics and Applied Human Factors*, pages 125–143. Berlin: Springer.

Mathur, R. M. and Varma, R. K. (2002). *Thyristor-based FACTS Controllers for Electrical Transmission Systems*. New York: Wiley.

Noroozian, M., Angquist, L., Ghandhari, M., and Andersson, G. (1997). Improving power system dynamics by series-connected FACTS devices. *IEEE Transactions on Power Delivery*, 12(4):1635–1641.

Reddy, S. S. (2018). Optimal placement of FACTS controllers for congestion management in the deregulated power system. *International Journal of Electrical and Computer Engineering*, 8(3):1336.

Samet, H. and Jarrahi, M. A. (2015). A comparison between SVC and statcom in flicker mitigation of electric arc furnace using practical recorded data. In *2015 30th International Power System Conference (PSC)*, pages 300–304. Piscataway, NJ: IEEE.

Sauer, P. W., Pai, M. A., and Chow, J. H. (2017). *Power System Dynamics and Stability: With Synchrophasor Measurement and Power System Toolbox*. New York: Wiley.

Sood, V. K. (2006). *HVDC and FACTS Controllers: Applications of Static Converters in Power Systems*. Berlin: Springer Science & Business Media.

Yazdani, A. and Iravani, R. (2010). *Voltage-sourced Converters in Power Systems: Modeling, Control, and Applications*. New York: Wiley.

Chapter 3

Metaheuristic optimization of electrical distribution systems with energy storage and reactive compensators

Ahmad Eid[1,2] and Pierluigi Siano[3,4]

Abstract

This study, using the dynamic arithmetic optimization algorithm (DAOA), provides a solution for the optimal integration of renewable energy sources, reactive power compensators (RPCs), and battery energy storage systems (BESS) into electric power systems. To reduce annual energy dissipation and RPC operating costs as much as possible, the DAOA optimizes the BESS and RPC levels. Uncertain residential, commercial, and industrial demand is also factored into the analysis. Annual energy loss and reactive power compensation costs are reduced significantly after implementing optimal active and reactive power compensation allocations. The yearly cost of energy loss decreases from 2631.77 k$ to 27.21 k$ for residential demand, 3995.48 k$ to 40.65 k$ for commercial demand, and from 4838.24 k$ to 48.54 k$ for industrial demand. In addition to determining critical performance measures, including the voltage deviation from unity, system stability, and voltage profile during the simulation horizon, the capability of the DAOA algorithm to minimize the different costs of units while satisfying the problem and system restrictions is also outlined. The results confirm the efficient methodology adopted using the DAOA in controlling the BESS units' charging and discharging modes according to the system operating conditions.

Keywords: Optimization algorithm; renewable energy source; energy storage system; reactive compensators; distribution systems

[1]Department of Electrical Engineering, College of Engineering, Qassim University, Saudi Arabia
[2]Department of Electrical Engineering, Faculty of Engineering, Aswan University, Egypt
[3]Department of Management and Innovation Systems, University of Salerno Via Giovanni Paolo II, Italy
[4]Department of Electrical and Electronic Engineering Science, University of Johannesburg, South Africa

3.1 Introduction

3.1.1 Background

Distribution networks (DNs) have different loss components due to the presence of cables or lines. The power losses can be categorized into real losses due to the copper losses in overhead lines or underground cables and reactive power losses. These losses are proportional to the squared current of conductors. At the same time, passing the current in the conductors initiates a voltage drop along these long lines, and consequently, the voltages reduce dramatically, especially at the far ends. The researchers always try to find solutions to improve the DNs in different aspects to save operation and component costs. Some researchers attempted to solve these problems by injecting real power into the system using distributed generation (DG) technology. The DG can be wind turbine (WT) power generation, photovoltaic (PV) power, or a combination of different renewable energy sources. Other researchers used capacitor banks or reactive power compensators (RPCs) to inject reactive power into the distribution system.

In contrast, battery energy storage systems (BESS) regulate the voltage profile, provide fast charging operations, and store electrical power to maintain grid stability [1] under fast-changing operating conditions such as connecting or disconnecting a vast demand. Intelligent grid stability, security, and cost-effectiveness can all be improved with the help of energy storage technology [2]. Peak shaving, making it easier to maintain outdated infrastructure, competitively lowering the cost of energy and auxiliary services, pollution reduction, and maximizing the penetration of renewable energy systems (RESs) are examples of BESS's outstanding prospects and applications. However, because of its shifting function as backup generation (discharging mode) or demand (charging mode) [3], the BESS integration follows a specific scenario other than DGs. In addition, the BESS state of charge (SoC) is considered for a long-life battery operation. Therefore, maintaining and using its SoC levels as needed may require optimum network management. In [4], the authors tried to minimize the studied system's battery-consumed power and stability index under a continually growing load of power using particle swarm optimization (PSO).

Researchers tried optimizing distribution systems using different techniques and methods while integrating BESS units and renewable energy sources. Non-optimal BESS deployments can lead to increases in system losses and associated costs. The authors of [5] recommended an optimal BESS size utilizing PSO with a frequency regulation technique to keep the microgrid stable after the loss of utility. In a similar study [6], the authors optimally integrated the RESs and BESS units using a combination of linear programming and genetic multi-objective solver. They have managed to diminish losses and environmental discharges. Another study [7] proposed optimal energy management by combining approximation dynamic programming and a hybrid big bang big crunch algorithm to achieve optimal scaling of RES and BESS for microgrid operation, but not optimal siting. Authors of [8] designed a hybrid system consisting of BESS and PV-WT to supply demand with the lowest investment cost using a deterministic approach. A multi-criteria design approach for a hybrid RES is provided in [9]. It considers factors, including the likelihood of a power supply failure, blackout, and power shortage between production and storage.

However, reactive loads may cause DNs to experience voltage fluctuations. Some delicate electrical equipment may suffer significant failure due to voltage disturbances, which can lower power supply efficiency. These problems can be partially remedied by installing RPCs. Many authors applied different RPC types, including shunt capacitors [10–15], SVC [16–19], or distribution static synchronous compensator (DSTATCOM) with DNs, to reach various goals [10,20]. Authors in [21] used a DSTATCOM with a transformer to boost the efficiency of a small DN. A technique of mixed integer non-linear programming (MINP) in conjunction with model predictive control (MPC) model has been proposed in [22] to allocate optimal capacitor banks, BESS, and PV units into a distribution system. The method was employed to lessen DN operational losses, active power curtailments of PV, and bus voltage swings. In another study [23], shunt capacitors are used to get the finest operation of DN with nonlinear demands using the jellyfish optimization algorithm. A similar study [24] used the whale optimization algorithm to deduct system losses and operating costs by optimal allocating capacitor banks into distribution systems. The improved golden ratio optimization algorithm (IGROA) was adopted in [24] to allocate shunt capacitors and DSTATCOM in DNs to boost network performance and lessen reactive investment costs. The IGROA solved the problem in a multiobjective scenario with a fuzzy-based function. Optimal integration of PV-BESS with a unified power quality controller device was proposed in [25] for peak load shaving of distribution systems. The BESS charges during PV peak hours and discharges during peak demand periods.

3.1.2 Problem and motivation

A careful review of the above studies revealed that the appropriate and concurrent placement of BESS, WT, and RPCs in DNs is a problem that has not been adequately addressed. Employing various equipment to balance reactive power in DNs interests designers and operators [26–30]. Using this kind of equipment to make up for the network and improve its features may be an exciting aspect. Additionally, the contribution of these types of equipment as compensation must be estimated ideally using a robust optimization algorithm. They should be placed at optimal network buses with optimal sizes. According to a literature survey, most active and reactive source optimization studies are based on determining the place and size of one or more of these devices without considering the cost of energy losses or the installation and operation of reactive compensators. However, most researchers are interested in the optimal allocation of one or more active or reactive sources that operate at full-load conditions, not considering the stochastic nature of RESs or system demand.

3.1.3 Proposed solution and contribution

In this chapter, the authors overcome the above drawbacks and apply an efficient methodology for integrating three active and reactive power compensation sources. The study controls the optimal integration of BESS, WTs, and RPCs into DNs, lessening the cost of energy loss per year and RPC installation and operation costs using the dynamic arithmetic optimization algorithm (DAOA). Different residential, commercial, and industrial demands are reflected. Moreover, the BESS units are optimized to operate in charging and discharging modes depending on the SoC of BESS units and the state action of DN. In addition, the methodology calculates essential indicators, including the transient voltage dip (TVD), SI, and voltage profile during the simulation horizon, to ensure the capability of the DAOA algorithm while satisfying the objectives and system constraints.

3.2 Device modeling

Integrating RESs and RPCs into distribution systems enhances the voltage profile, decreases power losses, improves power factors, and improves system operation. To clarify this idea, suppose we have a two-bus system, as shown in Figure 3.1. At first, the demand real and reactive powers (P_D, Q_D) are fed by the source where the apparent power ($P_s + jQ_s$) of the source is equal to the apparent power ($P_D + jQ_D$) of the demand and above the losses of the feeder. When DG and RPCs are connected to the load bus, they share their powers with the load. The required apparent power from the source decreases with the injection amounts from the DG and RPC devices. Hence, the current in the feeder is reduced after integrating DG and RPC ($I_2 < I_1$). As the supply voltage is assumed constant, the load voltage V_r must increase ($V_{r2} > V_{r1}$), as shown in the phasor diagram. Reducing the current results in a considerable reduction in the feeder losses as they are proportional to the current squared (I^2R, I^2X). Moreover, the power factor in the second case is more significant than in the first. Thus, integrating renewable energy sources and RPCs provides many advantages for the DN, including improving the bus voltage distribution, reducing losses, and improving the overall power factor. The following sections provide detailed mathematical equations for the performance parameters and a cost analysis approach.

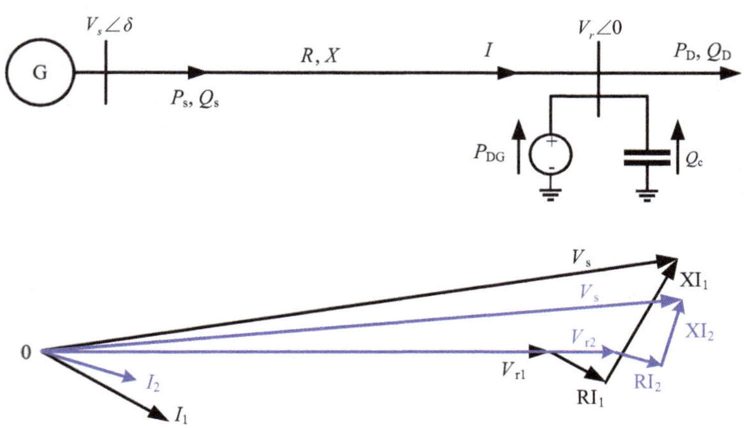

Figure 3.1 Two-bus network and its associated phasor diagram with active and reactive power compensations

3.2.1 BESS modeling

The BESS unit is normally connected to the DN through a DC/AC inverter with a specific efficiency. The BESS inverter is usually designed to work in the four quadrants over all periods of operation. Thus, the BESS can be considered a load with charging mode or a power source with discharging mode. Moreover, when properly controlled, the BESS inverter can work with any power factor (leading, unity, lagging).

The stored energy of BESS during the charging period (load) when $P_{BESS}(t) > 0$ is expressed as [31,32]

$$E_{BESS}(t) = E_{BESS}(t-1) + \Delta t \times P_{BESS}/\eta_c \tag{3.1}$$

where $E_{BESS}(t)$, $E_{BESS}(t-1)$ are the BESS energy during the current and previous time steps; P_{BESS} is its consumed power; η_c is the efficiency; Δt is the time step of the simulation.

Note that the energy storage of the BESS unit increases during charging mode. Hence, the SoC of the BESS unit rises with time as:

$$SoC(t) = SoC(t-1) + \frac{\Delta t \times P_{BESS}}{\eta_c \times C_{BESS}} \tag{3.2}$$

where $SoC(t)$, $SoC(t-1)$ are SoC during the current and previous time steps; C_{BESS} is the capacity of the BESS unit (kWh).

During the discharge mode (generating mode), when the $P_{BESS}(t) < 0$, the energy stored is:

$$E_{BESS}(t) = E_{BESS}(t-1) + \Delta t \times P_{BESS} \times \eta_d \tag{3.3}$$

where η_d is the discharge efficiency.

Here, the energy storage of the BESS unit decreases with time during the discharging mode. Hence, the SoC of the BESS unit decreases with time as:

$$SoC(t) = SoC(t-1) + \frac{\Delta t \times P_{BESS} \times \eta_d}{C_{BESS}} \tag{3.4}$$

When the SoC of the BESS is less than $SoC_{\min}$ or greater than $SoC_{\max}$, the P_{BESS} is considered zero with the BESS in the idle state (no charging nor discharging).

3.2.2 Wind turbine modeling

WT power generation is described in many pieces of literature [3,33,34] and will not be repeated here. The WT power generation is considered to be varying during the day. The WT power profile follows a stochastic pattern [3], as shown in Figure 3.2. The study includes the best allocation of three WTs based on the system's minimum loss at full-load conditions. Wind power generation varies from place to place and from season to season [35]. In this work, the WT profile of Figure 3.2 is considered. The seasonal WT profiles will be similar to daily ones, as the WT is considered a generator with a total capacity of the optimal value at each

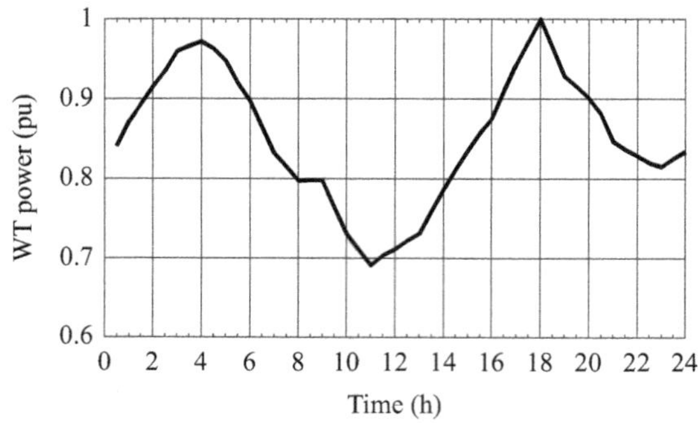

Figure 3.2 Normalized daily profile of WT power generation

best location. The WT power profile is similar to winter, spring, and fall seasonal profiles. During summer, the WT power is low due to low wind.

3.2.3 RPC modeling

Depending on the size, the RPC device maybe SC, static var compensator (SVC), or DSTATCOM [30,36]. This study uses the RPC term for any type of them. Any RPC's main task is injecting reactive power into the DN when needed. The injected RPC current is calculated from the optimized value of its reactive power and the voltage of the connected bus. The optimization algorithm determines the reactive power according to the distribution system operating conditions and applied constraints.

$$I_{RPC} = \frac{Q_{Ci}}{V_i} \tag{3.5}$$

where I_{RPC} is the current of the RPC unit; Q_{Ci} is the amount of the optimized reactive power; V_i is the ith bus voltage at which the unit is located. Three RPC units are located in the DN based on the full-load condition.

3.3 Method of analysis

3.3.1 Calculation of objective functions

Suppose a branch between buses i, j of a DN that has a current $I_{i,j}$ passing through it. Here, bus i, j are the sending and receiving end buses, respectively. The branch has an impedance of $Z_{i,j}$. The power loss of this branch is estimated as follows:

$$P_L^{Br} = |I_{i,j}|^2 \times R_{i,j} \tag{3.6}$$

$$Q_L^{Br} = |I_{i,j}|^2 \times X_{i,j} \tag{3.7}$$

where P_L^{Br} and Q_L^{Br} are the real and reactive losses of each branch, respectively; $R_{i,j}$, $X_{i,j}$ are impedance components of the branch.

The system's losses are the summation of all losses as follows:

$$P_L = \sum_{i=1}^{N_{Br}} P_L^{Br}, Q_L = \sum_{i=1}^{N_{Br}} Q_L^{Br} \tag{3.8}$$

The P_L and Q_L are calculated for every time step of the simulation of 24 h. If all loss values are recorded during the day, we can get the loss as a function $p_L(t)$ from which the energy can be calculated. The TVD is calculated from the summation of the differences between current voltages and the reference as follows:

$$TVD = \sum_{i=1}^{N_B} (1 - |V_i|)^2 \tag{3.9}$$

where N_B is the number of buses inside the distribution system; V_i is the ith bus voltage in per unit.

The TVD is important because it shows the status of the voltage profile for the entire system. Moreover, the squared differences consider whether the deviation is positive or negative. The best value of TVD is zero when all bus voltages are unity.

The stability of a DN can be expressed in a normalized index SI [37] as:

$$SI_{i,j} = 1 - \left[2\left(P_j R_{i,j} + Q_j X_{i,j}\right) - |V_i|^2\right]^2 - 4|Z_{i,j}|^2 \times \left(P_j^2 + Q_j^2\right) \tag{3.10}$$

$$SI = \max\left(SI_{i,j}\right) \tag{3.11}$$

where P_j, Q_j are the power components at bus j; V_i is the voltage at the sending side.

This study aims to reduce the yearly cost of energy loss and the cost of reactive power compensation. The cost of RPC includes the purchasing, installation, and operation of these units. The energy loss is calculated as:

$$E_L^d = \int_{t=1}^{24} p_L(t) \times dt \tag{3.12}$$

where $p_L(t)$ is the function of power loss; E_L^d is the daily energy loss, kW.

The study assumes 0.5 h per step of calculations (48 steps per day). Thus, the above integration can be transferred into a summation to calculate the energy per day:

$$E_L^d = \sum_{i=1}^{N_{steps}} p_{L,i}(t) \times \Delta t \tag{3.13}$$

where N_{steps} is the number of simulation steps per day; Δt is the time step of 0.5 h.

The cost of energy loss (C_{EL}) is formulated as:

$$C_{EL} = C_e \times E_L^d \times D_y \tag{3.14}$$

where C_e is the energy cost ($/kWh); D_y is the total days per year.

The cost of RPCs (C_{RPC}) includes the purchase, installation, and operation [23,38]:

$$C_{RPC} = D_f \left(C_p \times \sum_{i=1}^{N_{RPC}} Q_{c,i} + C_i \times N_{RPC} \right) + C_o \times N_{RPC} \quad (3.15)$$

where D_f is the depreciation factor; C_p is the RPC purchase cost; N_{RPC} is the RPC units; $Q_{c,i}$ is the capacity of each RPC in kVar; C_i is the installation cost ($); and C_o is the operation cost ($).

The depreciation factor is taken as 10% in [38], and 20% in [39], and the latter value is considered in this study. The cost of BESS includes the investment, operation and maintenance, and exchange power between BESS units and DN. The total cost can be expressed as follows [7,9]:

$$Cost_{BESS} = C_{invB} + C_{OMB} + C_{exB} \quad (3.16)$$

$$C_{invB} = (0.17 \times C_{BESS} + 3) \times 1000 \quad (3.17)$$

$$C_{OMB} = 0.07 \times C_{BESS} + 9 \quad (3.18)$$

$$C_{exB} = C_{exB}^{d} \times D_y \quad (3.19)$$

$$C_{exB}^{d} = B_{BESS} \times \left(\sum_{i=1}^{N_{steps}} \left[P_{BESS,i}^{ch}(t) - P_{BESS,i}^{disch}(t) \right] \times \Delta t \right) \quad (3.20)$$

where $Cost_{BESS}$ is the total cost per year; C_{invB}, C_{OMB}, C_{exB} are the BEESS investment, O&M, and power exchange costs per year, respectively; C_{BESS} is the battery capacity; B_{BESS} is the considered energy bid ($); $P_{BESS,i}^{ch}$, $P_{BESS,i}^{disch}$ are the charging/discharging of BESS power. The BESS investment and O&M costs are constant per year.

The WT cost includes the investment, O&M, and trade of power between the WT units and DN. The total WT cost can be given as follows [7,9]:

$$Cost_{WT} = C_{invWT} + C_{OMWT} + C_{exWT} \quad (3.21)$$

$$C_{invWT} = (1.24 \times P_{WTr} + 2) \times 1000 \quad (3.22)$$

$$C_{OMWT} = 0.25 \times P_{WTr} + 8.2 \quad (3.23)$$

$$C_{exWT} = C_{exWT}^{d} \times D_y \quad (3.24)$$

$$C_{exWT}^{d} = B_{WT} \times \left(\sum_{i=1}^{N_{steps}} P_{WT,i}(t) \times \Delta t \right) \quad (3.25)$$

where $Cost_{WT}$ is the total cost per year; C_{invWT}, C_{OMWT}, C_{exWT} are the BESS investment, O&M, and power exchange costs per year, respectively; P_{WTr} is the rated WT output power; B_{WT} is the considered energy bid ($); $P_{WT,i}$ is the output WT power at any interval of time. The investment and O&M costs of the WT are also constant per year. This research aims to determine the minimum costs for

Table 3.1 Used constant and parameters

Parameter	Value	Parameter	Value
C_e	0.07 \$/kWh	D_y	365 days
D_f	20 %	N_{RPC}	3.0
C_p	25 \$/kVar	N_{WT}	3.0
C_i	1600 \$	N_{BESS}	3.0
C_o	300 \$	N_{steps}	48.0
B_{BESS}	0.38 c/kWh	C_{BESS}	667 kWh
B_{WT}	1.073 c/kWh	P_{WTr}	527, 380, 1719 kW

energy loss and RPC units. The used parameters are shown in Table 3.1 [23,38,39]. The fitness function is:

$$f_{obj} = \min(C_{EL} + C_{RPC}) \tag{3.26}$$

3.3.2 Problem constraints

The proposed solution methodology solves the 69-bus DN while satisfying all problem constraints. The methodology adopts constraints on the voltage magnitude, the power transferred to the DN, the power balance, and the locations of the devices.

The voltage after the integration of the renewable sources must lie within standard limits:

$$|V_{\min,i}| \leq |V_i| \leq |V_{\max,i}| \tag{3.27}$$

where $V_{\min,i}$, $V_{\max,i}$ are the extremes of voltage V_i.

At every solution, the power balance equations are satisfied.

$$P_U \pm P_{BESS} + P_{WT} = P_D + P_L \tag{3.28}$$

$$Q_U + Q_{RPC} = Q_D + Q_L \tag{3.29}$$

where P_U, Q_U are the utility power components; P_D, Q_D are the demand power components; P_{WT} is the power of WT units; Q_{RPC} is the reactive power of RPC units; P_L, Q_L are the power losses; P_{BESS} is the BESS capacity.

The positive sign of P_{BESS} is considered for discharging mode, while the negative sign is used for charging mode. The reactive power of RPC units lies within acceptable limits:

$$Q_{c,\min} \leq \sum_{i=1}^{N_{RPC}} Q_{ci} \leq Q_{c,\max} \tag{3.30}$$

where $Q_{c,\min}$, $Q_{c,\max}$ are the reactive power levels fed into the DN; Q_{ci} is the reactive power per unit. When any constraint is violated, the problem is resolved once the parameters are updated. The optimization methodology of the research is shown in Figure 3.3, showing the simulation steps and solution sequences.

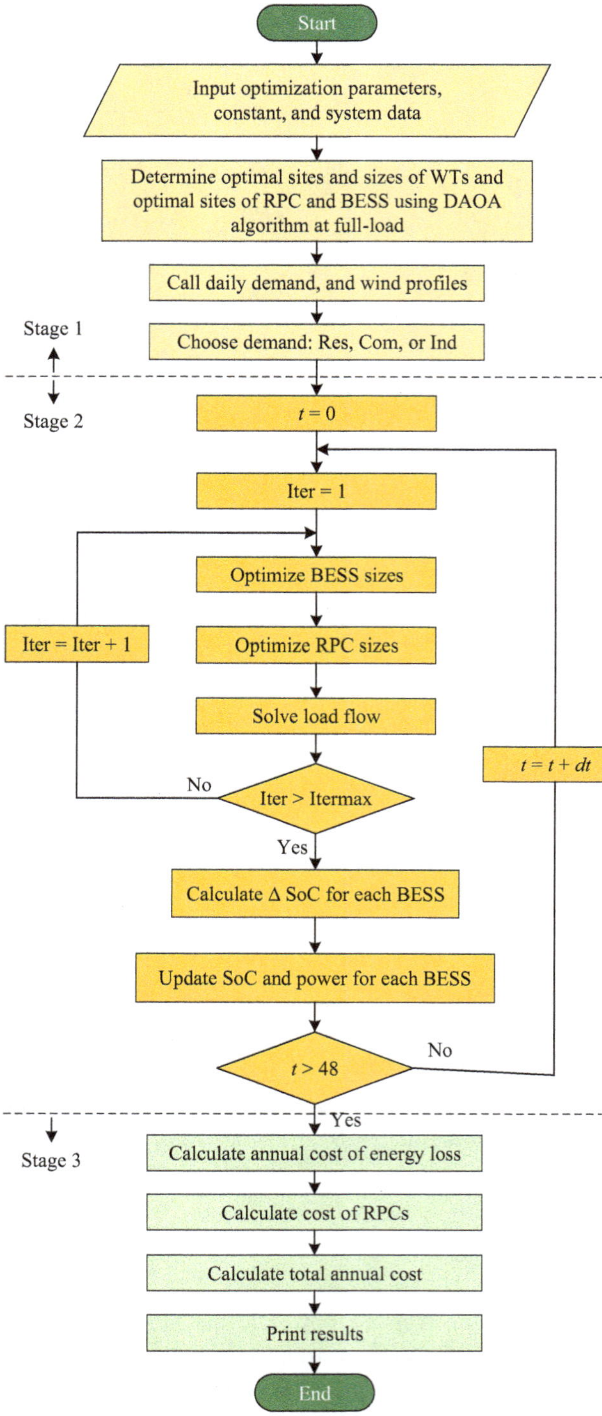

Figure 3.3 Flow chart of the proposed problem-solving methodology

There are three sequential steps for minimizing the cost of energy losses and RPC installation and operation. After collecting the optimization parameters and system data, the distribution system is optimized at full load conditions using the DAOA in the first stage. The DAOA determines the optimal sites and sizes of the WT units and the optimal locations for both BESS and RPS units. The optimal sizes of BESS and RPC units are decision variables. They are decided in the subsequent stage, which simulates a 24-h operation with variable demand. At the end of the first stage, the program calls the WT profile and a daily demand of residential, commercial, or industrial types. In the second stage, the DAOA algorithm optimizes the BESS and RPC sizes during the 24 h. With each time step, the SoC of the BESS units is updated based on the operating mode. As the considered time step is 0.5 h, this process is repeated 48 times. The cost analysis is executed in the last stage, where the total annual cost of energy loss and RPC devices are obtained.

3.4 Simulation results

The 69-bus DN, as shown in Figure 3.4, is considered for analysis to study the impact of BESS integration using the DAOA optimization algorithm [40]. The DAOA determines the best locations for the BESS, RPC, and WT at the full-load demand of the system. The BESS and WT units are optimally located at 11, 18, and 61, while RPC units are located at 17, 50, and 61. The DAOA determines its best

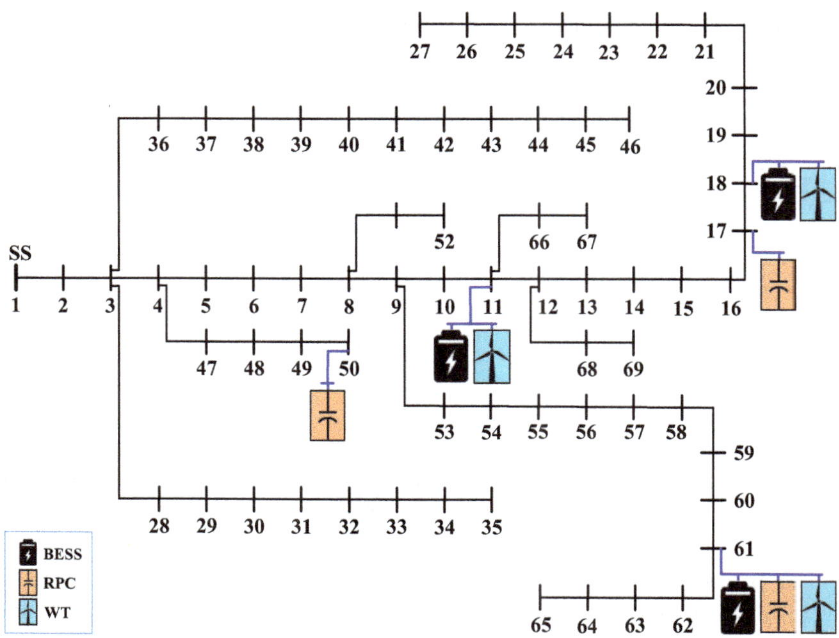

Figure 3.4 System layout with optimal locations of devices at full-load conditions

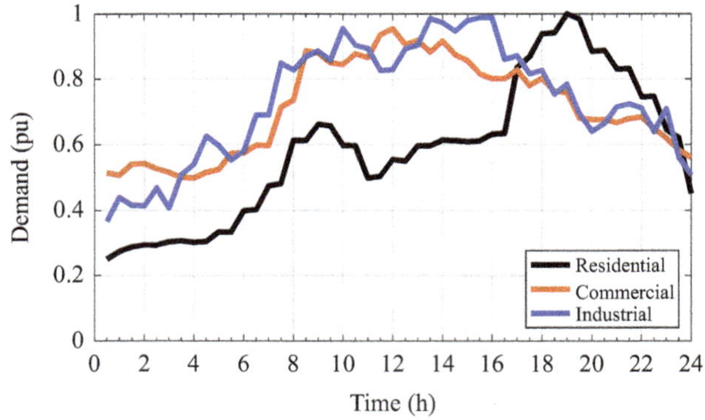

Figure 3.5 Normalized profiles of daily demand

locations while satisfying the objective of minimum losses for the system. The optimal sizes of BESS and RPC units are determined during the simulation process. At the same time, the WT power profile follows a stochastic pattern [3], as shown in Figure 3.2. The WT profile has two peaks at around 4:00 am and 6:00 pm and two valleys at 11:00 am and 11:00 pm. The optimal sizes at full-load conditions of WT units are 651.114 kW, 483.42 kW, and 2000.87 kW. The sizes of the WT vary during the day according to the wind profile. On the other side, the demand varies according to its type [41], as shown in Figure 3.5. This study considers the residential, commercial, and industrial demand profiles, which will be simulated in the following sections.

3.4.1 Case 1: residential demand

Residential or domestic demand is the quantity of electricity entering a home at any time. The amount of electricity a residence can obtain is often limited by the amount of service drop, and most residential loads are only connected for a few hours per day. The peak period of residential demand is from 4:00 pm until midnight, as shown in Figure 3.5.

The DAOA optimizes the BESS and RPC units throughout the day with the residential demand connected to the system. While the optimized three WTs and RPCs are in their optimal positions, the size of the WT power is determined according to its daily profile. For every step of 0.5 h, the SoC of each BESS is updated after calculating the change in its SoC. Suppose the optimized size of the BESS unit is positive at any particular time. In that case, the BESS unit charges and associated SoC increases, and vice versa. The DN is solved for every iteration using the forward–backward sweep method [42].

The BESS powers along the day are shown in Figure 3.6. When the BESS power is positive, the battery is charged, and as a result, its SoC increases, as shown in Figure 3.7. The charging rate depends on the amount of power and its increased rates. In contrast, when the power is reversed, the BESS discharges with a

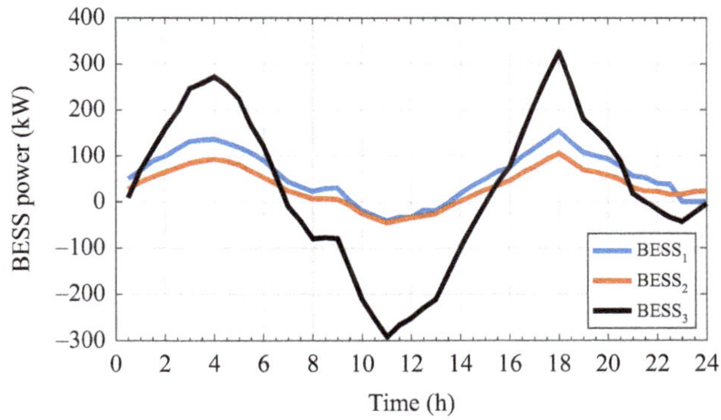

Figure 3.6 Daily BESS power for residential demand

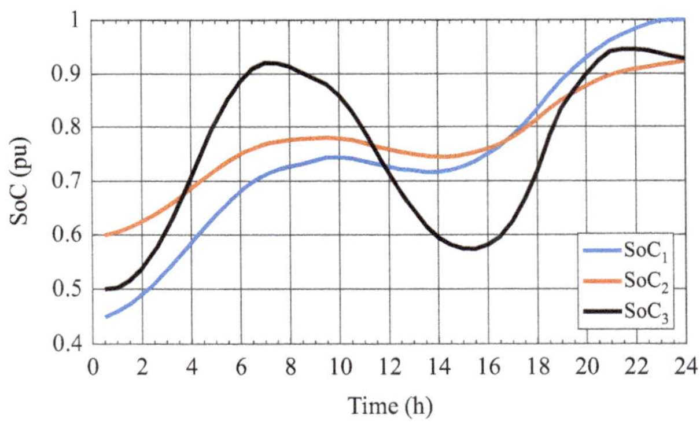

Figure 3.7 Daily SoC of BESS units for residential demand

diminishing SoC. The BESS units have started at random initial SoC. The faster changing of SoC_3 compared to SoC_2 and SoC_1 is due to the significant power of $BESS_3$ compared to the other two units. The high capacities of the WT (2000.87 kW) and RPC (1237.84 kVar) at bus number 61, to which $BESS_3$ is connected, supports the system, and provides enough power to charge the battery. The power of $BESS_2$ is always more significant than that of $BESS_1$ almost all the day. The SoC_2 reached 100% by the end of the day, although it starts at a lower initial value of 0.45.

The DAOA optimizes the BESS and RPC units with the help of WT to achieve minimum daily losses. Decreasing the losses results in decreasing the energy losses, which is the objective of the optimization. The active (P_{Loss}) and reactive power (Q_{Loss}) losses of the 69-bus DN are recorded during the simulation and shown in Figure 3.8. If we compare the residential demand to the losses, we can find a direct relationship between the demand and the losses. The maximum demand period from 4:00 pm to midnight also has maximum losses. Moreover, the reactive power losses exceed the active losses due to the available power from WT and BESS units. The active power loss profile is used to calculate the energy loss per day using the trapezoidal rule.

The corresponding profile of *TVD* is shown in Figure 3.8 along with *SI* index. The TVD shows the voltage deviation from the unity reference value. The results show that the values of the TVD are small, leading to an excellent voltage profile. Furthermore, the system is rigid due to the small daily SI values. High values of TVD and SI appear during the period of maximum demand. When DAOA optimizes the system, system constraints should be met, especially voltage constraints. The minimum (V_{min}) and maximum (V_{max}) voltages of DN are recorded during the daily simulation period to ensure that the voltage profile does not violate the voltage constraint. The profiles of V_{min}, and V_{max} are shown in Figure 3.8. The voltage does not exceed the unity nor goes below 0.995 pu. The RPC capacities Q_c are also shown. The reactive power for the three units are constants with slight variations, and they have an average of 352.25, 514.67, and 1237.69 kVar for Q_{c1}, Q_{c2}, and Q_{c3}, respectively.

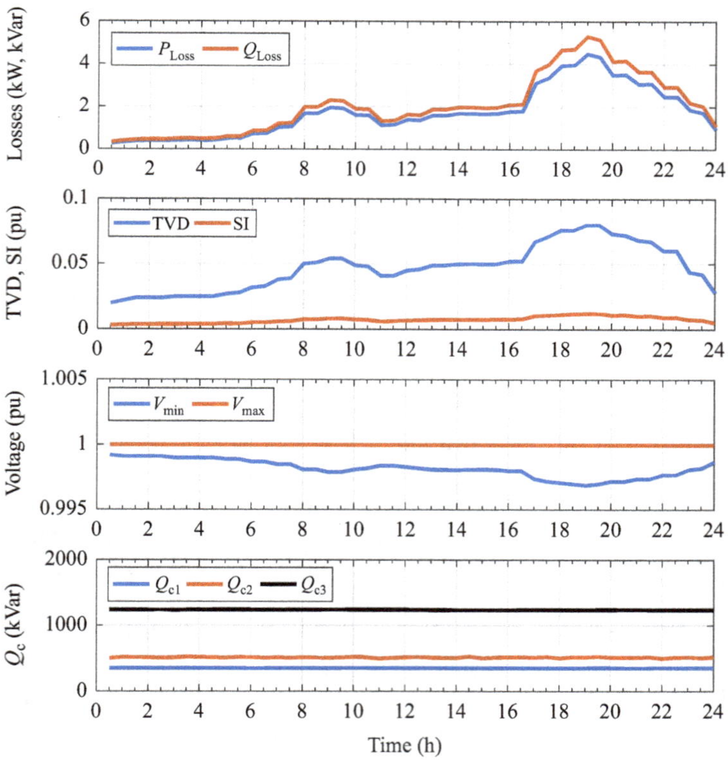

Figure 3.8 Active and reactive losses, TVD and SI, voltage profiles, and RPC capacities for residential demand

3.4.2 Case 2: commercial demand

Commercial demand is electrical loads intended for commercial usage such as those found in restaurants, shops, schools, colleges, and shopping malls. This load happens for a more extended period during the day than the residential demand. The switching of air-conditioning equipment is primarily responsible for commercial load fluctuations, and the season significantly impacts these demands.

The commercial demand has a flat peak centered at noon, as shown in Figure 3.5. Like residential demand, the DAOA optimizes the BESS and RPC powers to reduce the total system losses while integrating three WTs. The BESS powers of the three units are shown in Figure 3.9. The $BESS_1$ power starts to increase until 4:00 am and then decreases until 10:00 am. After that, $BESS_1$ starts to be negative power, meaning that the battery discharges and supports the system. Around 1:00 pm, $BESS_1$ turns into positive power again to charge until after 5:00 pm. After 5:00 pm, $BESS_1$ power is almost zero, which means it is in an idle state or has no power exchange. The sign of the battery power reflects on its SoC, as shown in Figure 3.10. The SoC1 of $BESS_1$ begins at a random value of 0.57 and

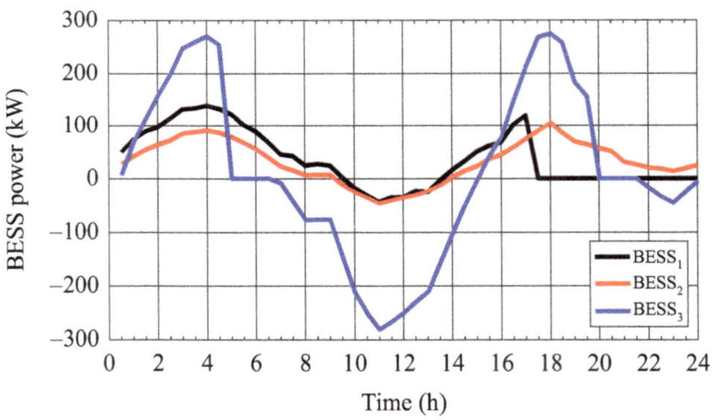

Figure 3.9 Daily BESS power for commercial demand

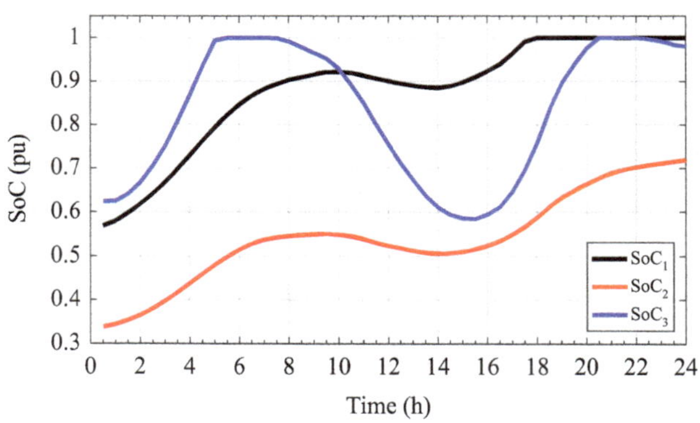

Figure 3.10 Daily SoC of BESS units for commercial demand

increases with positive power at varying rates depending on the magnitude of the optimal power. The SoC_1 remains fixed at 100% when the $BESS_1$ has zero power.

The behavior of the $BESS_2$ is similar to $BESS_1$ during the day. While the SoC_1 reaches 100% with zero power in the last 3 h, SoC_2 reaches a maximum of 71%. The low value of SoC_2 may refer to its low initial value of 0.34 at the start of the operation. The power of $BESS_3$ has a more significant value than $BESS_1$ or $BESS_2$. The value of the SoC_3 is impacted by the rapid rate of changing power of $BESS_3$. The power of $BESS_3$ has zero values during two periods; from 5:00 to 7:00 am and from 8:00 to 10:00 pm. Consequently, the SoC_3 reaches the maximum value of 100% during these periods, as shown in Figures 3.9 and 3.10.

The objective function (P_{Loss}) and other factors are shown in Figure 3.11 for the commercial demand case study. The losses follow the same pattern as the commercial demand profile. The active power loss always has a slightly higher value compared to the reactive power loss for the highly injected active power compared to the amount of injected reactive power. Moreover, it is the objective function, where the algorithm tries to reduce whatever the other parameters' values. The TVD also follows the pattern of commercial demand. When the demand increases, the losses increase; therefore, the voltage decreases at the system buses, resulting in higher TVD values. The SI

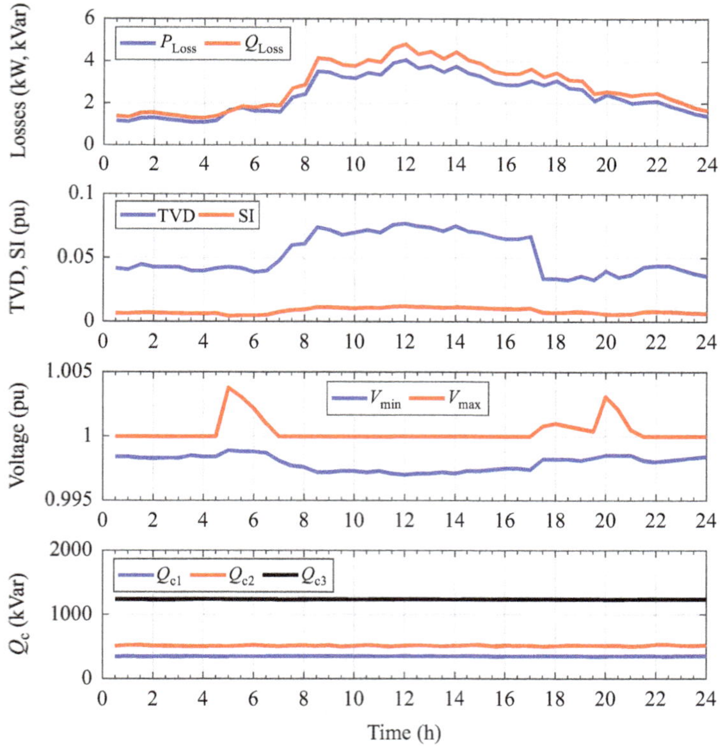

Figure 3.11 Active and reactive losses, TVD and SI, voltage profiles, and RPC capacities for commercial demand

values are very small, similar to the previous case, indicating that the system is stable. The voltage profiles are around unity. Optimizing the BESS units in the presence of the WT and RPC enhances the voltage profiles and remains within their constraint values. The RPC capacities Q_c are also shown. The reactive power for the three units are constants with slight variations, and they have an average of 352.42, 516.15, and 1237.67 kVar for Q_{c1}, Q_{c2}, and Q_{c3}, respectively.

3.4.3 Case 3: industrial demand

Industrial demand is made up of load demand from numerous industries, and it comprises all electrical loads utilized from small-scale to heavy industries and the machinery used. Industrial demand can be connected throughout the day, similar to commercial demand, as shown in Figure 3.5. The DAOA optimizes the BESS and RPC units for minimum daily loss while incorporating WTs into DN with a particular profile, as explained before.

The daily BESS power is shown in Figure 3.12, while the corresponding *SoC* is shown in Figure 3.13. Because each battery's starting *SoC* varies, distinct changes

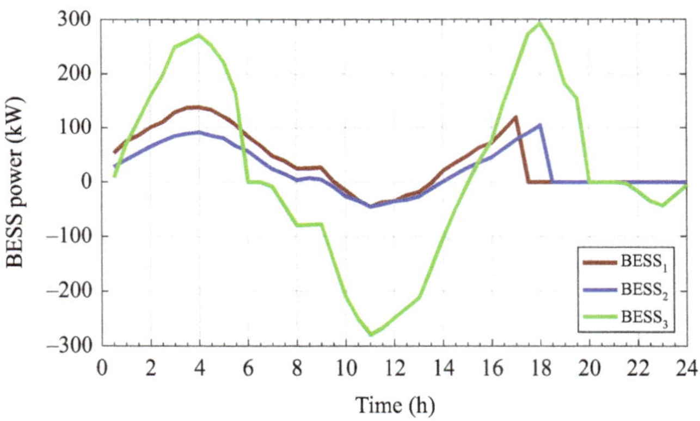

Figure 3.12 Daily BESS power for industrial demand

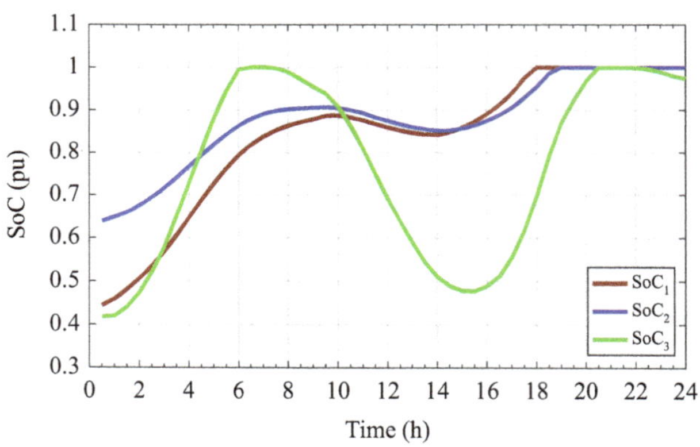

Figure 3.13 Daily SoC of BESS units for industrial demand

occur from one battery to the other based on the charging or discharging power. We also notice that some have *SoC* values reaching 100%. As previously stated, the battery power is zero at that moment. The performance parameters are shown in Figure 3.14, and they have excellent behavior for system operation. All constraints are met, and the voltage profile is improved. The RPC capacities Q_c are also shown. The reactive power for the three units are constants with slight variations, and they have an average of 352.21, 516.98, and 1237.83 kVar for Q_{c1}, Q_{c2}, and Q_{c3}, respectively.

The cost analysis for the three case studies, including residential, commercial, and industrial demands, is listed in Table 3.2. A careful examination of

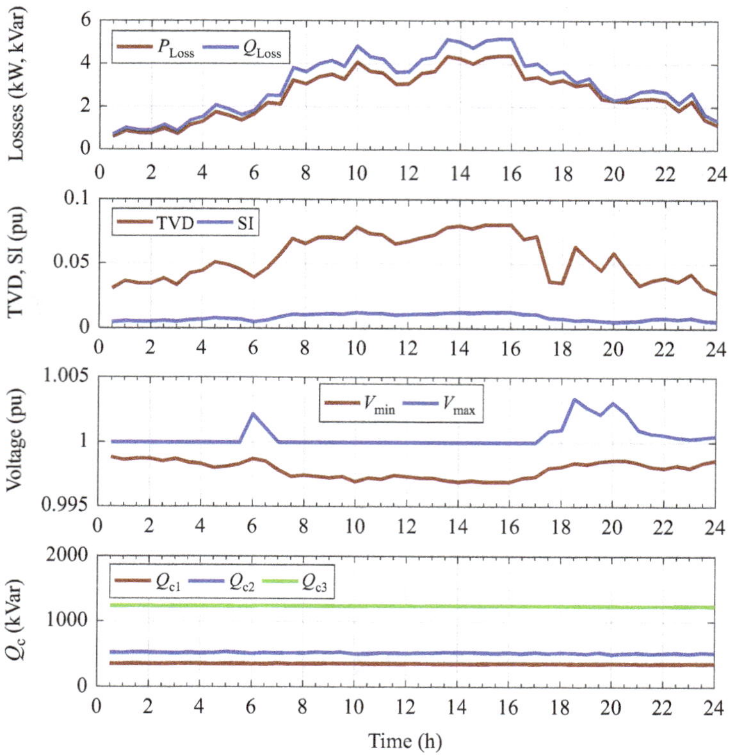

Figure 3.14 Active and reactive losses, TVD and SI, voltage profiles, and RPC capacities for industrial demand

Table 3.2 Cost analysis of energy loss, RPC, BESS, and WT devices

Demand	Energy loss (k$)	RPC (k$)	BESS (k$)	WT (k$)	Total cost (k$)
Res	27.214	12.383	120.443	209.897	369.937
Com	40.653	12.391	119.134	209.897	382.075
Ind	48.543	12.395	119.098	209.897	389.933

demand costs reveals that industrial demand has the highest cost, whereas residential demand has the lowest cost. The cost of RPC units is constant for all demands due to their similar magnitudes as Q_{c1}, Q_{c2}, and Q_{c3}. Moreover, the cost of BESS units during the three demands is very similar, with slight changes. The WT costs are also constant due to the constant power generation of wind at every demand.

3.4.4 Effect of BESS and RPC integration

The integration of RESs and RPC into DNs enhances the system's performance from different sides. One of these crucial sides is the daily voltage profile, where the customer needs a specified voltage that matches the devices to work correctly. In this study, BESS and WT units inject active power into the system, and the RPC units supply reactive power. This power will cover a part of the demand; hence, a smaller amount of power will be needed from the utility. Consequently, smaller currents run in the system's feeders, which leads to smaller voltage drops and a better voltage profile. The minimum voltage during the day ($V^d_{\min}$) is recorded for the three types of demand before and after integrating BESS, WT, and RPC devices and shown in Figure 3.15. The $V^d_{\min}$ is improved for all types of demand and achieves increasing amounts of 9.67%, 9.13%, and 9.52% for *Res*, *Com*, and *Ind* demands, respectively. For the *Res* demand, the voltage during the day rises from 0.9092 pu to 0.9971 pu. A similar achievement has occurred for the other two demand types. The new voltages are approaching unity.

Tremendous improvements have been achieved for the distribution system in terms of the other performance parameters such as P_{Loss}, Q_{Loss}, TVD, SI, input utility power (P_{Grid}), and the cost of energy loss per year (C_{EL}), as listed in Table 3.3. All performance parameters are decreased for all demand types after integrating optimized BESS, RPC, and WT units into the distribution system. Except for the last row of energy loss cost, all the figures in Table 3.3 indicate the maximum values of the parameters per day.

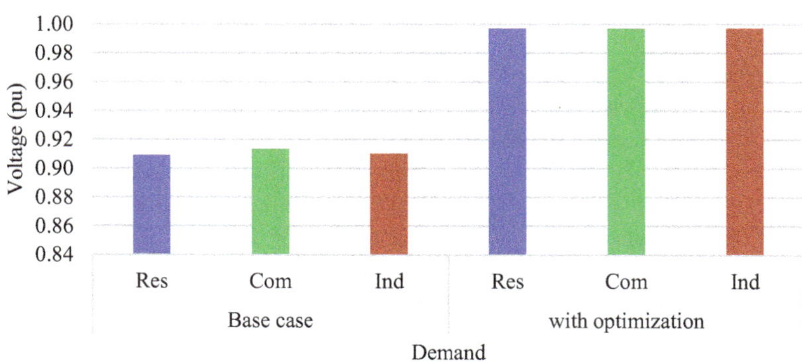

Figure 3.15 Minimum voltage before and after optimization of demand types

Table 3.3 System performance before and after optimization of all demands

Parameter	Base case			With optimization		
	Res	*Com*	*Ind*	*Res*	*Com*	*Ind*
P_{Loss} (kW)	224.94	203.08	219.38	4.61	4.06	4.36
Q_{Loss} (kVar)	102.14	92.26	99.63	5.30	4.81	5.16
TVD (pu)	1.836	1.745	1.813	0.070	0.071	0.076
SI (pu)	0.317	0.303	0.313	0.012	0.012	0.012
P_{Grid} (kW)	4016.83	3821.29	3968.04	1168.85	1190.09	1235.03
C_{EL} (k$)	2631.77	3995.48	4838.24	27.21	40.65	48.54

3.5 Conclusion

The work provides a competent technique for the integration of renewable energy sources, RPCs, and BESS utilizing the recently published DAOA. In the first stage of the study, the optimal sizes of the WTs and sites of the BESS and RPC units are determined. In the second stage, the DAOA optimizes the levels of the BESS and RPCs to get the lowest annual energy loss cost and the cost of RPC installation and operation. Furthermore, the study considers the unpredictable demand for residential, commercial, and industrial types. Significant cost savings demonstrate the benefit of efficient allocations of active/reactive power compensations in yearly energy loss and reactive power compensation. The power losses of the systems decreased from 224.94 kW, 203.08 kW, and 219.38 kW to 4.61 kW, 4.06 kW, and 4.36 kW for residential, commercial, and industrial demands, respectively. Additionally, there are significant reductions in other performance parameters. The results show average reductions in power losses, TVD, SI, and utility input power of 98%, 95 %, 96 %, 96%, and 70%, respectively, in addition to decreasing the annual costs of both energy loss and reactive power compensation to vast values.

References

[1] F. Qiao and J. Ma, "Voltage/var control for hybrid distribution networks using decomposition-based multiobjective evolutionary algorithm," *IEEE Access*, vol. 8, pp. 12015–12025, 2020.

[2] A. Kumar, N. K. Meena, A. R. Singh, *et al.*, "Strategic integration of battery energy storage systems with the provision of distributed ancillary services in active distribution systems," *Appl. Energy*, vol. 253, p. 113503, 2019.

[3] A. Eid, O. Mohammed, and H. El-Kishky, "Efficient operation of battery energy storage systems, electric-vehicle charging stations and renewable energy sources linked to distribution systems," *J. Energy Storage*, vol. 55, p. 105644, 2022.

[4] O. B. Adewuyi, R. Shigenobu, K. Ooya, T. Senjyu, and A. M. Howlader, "Static voltage stability improvement with battery energy storage

considering optimal control of active and reactive power injection," *Electr. Power Syst. Res.*, vol. 172, pp. 303–312, 2019.

[5] T. Kerdphol, K. Fuji, Y. Mitani, M. Watanabe, and Y. Qudaih, "Optimization of a battery energy storage system using particle swarm optimization for stand-alone microgrids," *Int. J. Electr. Power Energy Syst.*, vol. 81, pp. 32–39, 2016.

[6] M. R. Jannesar, A. Sedighi, M. Savaghebi, A. Anvari-Moghadam, and J. M. Guerrero, "Optimal multi-objective integration of photovoltaic, wind turbine, and battery energy storage in distribution networks," *J. Energy Manag. Technol.*, vol. 4, no. 4, pp. 76–83, 2020.

[7] M. Sedighizadeh, A. H. Mohammadpour, and S. M. M. Alavi, "A two-stage optimal energy management by using ADP and HBB-BC algorithms for microgrids with renewable energy sources and storages," *J. Energy Storage*, vol. 21, pp. 460–480, 2019.

[8] K. Anoune, A. Laknizi, M. Bouya, A. Astito, and A. Ben Abdellah, "Sizing a PV-wind based hybrid system using deterministic approach," *Energy Convers. Manag.*, vol. 169, pp. 137–148, 2018.

[9] H. Bakhtiari and R. A. Naghizadeh, "Multi-criteria optimal sizing of hybrid renewable energy systems including wind, photovoltaic, battery, and hydrogen storage with ϵ-constraint method," *IET Renew. Power Gener.*, vol. 12, no. 8, pp. 883–892, 2018.

[10] S. R. Arya, R. Niwas, K. K. Bhalla, B. Singh, A. Chandra, and K. Al-Haddad, "Power quality improvement in isolated distributed power generating system using DSTATCOM," *IEEE Trans. Ind. Appl.*, vol. 51, no. 6, pp. 4766–4774, 2015.

[11] Y. W. Liu, S. H. Rau, C. J. Wu, and W. J. Lee, "Improvement of power quality by using advanced reactive power compensation," *IEEE Trans. Ind. Appl.*, vol. 54, no. 1, pp. 18–24, 2018.

[12] H. S. Ramadan, A. F. Bendary, and S. Nagy, "Particle swarm optimization algorithm for capacitor allocation problem in distribution systems with wind turbine generators," *Int. J. Electr. Power Energy Syst.*, vol. 84, pp. 143–152, 2017.

[13] J. Cabral Leite, I. Pérez Abril, and M. S. Santos Azevedo, "Capacitor and passive filter placement in distribution systems by nondominated sorting genetic algorithm-II," *Electr. Power Syst. Res.*, vol. 143, pp. 482–489, 2017.

[14] A. Eid, S. Kamel, and L. Abualigah, "Marine predators algorithm for optimal allocation of active and reactive power resources in distribution networks," *Neural Comput. Appl.*, vol. 33, pp. 14327–14355, 2021.

[15] A. Y. Abdelaziz, E. S. Ali, and S. M. Abd Elazim, "Optimal sizing and locations of capacitors in radial distribution systems via flower pollination optimization algorithm and power loss index," *Eng. Sci. Technol. Int. J.*, vol. 19, no. 1, pp. 610–618, 2016.

[16] M. Čerňan, Z. Müller, J. Tlustý, and V. Valouch, "An improved SVC control for electric arc furnace voltage flicker mitigation," *Int. J. Electr. Power Energy Syst.*, vol. 129, p. 106831, 2021.

[17] A. Sode-Yome and N. Mithulananthan, "Comparison of shunt capacitor, SVC and STATCOM in static voltage stability margin enhancement," *Int. J. Electr. Eng. Educ.*, vol. 41, no. 2, pp. 158–171, 2004.

[18] B. Mahdad and K. Srairi, "Adaptive differential search algorithm for optimal location of distributed generation in the presence of SVC for power loss reduction in distribution system," *Eng. Sci. Technol. Int. J.*, vol. 19, no. 3, pp. 1266–1282, 2016.

[19] J. G. Jamnani and M. Pandya, "Coordination of SVC and TCSC for management of power flow by particle swarm optimization," *Energy Procedia*, vol. 156, pp. 321–326, 2019.

[20] W. Rohouma, R. S. Balog, A. A. Peerzada, and M. M. Begovic, "D-STATCOM for harmonic mitigation in low voltage distribution network with high penetration of nonlinear loads," *Renew. Energy*, vol. 145, pp. 1449–1464, 2020.

[21] Z. Wang, X. Yin, Y. Chen, J. Lai, L. Li, and Z. Qi, "DSTATCOM integrated with Y-y connection transformer for reactive power compensation," *Int. J. Electr. Power Energy Syst.*, vol. 117, p. 105721, 2020.

[22] R. Hu, W. Wang, X. Wu, *et al.*, "Coordinated active and reactive power control for distribution networks with high penetrations of photovoltaic systems," *Sol. Energy*, vol. 231, pp. 809–827, 2022.

[23] A. Eid, "Cost-based analysis and optimization of distributed generations and shunt capacitors incorporated into distribution systems with nonlinear demand modeling," *Expert Syst. Appl.*, vol. 198, p. 116844, 2022.

[24] A. Noori, Y. Zhang, N. Nouri, and M. Hajivand, "Hybrid allocation of capacitor and distributed static compensator in radial distribution networks using multi-objective improved golden ratio optimization based on fuzzy decision making," *IEEE Access*, vol. 8, pp. 162180–162195, 2020.

[25] S. Lakshmi and S. Ganguly, "Multi-objective planning for the allocation of PV-BESS integrated open UPQC for peak load shaving of radial distribution networks," *J. Energy Storage*, vol. 22, pp. 208–218, 2019.

[26] A. Eid and H. El-Kishky, "Renewable energy integration amid the 4th industrial revolution," in D.-E. A. Mansour, M. G. Hussien, P. S. Subudhi, A. R. Kumar, and S. Padmanaban (eds.) *Power Systems Amid the 4th Industrial Revolution*. River Publishers, Gistrup, 2024, pp. 111–144.

[27] A. Eid, "The role of battery energy storage in mitigating demand fluctuations of distribution networks caused by uncertain weather conditions using multi-objective Bonobo Optimizer," *Neural Comput. Appl.*, vol. 0123456789, pp. 1–18, 2024.

[28] A. Eid, "Power quality enhancement of balanced/unbalanced distribution systems using metaheuristic optimizations," in A. F. Zobaa and S. H. Abdel Aleem (eds.) *Modernization of Electric Power Systems: Energy Efficiency and Power Quality*. Springer, Cham, 2023, pp. 235–268.

[29] A. Eid, S. Kamel, and H. Abdel-mawgoud, "Shunt reactive compensations for distribution network optimization," in K. Pal, S. M. Tripathi, and S. Pandey (eds.) *Advanced Control & Optimization Paradigms for Energy*

System Operation and Management, 1st ed. River Publishers, Gistrup, 2023, pp. 109–135.

[30] A. Eid and M. Abdel-Salam, "Management of electric vehicle charging stations in low-voltage distribution networks integrated with wind turbine–battery energy storage systems using metaheuristic optimization," *Eng. Optim.*, vol. 56, no. 9, pp. 1335–1360, 2023.

[31] A. Eid, S. Kamel, and E. H. Houssein, "An enhanced equilibrium optimizer for strategic planning of PV-BES units in radial distribution systems considering time-varying demand," *Neural Comput. Appl.*, vol. 5, pp. 17145–17173, 2022.

[32] D. Q. Hung, N. Mithulananthan, and R. C. Bansal, "Integration of PV and BES units in commercial distribution systems considering energy loss and voltage stability," *Appl. Energy*, vol. 113, pp. 1162–1170, 2014.

[33] M. Kharrich, S. Kamel, A. S. Alghamdi, *et al.*, "Optimal design of an isolated hybrid microgrid for enhanced deployment of renewable energy sources in Saudi Arabia," *Sustainability*, vol. 13, no. 9, pp. 1–26, 2021.

[34] M. Kharrich, S. Kamel, R. Ellaia, *et al.*, "Economic and ecological design of hybrid renewable energy systems based on a developed IWO/BSA algorithm," *Electronics*, vol. 10, no. 6, pp. 1–31, 2021.

[35] Y.-H. Wan, "Wind power plant behaviors: analyses of long-term wind power data," *Tech. Report-NREL/TP-500-36551*, pp. 1–59, 2004.

[36] A. Eid, S. Kamel, A. Korashy, and T. Khurshaid, "An enhanced artificial ecosystem-based optimization for optimal allocation of multiple distributed generations," *IEEE Access*, vol. 8, pp. 178493–178513, 2020.

[37] A. Eid, "Allocation of distributed generations in radial distribution systems using adaptive PSO and modified GSA multi-objective optimizations," *Alexandria Eng. J.*, vol. 59, no. 6, pp. 4771–4786, 2020.

[38] A. M. Shaheen, E. E. Elattar, R. A. El-Sehiemy, and A. M. Elsayed, "An improved sunflower optimization algorithm-based Monte Carlo simulation for efficiency improvement of radial distribution systems considering wind power uncertainty," *IEEE Access*, vol. 9, pp. 2332–2344, 2021.

[39] A. El-Fergany and A. Y. Abdelaziz, "Multi-objective capacitor allocations in distribution networks using artificial bee colony algorithm," *J. Electr. Eng. Technol.*, vol. 9, no. 2, pp. 441–451, 2014.

[40] N. Khodadadi, V. Snasel, and S. Mirjalili, "Dynamic arithmetic optimization algorithm for Truss optimization under natural frequency constraints," *IEEE Access*, vol. 10, pp. 16188–16208, 2022.

[41] K. Qian, C. Zhou, M. Allan, and Y. Yuan, "Modeling of load demand due to EV battery charging in distribution systems," *IEEE Trans. Power Syst.*, vol. 26, no. 2, pp. 802–810, 2011.

[42] S. Sunisith and K. Meena, "Backward/forward sweep based distribution load flow method," *Int. Electr. Eng. J.*, vol. 5, no. 9, pp. 1539–1544, 2014.

Chapter 4

Classification and hidden neuron count effect on renewable microgrid power management

Ismail Elabbassi[1], Mohamed Khala[1], Naima El yanboiy[1], Omar Eloutassi[2] and Youssef El hassouani[3]

Abstract

The chapter presents a proposed energy management system using Neural Network Time Series model, with Levenberg–Marquardt (LM) training method, compared with the author's previous works, which incorporated classification methods. The study explores the effect of neuron count in a neural network time series model to enhance energy management, revealing that 10 neurons optimize validation performance, achieving a 96.2196% determination coefficient using the LM algorithm and with an error of 0.0379, outperforming all other simulations. The proposed method excelled in simulating energy systems, achieving 99.864% accuracy and surpassing previous benchmarks (99.747%–99.81%).The research underscores the neural network approach based on neuron count determined value to enhance energy storage management and promoting sustainable energy integration.

Keywords: Power storage management; neural network; Levenberg–Marquardt; microgrid; fuel cell; determination factor

4.1 Introduction

The advent of machine learning techniques has brought dynamic changes to the management of hybrid microgrids and energy storage systems. These technologies leverage advanced computational methods to optimize energy distribution, improve efficiency, and integrate renewable energy sources more effectively. Currently, the literature research works in energy management in microgrids field focus on

[1]Physics Department, Faculty of Science and Techniques Errachidia, Moulay Ismail University of Meknes, Morocco
[2]Materials and Modelling Laboratory, Department of Physics, Faculty of Science, Moulay Ismail University, Morocco
[3]New Energies and Materials Engineering, Faculty of Science and Techniques Errachidia, Moulay Ismail University of Meknes, Morocco

changing methods of machine learning, but there is not much attention on the architecture of the proper system and of the approach used. The novel approaches proposed to overcome the energy management of microgrid systems include artificial neural network (ANN) with Levenberg–Marquardt (LM) training and neurons count changing. The comprehensive state of art of the latest advancements in machine learning applications focuses on three key methods: the adaptive neural fuzzy inference system (ANFIS), AdaBoost, and random forest (RF). We will also examine their roles in fuel cell electric vehicle (FCEV) systems and their integration with emerging technologies such as the Internet of Things (IoT) and deep learning. Ramadan *et al.* [1] demonstrated that the ANFIS controller can improve the dynamic response of the SOFC model. RF is employed for data transformation. RF is utilized for data transformation to enhance fault detection and classification in PV arrays, leveraging the Stockwell transform and data mining techniques [2]. It integrates photovoltaic (PV) panels, batteries, supercapacitors, and fuel cells to manage power balance and ensure reliable energy delivery. Direct current microgrids that utilize renewable sources offer high energy efficiency and simpler design, but balancing supply and demand remains challenging [3].

The foundational aim of this research is to add value to scientific studies by underscoring the critical role of neuron count in optimizing energy storage management systems, facilitating the efficient utilization of renewable energy sources across diverse engineering applications. Hybrid microgrid systems face significant electrical challenges, particularly with PV generators, diesel generators, and lead-acid batteries [4]. The results explore the versatility of fuel cells in sustainable cities, where they provide backup power and enhance energy efficiency [5] compared to [6–8] that they achieved a value of training in order of 99.747%, 99.81%, and mean correlation between solar PV system and battery pack value of 99.765% respectively, but in the study we achieved 99.864% in a system that includes a lot of renewable energy resources.

The remainder of the chapter has been organized as follows: Section 4.2 proposes system and problematic study, Section 4.3 outlines review of machine learning applications and work methodology, Section 4.4 presents work methodology, Section 4.5 presents results and discussions, and Section 4.6 presents conclusion and future outlook.

4.2 System and problematic study

The referenced literature indicates that neural network time series (NNTS) using LM has been implemented in microgrids as shown in Figure 4.1. The major contribution of this work is the integration of a hydrogen reservoir as an alternative source in the storage system. The control method is based on the variation of the number of neurons, which will frankly optimize the performance of the control model, as we will show in Section 4.6, vis-à-vis other research in the field. The operational principles of PV cells and diodes are fundamentally similar, leading to equivalent modeling as p-n junctions [9]. The PV cell is shown; proton exchange membrane fuel cells (PEMFCs) and solid oxide fuel cells (SOFCs) are crucial for

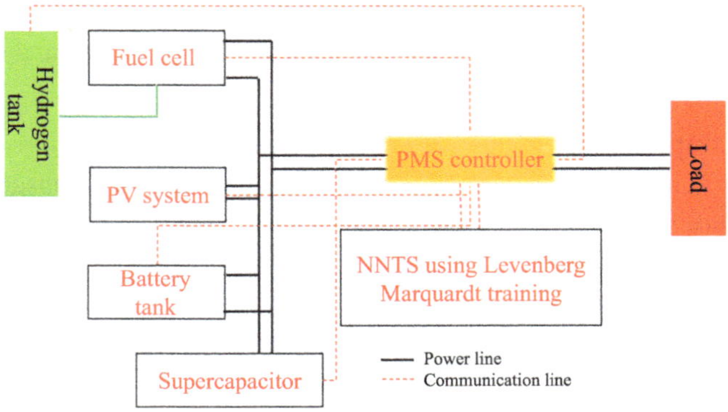

Figure 4.1 *System framework*

microgrid systems due to their efficiency, reliability, and environmental benefits [10]. PEMFCs are well suited for compact, low-temperature applications, whereas SOFCs provide high efficiency and durability at elevated temperatures [11]. Both types of fuel cells contribute to the stability and sustainability of microgrids by integrating effectively with renewable energy sources. Although fuel cells can operate on various fuels and oxidants, hydrogen is the preferred fuel due to its high reactivity and low catalyst cost, and oxygen is the preferred oxidant due to its abundance in atmosphere [12].

4.3 Review of machine learning applications and work methodology

4.3.1 Adaptive neural fuzzy inference system

The ANFIS is a hybrid model that integrates fuzzy logic (FL) with ANN to manage energy storage systems in microgrids. ANFIS has shown excellent performance and potential use among AI methods [13]. ANFIS is designed to handle the complexities of variable energy sources and fluctuating demands, providing several key advantages.

4.3.1.1 Efficient battery management

ANFIS is highly effective in maintaining the state of charge (SoC) of batteries within an optimal range. This capability ensures a stable and reliable energy supply, crucial for balancing the intermittent nature of renewable energy sources like solar and wind power. The study highlights the crucial role of artificial intelligence in battery systems for reliable power supply and energy management, showing significant improvements and promising results.

4.3.1.2 Fuel cell integration

ANFIS facilitates the integration of fuel cells into microgrid systems. During periods of energy shortage or when the battery SoC is low, the fuel cell can supply power to critical loads and simultaneously recharge the battery when conditions are favorable. This integration helps reduce greenhouse gas emissions and enhance the overall efficiency of the energy system [14].

4.3.1.3 Adaptability to variable conditions

ANFIS excels in managing energy across various production, consumption, and storage phases, even under variable solar irradiation conditions ranging from 0 W/m^2 to 1000 W/m^2. Its ability to adapt to changing conditions makes it a valuable tool for advancing microgrid technology [15].

Overall, ANFIS's combination of fuzzy logic and neural networks provides a powerful mechanism for optimizing energy management and integrating renewable energy sources effectively.

4.3.2 AdaBoost algorithm

AdaBoost, short for Adaptive Boosting, is a powerful ensemble learning algorithm that has demonstrated exceptional performance in managing energy systems, particularly within FCEVs. Key features of AdaBoost are mentioned below [16].

4.3.2.1 Exceptional accuracy

AdaBoost has achieved remarkable F1-score accuracies of 0.98 for fuel cell states and 0.90 for battery states. This high level of accuracy is crucial for effective decision-making in energy management and ensures that the system operates efficiently and reliably.

4.3.2.2 High recall

The algorithm's recall macro averages are 0.95 for the fuel cell and 0.84 for the battery. These metrics reflect AdaBoost's robustness in correctly identifying and managing different states of the energy system, further enhancing its suitability for energy management tasks.

4.3.2.3 Effective classification

AdaBoost's superior performance in classification tasks underscores its ability to handle complex decision-making processes, making it well suited for optimizing power management in microgrids and vehicle-grid integration scenarios. AdaBoost's strengths in classification and management make it a valuable tool for optimizing energy systems, particularly in dynamic and complex environments.

4.3.3 Random forest method

The RF method is renowned for its robustness and reliability in decision-making tasks, particularly within microgrid systems. Its advantages are mentioned below.

4.3.3.1 Precision and F1-scores

RF consistently delivers high macro-average precision factors of 90% for fuel cells, 86% for batteries, 84% for supercapacitors, and 82% for grid systems. Its macro-average F1-scores are similarly high, indicating effective management across various microgrid components.

4.3.3.2 Handling complex datasets

RF's ability to manage complex and imbalanced datasets is a key strength. The ensemble learning approach of RF, which aggregates predictions from multiple decision trees, enhances its accuracy and reliability.

4.3.3.3 Versatility and robustness

RF's robustness in handling diverse and complex datasets makes it a versatile tool for sustainable energy management. Its effectiveness in managing different microgrid components underscores its value in optimizing energy systems.

The RF method's combination of precision, versatility, and robustness makes it a leading choice for decision-making in microgrid management [17].

4.3.4 Neural fitting techniques and training algorithms

Neural fitting techniques are critical for simulating and optimizing energy management systems. The LM algorithm has emerged as a leading method in this domain due to the below mentioned reasons.

4.3.4.1 Superior performance

The LM algorithm demonstrates a training error of 0.00073 and a high correlation rate of 99.907%. Its performance during validation and testing is also exceptional, maintaining high correlation rates with minimal errors.

4.3.4.2 Comparison with other methods

The LM algorithm outperforms Bayesian regularization and scaled conjugate gradient methods, offering better performance in simulation accuracy and efficiency.

4.3.4.3 Potential for advanced integration

The algorithm's effectiveness in neural fitting highlights its potential for integrating advanced technologies such as FCEV to grid. Its high accuracy and efficiency make it a valuable tool for advancing energy management simulations.

The LM algorithm's superior performance in neural fitting underscores its potential for enhancing simulation accuracy and integrating advanced energy management technologies [18].

4.4 Methodology

4.4.1 Neural network time series

The time neural network is an advanced time-series forecasting method that enhances accuracy through kernel filters, time attention mechanisms, and adaptive weight adjustments, outperforming traditional models like recurrent neural network and long–short-term memory in terms of AE, and R^2, even with noisy data [19]. Figure 4.2 shows the proposed methodology framework.

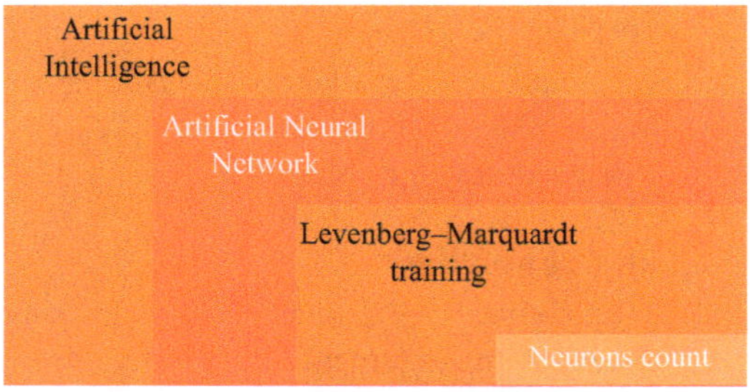

Figure 4.2 Proposed methodology framework

Table 4.1 Selected Levenberg–Marquardt model [23]

Parameters	Values
No of layers	16 (input layers, hidden layers, output layers)
Hidden layer neurons number	Variable
Input layer neurons	3
Output layer neurons	3
Number of hidden layers	10
Activation function	Sigmoid function
Training algorithm	LM
Number of epochs	1,000 (default)

In the model, $y(t)$ represents the target values, while $x(t)$ denotes the input data. The sigmoid function is applied in the hidden layer. The NNTS algorithm is trained using the LM method as shown in Figure 4.2, which is memory-intensive but efficient in terms of training time [20].

4.4.2 Levenberg-Marquardt algorithm

The LM algorithm is an advanced optimization technique that blends gradient descent with Gauss–Newton methods, typically falling under the category of trust-region approaches. It efficiently optimizes multilayer perception weights by minimizing errors effectively, even from distant starting points [21]. Bilski *et al.* [22] explores the influence of activation functions and LM parameters on neural network convergence, introducing the adaptive LM algorithm, which enhances both accuracy and efficiency by dynamically adjusting the descent direction and step size. In energy management simulations, the LM algorithm has demonstrated exceptional performance, achieving high determination rates and minimal errors, with a best mean squared error of 0.00081724. Despite some unresolved issues, it remains highly effective for small- to medium-sized networks and datasets, making it a recommended method for neural network training. Table 4.1 presents the values of the selected model parameters.

4.5 Simulation factors

4.5.1 Determination factor (R^2)

The determination coefficient, often denoted as R^2, measures both the strength and direction of the linear relationship between two variables. It ranges from −1 to 1, with the following interpretations:

✓ 1 indicates a perfect positive linear relationship,
✓ 0 indicates no linear relationship (the variables are independent),
✓ −1 indicates a perfect negative linear relationship.

In model evaluation, R^2 typically represents the square degree of agreement between predicted values and the actual observed values in the dataset. A determination coefficient close to 1 suggests that the model's predictions are closely aligned with the observed values, reflecting high accuracy and reliability. The formula for calculating the determination coefficient is as follows [24]:

$$R^2 = \frac{\sum_{i=1}^{n} [(x_i - \bar{x})(y_i - \bar{y})]^2}{\sum (x_i - \bar{x})^2 * \sum (y_i - \bar{y})^2} \quad (4.1)$$

4.5.2 Absolute error

Absolute error (AE), also known as root mean square deviation, is a widely used metric for assessing prediction accuracy. It measures the average deviation between predicted values and actual measurements by calculating the Euclidean distance of the discrepancies. The formula for AE is given in (4.2) [25], where:

✓ $y(i)$ denotes the ith actual measurement,
✓ $\widehat{y}_i$ denotes the corresponding predicted value.

$$\text{Absolute error} = y(i) - \widehat{y}(i) \quad (4.2)$$

4.6 Results and discussions

The PV system is the main energy source for the load, with its output varying directly with irradiation levels. To ensure uninterrupted production at the plant, maintaining a continuous power supply during sunset is critical. The load supply relies on the combined power contributions from the PV system: battery, fuel cell, and supercapacitor as shown in Figure 4.3. In keeping with environmental sustainability, it is essential that alternative power sources are green, eco-friendly, and silent. Hydrogen is a prominent clean energy option, while Li-ion batteries have shown robust performance in various applications, reflecting a significant technological trend.

The method achieved an R^2 value of 96.2126 and an AE of approximately 0.0379 during simulations with ten neurons. With 20 neurons, these values changed

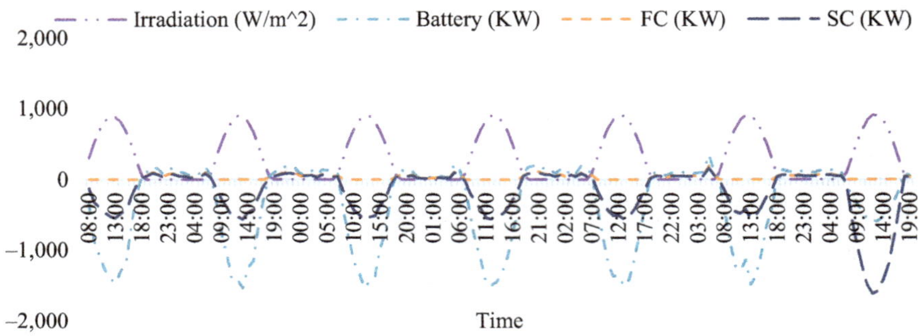

Figure 4.3 System elements powers behavior

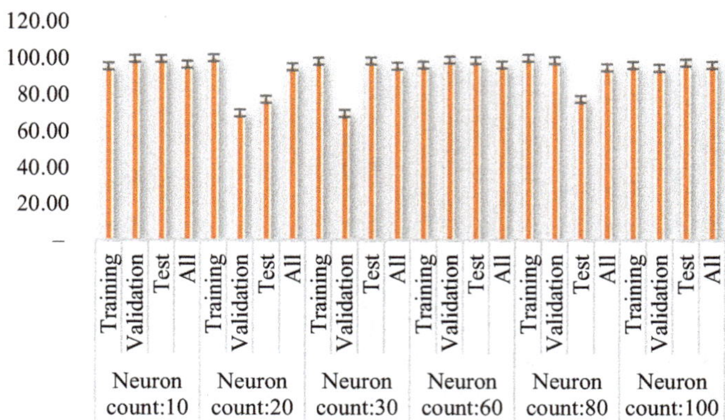

Figure 4.4 Regression analysis was conducted on the predicted and targeted outputs from the trained NNTS network, revealing the determined R values for each phase corresponding to varying numbers of neurons

to 94.9280 for R^2 and 0.0507 for AE. When the number of neurons increased to 30, the coefficient of determination reached 95.3045, with an AE of 0.0470. At 60 neurons, the R^2 value improved to 96.0988, and the AE decreased to 0.0390. Finally, for 80 neurons, the values were 94.5601 for R^2 and 0.0544 for AE, while at 100 neurons, the values were 95.8676 for R^2 and 0.0413 for AE, respectively. Figure 4.4 shows all regression analysis found in the study.

The fluctuation in the number of neurons improves the performance of the suggested approach over those containing a fixed number of neurons, comparable to the recent findings. With an accuracy of 99.864%, our study shows a significant improvement over the results shown in [6–8], where training accuracy values of 99.747%, 99.81%, and a mean correlation of 99.765% between the solar PV system and the battery pack were attained. This was achieved in a system that incorporates a wide range of renewable energy sources, demonstrating the scalability and resilience of our methodology in managing intricate, multi-source energy systems.

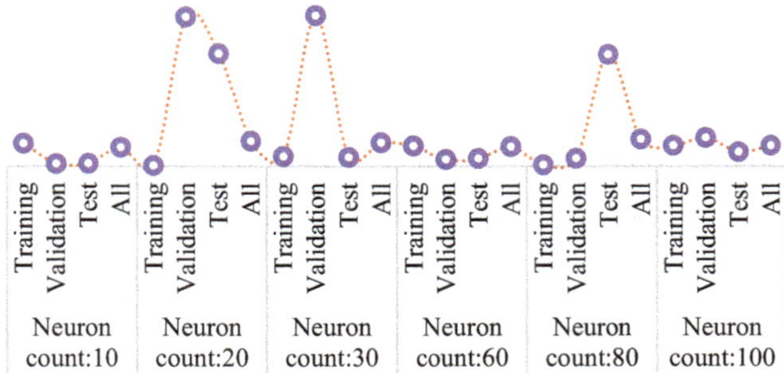

Figure 4.5 Training, validation, and test errors of NNTS strategy

The AE values obtained during the simulations exhibited significant variability across different neuron configurations. For a configuration of 10 neurons, the AE was measured at 0.0457. This value improved markedly to 0.0048 with 20 neurons and further decreased to 0.0056 at 30 neurons. However, at 60 neurons, the AE increased to 0.0379, before reaching a minimal value of 0.0014 at 80 neurons. At 100 neurons, a notable increase in AE was observed, rising to 0.3021. Additional values recorded included 0.2279 for 20 neurons, 0.0507 for 30 neurons, and 0.0197 for 60 neurons. Overall, the AEs ranged from a low of 0.0014 at 80 neurons to values of 0.0544 and 0.0412 at higher neuron counts, highlighting the intricate relationship between neuron configuration and model performance as shown in Figure 4.5.

4.7 Conclusion and future outlook

In conclusion, the proposed method excelled in simulating energy systems, achieving 99.864% accuracy and surpassing previous benchmarks (99.747%–99.81%). It demonstrated strong adaptability, robustness, and scalability for managing complex, multi-source energy systems. The AE varied across neuron configurations, from 0.0457 at 10 neurons to 0.3021 at 100 neurons, with a low of 0.0014 at 80 neurons, highlighting the complex relationship between neuron count and model performance.

The future prospects aim to implement an advanced control algorithm to stabilize, optimize, and secure energy storage systems.

References

[1] S. Ramadan, M. Al-Gabalaw, M. EL-Shimy, and A. Emarah, "Solid oxide fuel cell (SOFC) control strategy enhancement by adaptive neuro-fuzzy inference system (ANFIS)," *Sci. Rep.* 14, 17911 (2024).

[2] C. Saiprakash, S. R. Kumar Joga, A. Mohapatra, and B. Nayak, "Improved fault detection and classification in PV arrays using Stockwell transform and data mining techniques," *Results Eng.* 23, 102808 (2024).

[3] S. Sarwar, H. Zafar, M. Merlin, A. Arsalan, B. Papari, and A. Kiprakis, "Enhanced energy balancing and optimal load curtailment strategy for DC microgrid integration in hybrid AC/DC distribution networks," *IET Smart Grid* 7, 400–411 (2024).

[4] W. Chine, A. Mellit, A. M. Pavan, and S. A. Kalogirou, "Fault detection method for grid-connected photovoltaic plants," *Renew. Energy* 66, 99–110 (2014).

[5] H. Rezk and A. Fathy, "Hydrogen reduction-based energy management strategy of hybrid fuel cell/PV/battery/supercapacitor renewable energy system," *J. Energy Storage* 86, 111316 (2024).

[6] U. E. Ammara, S. S. Zehra, S. Nazir, and I. Ahmad, "Artificial neural network-based nonlinear control and modeling of a DC microgrid incorporating regenerative FC/HPEV and energy storage system," *Renew. Energy Focus* 49, 100565 (2024).

[7] M. A. Saddriwala and M. Alam, "Artificial neural network based decentralized current-sharing control for parallel connected DC-DC converters in DC microgrid application," *Comput. Electr. Eng.* 120, 109731 (2024).

[8] C. Ghenai, F. F. Ahmad, and O. Rejeb, "Artificial neural network-based models for short term forecasting of solar PV power output and battery state of charge of solar electric vehicle charging station," *Case Stud. Therm. Eng.* 61, 105152 (2024).

[9] K. I. Baradieh, M. A. A. M. Zainuri, N. A. Mohamed Kamari, *et al.*, "A study on the impact of different PV model parameters and various DC faults on the characteristics and performance of the photovoltaic arrays," *Inventions* 9, 93 (2024).

[10] B. Kanani and A. Zahedi, "Step toward sustainable development through the integration of renewable energy systems with fuel cells: A review," *Sustain. Energy Technol. Assess.* 70, 103935 (2024).

[11] K. Gunawardane, N. Padmawansa, and H. Jayasinghe, "A review: Compatibility of fuel cells as promising technology for DC-microgrids," *Renew. Energy Environ. Sustain.* 9, 7 (2024).

[12] X. Huang, Z. Zhang, and J. Jiang, "Fuel cell technology for distributed generation: An overview," in *2006 IEEE International Symposium on Industrial Electronics* (IEEE, 2006), pp. 1613–1618.

[13] E. N. Odonkor, P. M. Moses, and A. O Akumu, "Intelligent ANFIS-based distributed generators energy control and power dispatch of grid-connected microgrids integrated into distribution network," *Int. J. Electr. Electron. Eng. Telecommun.* 13, 112–124 (2024).

[14] I. Elabbassi, M. Khala, N. E. Yanboiy, O. Eloutassi, and Y. E. Hassouani, "Neural network for FCEVs and RM power management using V2G technology," in *2024 International Conference on Circuit, Systems and Communication (ICCSC)* (IEEE, 2024), pp. 1–8.

[15] I. Elabbassi, N. Elyanboiy, M. Khala, *et al.*, "Adaptive neural fuzzy inference system (ANFIS) in a grid connected-fuel cell-electrolyser-solar PV-battery-super capacitor energy storage system management," in *Artificial Intelligence and Smart Environment*, Y. Farhaoui, A. Rocha, Z. Brahmia, and B. Bhushab (eds.), Lecture Notes in Networks and Systems (Berlin: Springer International Publishing, 2023), Vol. 635, pp. 138–143.

[16] I. Elabbassi, N. Elyanboiy, M. Khala, Y. El Hassouani, O. Eloutassi, and C. Messaoudi, "Comparative study of machine learning for managing EV energy storage with battery-hydrogen tank," in *Advances in Electrical Systems and Innovative Renewable Energy Techniques*, M. Bendaoud, A. El Fathi, F. I. Bakhsh, and S. Pierluigi (eds.), Advances in Science, Technology & Innovation (Cham: Springer Nature, 2024), pp. 215–221.

[17] I. Elabbassi, M. Khala, N. El Yanboiy, Y. E. Hassouani, O. Eloutassi, and C. Messaoudi, "Comparative study on machine learning based decision-making for microgrid component operation," in *Artificial Intelligence, Big Data, IOT and Block Chain in Healthcare: From Concepts to Applications*, Y. Farhaoui (ed.), Information Systems Engineering and Management (Switzerland: Springer Nature, 2024), Vol. 5, pp. 412–417.

[18] I. Elabbassi, M. Khala, N. El Yanboiy, O. Eloutassi, and Y. El Hassouani, "Evaluating and comparing machine learning approaches for effective decision making in renewable microgrid systems," *Results Eng.* 21, 101888 (2024).

[19] L. Zhang, R. Wang, Z. Li, *et al.*, "Time-series neural network: A high-accuracy time-series forecasting method based on Kernel filter and time attention," *Information* 14, 500 (2023).

[20] M. Kulisz, J. Kujawska, M. Cioch, W. Cel, and J. Pizoń, "Comparative analysis of machine learning methods for predicting energy recovery from waste," *Appl. Sci.* 14, 2997 (2024).

[21] A. Alshawabkeh, M. Matar, and F. Almutairy, "Parameters identification for lithium-ion battery models using the Levenberg–Marquardt algorithm," *World Electr. Veh. J.* 15(9), 406 (2024).

[22] J. Bilski, J. Smoląg, B. Kowalczyk, K. Grzanek, and I. Izonin, "Fast computational approach to the Levenberg-Marquardt algorithm for training feedforward neural networks," *J. Artif. Intell. Soft Comput. Res.* 13, 45–61 (2023).

[23] S. Ali, S. Riaz, Safoora, X. Liu, and G. Wang, "A Levenberg–Marquardt based neural network for short-term load forecasting," *Comput. Mater. Contin.* 75, 1783–1800 (2023).

[24] M. Kirisci, "Measures of distance and entropy based on the Fermatean fuzzy-type soft sets approach," *Univers. J. Math. Appl.* 7, 12–29 (2024).

[25] F. F. Fadoul, A. A. Hassan, and R. Çağlar, "Integrating autoencoder and decision tree models for enhanced energy consumption forecasting in microgrids: A meteorological data-driven approach in Djibouti," *Results Eng.* 24, 103033 (2024).

Chapter 5

Energy management and voltage control strategy in microgrid using PV and battery

Ouadiâ Chekira[1], Hassan El Moussaoui[1] and Tijani Lamhamdi[1]

Abstract

This work proposes an efficient energy management strategy for a hybrid microgrid system including photovoltaic (PV) arrays and battery storage units, aimed at maintaining constant DC bus voltage and ensuring a reliable energy supply for isolated off-grid DC loads. The EMS uses a proportional–integral–derivative (PID) controller to regulate DC bus voltage and maintain stable power levels under fluctuating solar irradiation conditions. The system uses a perturb and observe (P&O) algorithm to maximize power point tracking (MPPT) by regulating the duty cycle of the bidirectional converter that connects the lithium-ion battery on the DC bus.

This dual-mode EMS is capable of controlling microgrid voltage while vectoring the power flow through batteries such that it can achieve superior stability on the DC bus, better cycle life for batteries, and lower operating costs. Simulation findings in MATLAB®/Simulink® confirm that the proposed method improves energy management, prolongs battery service life, and keeps the voltage stable under different level of irradiances.

Keywords: Microgrids; energy management system (EMS); battery management system (BMS); state of charge (SOC); voltage control strategy

5.1 Introduction

Microgrids have recently sparked significant attention in electrical energy distribution processes due to their extensive capabilities for promoting and strengthening the integration of renewable sources. They help to increase both quality and system efficiency. Distributed energy resources allow that consumers can utilize energy in a flexible manner while the system satisfies load demands at all times [1].

[1]Laboratory of Intelligent Systems, Geo-resources and Renewable Energies (SIGER), Sidi Mohamed ben Abdellah University, Morocco

A microgrid's electrical design is mostly dictated by its application, infrastructure, and user requirements. In general, microgrids are classified into three types: AC microgrids, DC microgrids, and hybrid microgrid. They provide numerous benefits such as allowing end-users to actively participate in demand response initiatives. An microgrid central controller for a microgrid can find optimal technologies and resource plans that save the overall costs while maintaining high dependability and security levels [2]. Supply-demand balance is a fundamental principle in electrical systems. Because power has very limited storage capacity, its availability in real time is critical, regardless of season or time of day. Production must always be balanced by consumption [3]. For example, when consumption exceeds production, the power grid turns off production sites due to its incapacity to accommodate increased demand and avoid a blackout [4]. As a result, an energy management system (EMS) typically incorporates an objective function designed to enhance the efficiency of hybrid renewable energy systems while minimizing energy consumption, hence elevating service quality for end-users. An EMS control plan can be established that prioritizes power availability, facilitating transitions among renewable energy sources (RES), storage devices, and the utility grid as required, without regard to electricity pricing or system profitability [5]. Alternatively, several EMS control mechanisms are developed to interact directly with power suppliers by placing limitations on power production to improve resource distribution and sustain system balance [6]. EMS is even more important, particularly when numerous resources must be managed, for improved power flow management and fulfilling load demand with lowest operational expenses [7]. To enhance electric energy utilization in outlying regions, the multi-objective particle swarm optimization technique is employed [8]. Ghaffari and Askarzadeh [9] optimized and sized a hybrid system using the Crow search technique. The probability of power outage and the proportion of RES are two limitations that are taken into account in order to reduce the overall net cost. Dash ad Bajpai [10] presents a power management system that coordinates the use of RES, fuel cells (FCs), and battery management systems (BMS) in conjunction with electrolyzer-supplied hydrogen; however, DC bus stabilization is not integrated into this system.

However, voltage is one of the most critical elements for ensuring the safety, stability, and proper operation of power systems. A good control method would enhance power quality while keeping voltage variations to acceptable levels. Various ways have been explored to ensure the stability of microgrid systems [11]. The Li-ion batteries can be connected to the DC bus via a bidirectional DC/DC converter, which allows charging and discharging to regulate the DC bus voltage [12]. As a result, ESS contributes significantly to system stability and availability. In this regard, a voltage control strategy is proposed that addresses energy management in an isolated microgrid equipped with PV panels and battery energy storage to meet load demand [13]. This strategy uses MPPT to run all of the power producing units at full capacity. A PI controller is used to control energy converters in a microgrid based on solar PV. A new energy management method, inspired by PI control with reference current, is intended to withstand weather variations. The chapter is organized as follows. Section 5.2 details the suggested system, Section 5.3 displays the energy management technique used, Section 5.4 discusses the outcomes, and Section 5.5 ends the work.

5.2 Proposed configuration

Figure 5.1 depicts the proposed design for a freestanding microgrid. To meet the load requirements, the proposed microgrid consists of a solar power generator and battery energy storage of batteries connected to the DC bus via a bidirectional converter and main grid from point of common coupling) when needed to connect equipment to a utility power grid.

5.2.1 System modeling

5.2.1.1 Modeling of PV panel

The PV generator used in this study is built up of interconnected modules that combine to form a high-constant energy-production unit that is compatible with standard electrical equipment [14]. The circuit model employed in this study consists of a current source that generates the current I_{ph} due to light incidence, two diodes, and two resistances, as illustrated in Figure 5.2, R_{sh} symbolizes the leakage current generated by ground losses, whereas R_{s} represents internal losses induced by solar cell current flow.

The two-diode model proposed in this study can better depict the non-idealities of a solar panel's behavior under varied weather circumstances [15].

The output current of the PV system can be expressed as

$$I_{PV} = I_{ph} - I_d - I_r \tag{5.1}$$

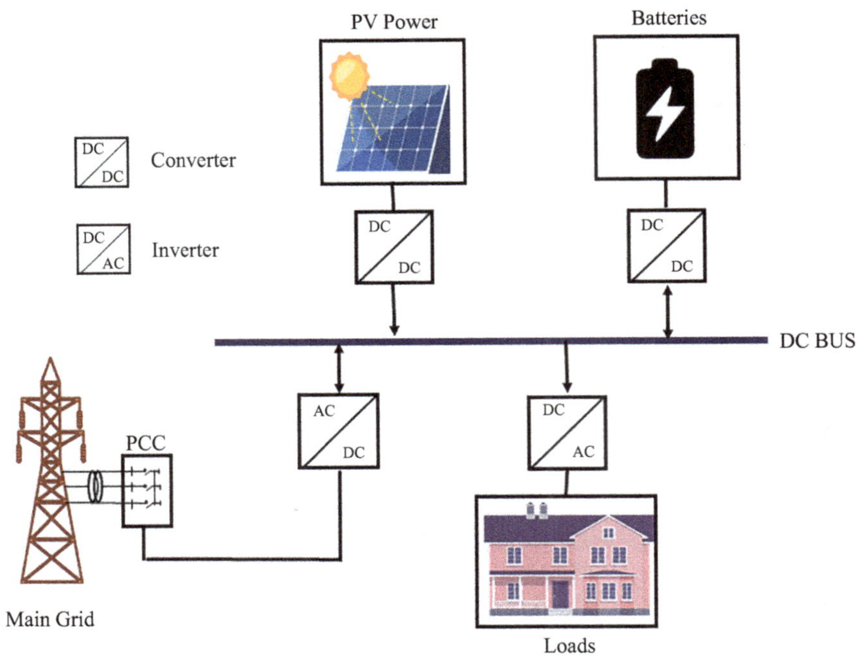

Figure 5.1 Proposed microgrid structure

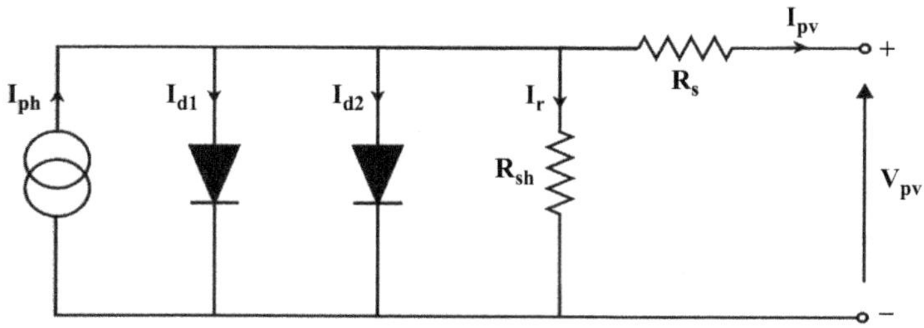

Figure 5.2 The solar cell's equivalent circuit with two diodes [15]

with

$$I_d = I_{d1} + I_{d2} = I_{01}\left(e^{\frac{V_{PV}+R_S I_{PV}}{n1V_T}} - 1\right) + I_{02}\left(e^{\frac{V_{PV}+R_S I_{PV}}{n2V_T}} - 1\right) \tag{5.2}$$

$$I_r = \frac{V_{PV} + R_s I_{PV}}{R_{sh}} \text{ and } V_T = \frac{K \times T}{q} \tag{5.3}$$

V_T: The thermal voltage;
n: The diode ideality constant;
T: The temperature of the p-n junction;
q: The charge of electron;
k: Boltzmann constant;
I_{01}: Reverse saturation currents of diode 1;
I_{02}: Reverse saturation currents of diode 2; and
I_{ph}: The photocurrent.

5.2.1.2 BES modelling

The BES discharges when RES are insufficient to power the load (based on SOC) and recharges from PV when there is extra generation. The modulation of dynamic batteries is [16]

$$SOC_{bat} = 100\left[1 - \left(\frac{1}{Q_{bat}} \cdot \int_0^t i_{bat}(t)\, dt\right)\right] \tag{5.4}$$

where Q_{bat} is the maximum battery capacity (Ah), i_{bat} is the battery current (A), and SOC_{bat} is the battery state of charge (SOC) (%).

5.2.1.3 Bidirectional DC–DC converter

Figure 5.3 shows an IGBT half-bridge battery-fed bidirectional DC–DC converter. In CCM, the switches function. There are two modes of operation: **boost mode** involves turning on switches S2 and diode D1 to drain the battery and feed current to the DC link; **buck mode** involves turning on switches S1 and diode D2 to reverse the power flow and charge the battery [17–19]. The battery can be charged and discharged using the buck-boost DC–DC converter utilized in this project [20].

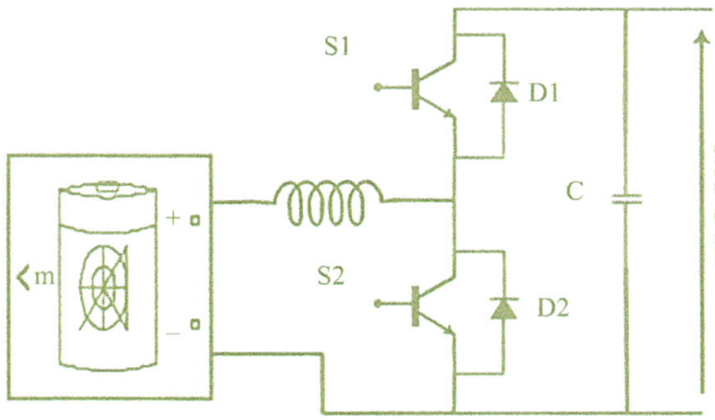

Figure 5.3 The topology of DC–DC converter

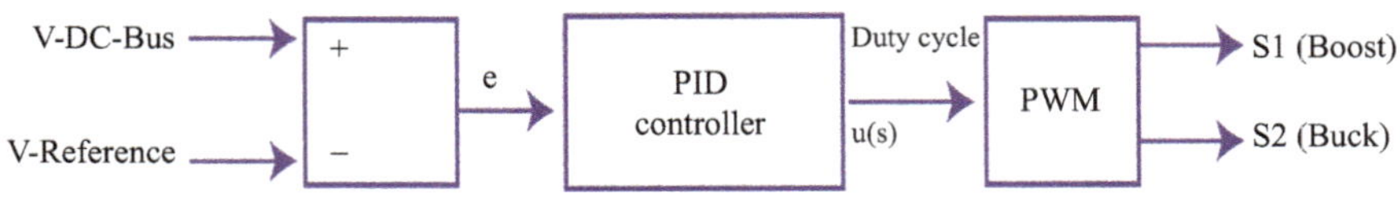

Figure 5.4 Implementation schematic for the PID controller [22]

5.3 Proposed control for energy management

With this method, the battery and supercapacitor share control over the DC bus voltage. The first step is to compare the reference DC bus voltage $V_{Reference}$ with the observed DC bus voltage $V_{DC\text{-}Bus}$ [21]. The reference current is obtained by feeding; the PID controller reduces the error signal by comparing the measured output signal with their intended references to determine the error signal, which is their input as shown in Figure 5.4.

The controller generates the error signal (*e*) by amplifying the difference between the measured voltage $V_{DC\text{-}Bus}$ and the desired voltage $V_{Reference}$. The PID controllers reduce the error signal by comparing the measured output signal with their intended references to determine the error signal, which is their input. The controller generates the error signal (*e*) by amplifying the difference between the measured voltage $V_{DC\text{-}Bus}$ and the desired voltage $V_{Reference}$ [22].

The duty cycle of each direction was produced by processing the difference between these voltage values in PID blocks, based on the voltage and battery state of charge.

The operational modes of the SOC-based and DC bus voltage-based EMS, which include surplus mode, deficit mode, and balance mode, are depicted in the flowchart below. These three modes specify whether the system should stay in standby mode or whether battery charging or discharging is required. The flowchart's subsequent sections match up with these modes as shown in Figure 5.5.

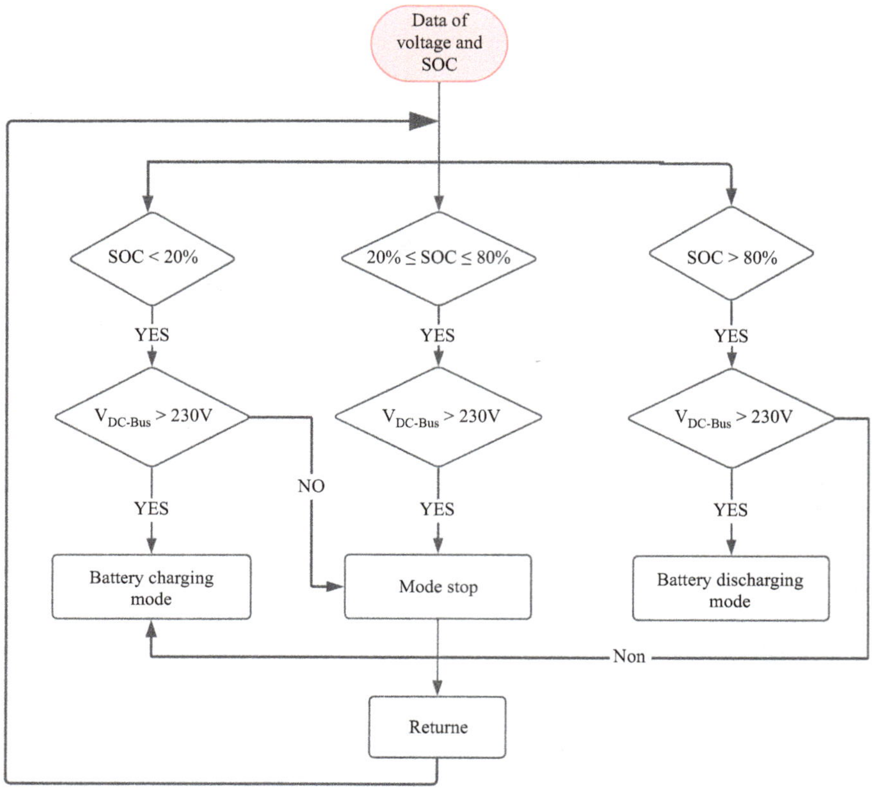

Figure 5.5 Flowchart of energy management strategy

(a) **Surplus mode (battery charging mode)**
When there is power that is not being used and the battery's charging capabilities need to be engaged, the system transitions to this mode. When the system's bus voltage is sufficiently high and its SOC is low, this scenario may occur.

$$Condition : (SOC < 20\%, V\_DC - Bus > 220V)$$

Action: The battery enters charging mode, in which surplus power is used to recharge the battery. In these cases, there is an endeavor to gather as much energy as possible within the operating system because SOC is low and there is enough bus voltage to continue charging.

Scenario: The system generates more energy than it consumes, allowing the battery to be charged.

(b) **Deficit mode (battery discharging mode)**
This mode is used when there is a power outage and the battery needs to be discharged to meet the system's power demands. This, however, is most

common when the SOC is quite high and the voltage exceeds a particular value, indicating energy consumption.

$$condition\ SOC > 80\%, V\_DC - Bus > 220V$$

Action: The battery is regarded to be in a battery discharging mode, which means that the load or voltage levels are sustained utilizing the battery's stored energy. This happens when the battery is almost completely charged and there is a need for energy from the storage.

Scenario: The system has a power shortage and requires energy from the battery to meet the load.

(c) **Balance mode (mode stop)**

This is the mode in which no volume of charge or discharge is necessary. It occurs when the SOC is in an intermediate condition (not too low, nor too high) and the DC bus voltage has achieved a level that can preserve system stability without the need of the battery.

$$Condition : 20\% \leq SOC \leq 80\%\ and\ V\_DC - Bus > 220V$$

Action: The system enters mode stop, which interrupts all charging and discharging processes. There is a steady mode in which the energy supply balances the demand, eliminating the need for the battery to be used.

Scenario: The generation and consumption are about equal, and the battery is not required.

$$Condition : V\_DC - Bus \leq 220V\ but\ 20\% \leq SOC \leq 80\%$$

Action: Without carrying out any battery activities, the system resumes monitoring; this indicates that the voltage is too low for balance but that neither charging nor discharging is necessary.

Scenario: The system is in balance mode, but no more battery operations are being done, and the voltage is not high enough for complete stability.

5.4 Results and discussions

A power converter links the PV energy source to the microgrid's DC bus to enhance power generation. A 10 kW resistive load is powered by the MG system, which harnesses energy from the sun's fluctuating rays. To elevate the operational voltage, 36 solar panels linked in series produce a maximum of 14 kW of power under optimal irradiation conditions. A bidirectional DC/DC converter links lithium-ion battery packs to the system, improving the reliability of the supplied electricity.

Simulations in MATLAB®/Simulink® were used to test how well the system worked in changing situations and to get the most power out of each source. Each source of energy has a boost converter built in. The P&O MPPT method is used to find the PV panel's highest power point. Table 5.1 lists the properties of the

Table 5.1 System component specification [17]

System component	Specification
PV power	V_{PV} = 14 kW
Load	P_{Load} =10 kW
Battery voltage	V_B = 215 V
DC bus voltage	$VV_{DC\text{-}Bus}$ = 230 V

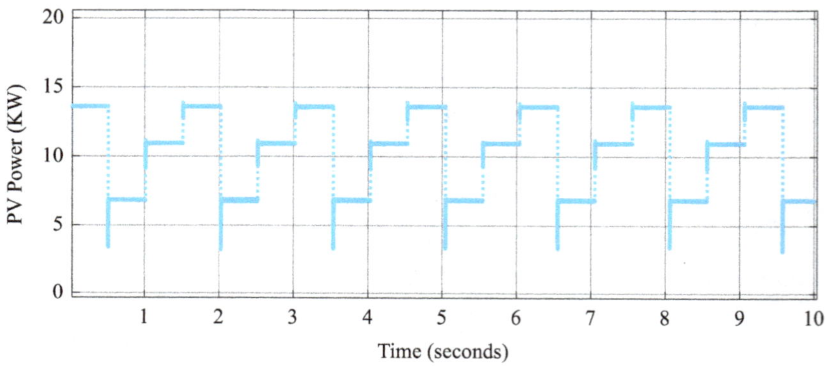

Figure 5.6 Radiation from the sun (W/m²)

system's parts, and Figure 5.6 shows how the power output changed at different amounts of irradiance over a 10-s modeling period with a sampling time of 125e−6. When the PV generator is in the MPPT mode, it keeps its power efficiency even when the amount of sunlight hitting it changes. Temperature doesn't have much of an effect on PV power, so the system is set to 25 °C and the amount of sunlight changes throughout the day.

A comparison was made with other recent research on microgrid stability and efficiency to elucidate the importance of these results. The performance of our PV-based microgrid system is demonstrated to be superior for stability and dependability under varying irradiance circumstances, showcasing its improved energy management. In contrast to traditional methods, our system utilizes a bidirectional DC/DC converter, which guarantees that excess power at the DC bus is regulated by the battery, as seen in Figure 5.7. This method mitigates instability by allowing the battery to provide the necessary power at the DC bus, so providing a consistent power output.

Since temperature has little effect on PV power, we set the system to 25 °C and a range of solar radiation. However, fluctuations in PV power are directly impacted by solar radiation. As shown in Figure 5.7, the power generated is unstable. At the start of the simulation, the power produced from PV generator is equal to 14 kW which means more than 10 kW (power demand from load), so there is a surplus of the power at the DC bus. The battery will therefore supply the remaining power at the DC bus.

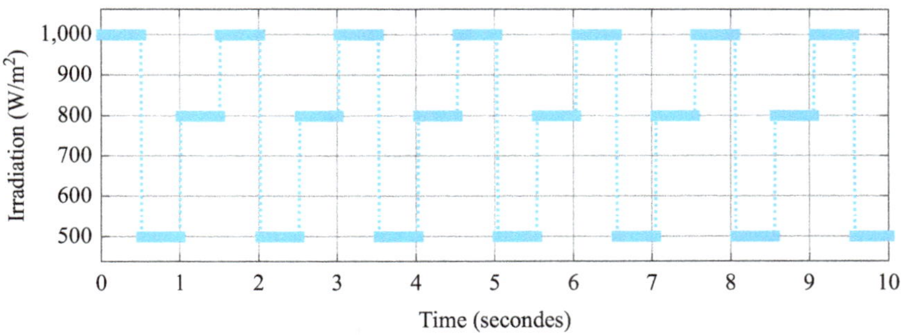

Figure 5.7 PV power (W)

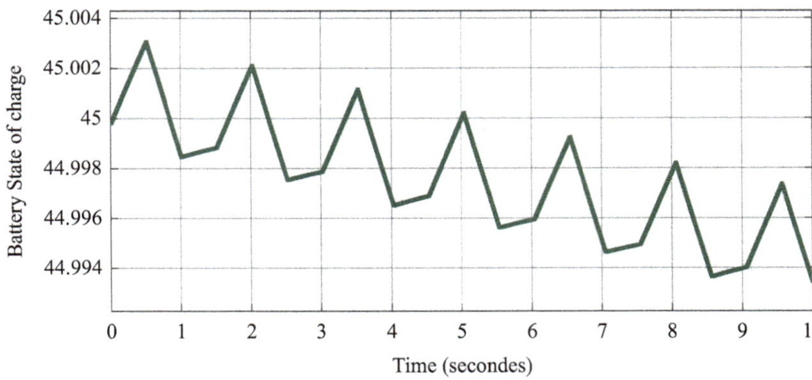

Figure 5.8 Battery state of charge (%)

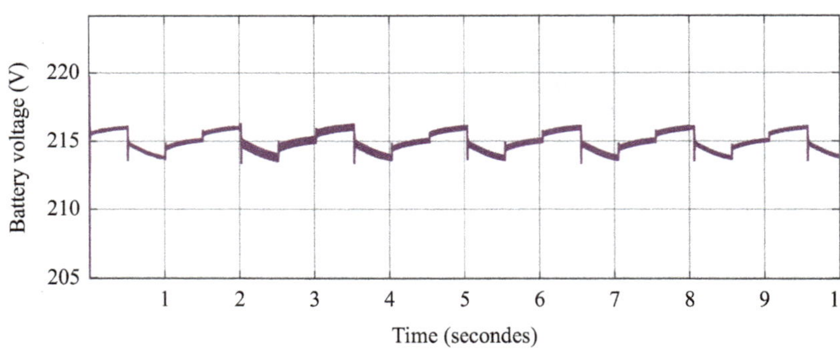

Figure 5.9 Battery voltage (V)

In order to store the excess power, the battery state of charge in Figure 5.8 rises. At t = 0.5 s, the solar panel system's power drops to 7 kW, indicating that more energy is required at the DC bus to meet the load demand. Figure 5.9 depicts the battery voltage change at charging and discharging modes.

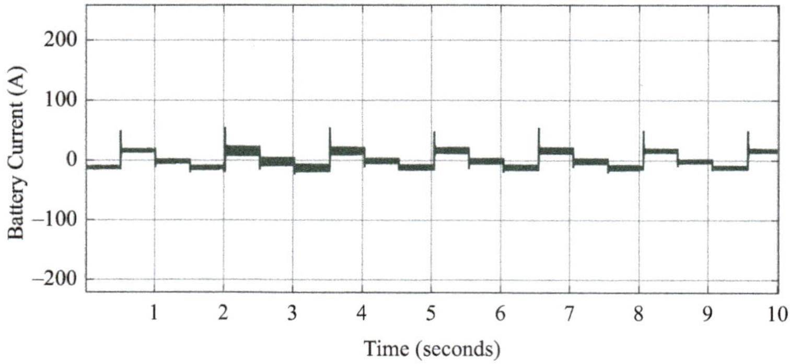

Figure 5.10 Battery current (A)

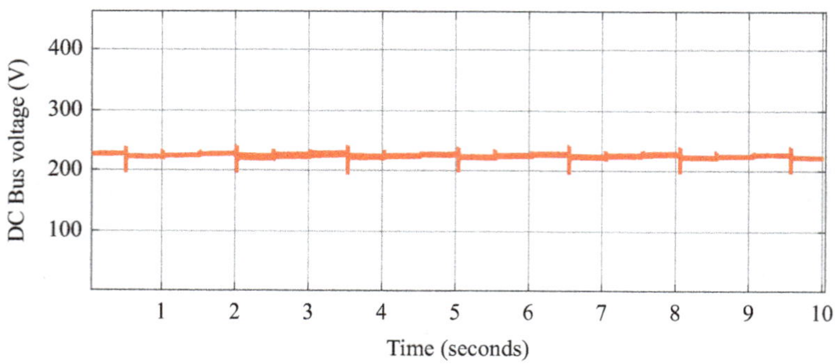

Figure 5.11 Voltage of DC bus

The PV output immediately decreases as a result of a drop in sun irradiation from t = 0.5 s to t = 1 s, as illustrated in Figure 5.10. As a result, there will be a minor decrease of the VB since the battery will be covered with energy to meet the changing needs. The current IB is negative when the battery is charged. However, the current IB is positive when the battery is in the state of discharge.

For the purpose of representing the change in voltage of the DC bus, a PI control approach is utilized, as can be seen in Figure 5.11. The results of the simulation demonstrate that the method is able to keep the DC bus voltage stable at a constant 230 V. This indicates that even if there are a few oscillations throughout the process, the voltage quickly returns to normal without putting the DC bus in danger.

5.5 Conclusion

When used properly, renewable technologies minimize environmental effects and generate the least amount of sustainable and secondary waste, making them clean energy sources. They provide enormous promise for lowering global warming and greenhouse gas emissions as well as transmission and distribution losses. This study offered an energy management plan based on RES. The main goal of this effort was to balance the power production and load demand. We used a control technique using PID for energy management in this contribution. The control strategy is predicated on optimizing the energy generated by renewable resource, managing battery charging and discharging to maintain DC bus voltage stability, and then safeguarding the battery by keeping its charge level between 20% and 80%.

References

[1] Habib HU, Waqar A, Junejo AK, *et al.* Optimal planning and EMS design of PV based standalone rural microgrids. *IEEE Access*. 2021;9:32908–30.

[2] Heidary J, Gheisarnejad M, and Khooban MH. Stability enhancement and energy management of AC–DC microgrid based on active disturbance rejection control. *Electric Power Systems Research*. 2023;217:109105.

[3] Chekira O, Boharb A, Lamhamdi T, El Moussaoui H, El Markhi H, and Beniss MA. Frequency control and energy management of microgrid with distributed energy storage. *E3S Web of Conferences* 2022;351:01047.

[4] El-Hendawi M, Gabbar HA, El-Saady G, and Ibrahim EN. Control and EMS of a grid-connected microgrid with economical analysis. *Energies*. 2018;11 (1):129.

[5] Prasad MBS and Chennaiah PB. Power management and control for PV integrated microgrid with battery energy storage system. In *IEEE 5th International Conference on Electronics and Sustainable Communication Systems (ICESC)* 2024 (pp. 1–6).

[6] Boutros F, Doumiati M, Olivier JC, Mougharbel I, and Kanaan HY. Optimal sizing assessment of an islanded microgrid based on different EMS strategies. *In 2023 6th International Conference on Renewable Energy for Developing Countries (REDEC)* 2023 (pp. 42–47). IEEE.

[7] Alhasnawi BN, Jasim BH, Sedhom BE, and Guerrero JM. A new communication platform for smart EMS using a mixed-integer-linear-programming. *Energy Systems*. 2023. https://doi.org/10.1007/s12667-023-00591-2.

[8] Indragandhi V, Logesh R, Subramaniyaswamy V, Vijayakumar V, Siarry P, and Uden L. Multi-objective optimization and energy management in renewable based AC/DC microgrid. *Computers & Electrical Engineering*. 2018;70:179–98.

[9] Ghaffari A and Askarzadeh A. Design optimization of a hybrid system subject to reliability level and renewable energy penetration. *Energy*. 2020; 193:116754.

[10] Dash V and Bajpai P. Power management control strategy for a stand-alone solar photovoltaic-fuel cell–battery hybrid system. *Sustainable Energy Technologies and Assessments*. 2015;9:68–80.

[11] Stanchev P, Vacheva G, and Hinov N. Evaluation of voltage stability in microgrid-tied photovoltaic systems. *Energies*. 2023;16(13):4895.

[12] Montegiglio P, Acciani G, Dicorato M, Forte G, and Marasciuolo F. A decentralized power and bus voltage regulation approach for DC microgrids. *IEEE Transactions on Industry Applications*. 2023;59(4):4773–85.

[13] Zhao C, Andersen PB, Træholt C, and Hashemi S. Grid-connected battery energy storage system: A review on application and integration. *Renewable and Sustainable Energy Reviews*. 2023;182:113400.

[14] Maka AO and Chaudhary TN. Performance investigation of solar photovoltaic systems integrated with battery energy storage. *Journal of Energy Storage*. 2024;84:110784.

[15] Chekira O, Boujoudar Y, El Moussaoui H, Boharb A, Lamhamdi T, and El Markhi H. An improved microgrid energy management system based on hybrid energy storage system using ANN NARMA-L2 controller. *Journal of Energy Storage*. 2024;98:113096.

[16] Chekira O, Boharb A, Boujoudar Y, El Moussaoui H, Lamhamdi T, and El Markhi H. An improved energy management control strategy for a standalone solar photovoltaic/battery system. *Indonesian Journal of Electrical Engineering and Computer Science*. 2022;27(2):647–58.

[17] Veerachary M and Khuntia MR. Design and analysis of two-switch-based enhanced gain buck–boost converters. *IEEE Transactions on Industrial Electronics*. 2021;69(4):3577–87.

[18] Elottri A, Mzouz L, Kouzou A, Teta A, Nekbil N, and Boutouta F. Enhancing grid performance of single-stage PV systems using modified SRF theory and SAPF with Quasi-Z-Source inverter. *Electrotehnica, Electronica, Automatica*. 2024;72(3):3–15.

[19] Abdolahi M, Adabi J, and Mousazadeh Mousavi SY. Implementation and control of a buck–boost converter connected to a constant power load in a DC microgrid. *Electrical Engineering*. 2025;107:1483–92.

[20] Ma Z, Gao F, Li W, *et al.* Two-mode balancing control scheme for modular DC/DC converter interfaced battery swapping stations. *IEEE Transactions on Transportation Electrification*. 2025;11(1):1640–51.

[21] Al Sumarmad KA, Sulaiman N, Wahab NI, and Hizam H. Energy management and voltage control in microgrids using artificial neural networks, PID, and fuzzy logic controllers. *Energies*. 2022;15(1):303.

[22] Ullah Z, Wang S, Lai J, *et al.* Implementation of various control methods for the efficient energy management in hybrid microgrid system. *Ain Shams Engineering Journal*. 2023;14(5):101961.

Chapter 6

Small scale renewable energies and storage for microgrids

Amanjot Kaur[1], *Pankaj Kumar*[1] *and Jasmeet Kaur*[1]

Abstract

Small-scale renewable energy systems, combined with advanced energy storage solutions, are transforming the sustainability and reliability of microgrids. With microgrids playing a vital role in decentralized power generation, incorporating renewable sources like solar, wind, and biomass helps minimize carbon emissions and boost energy security. These small-scale renewable solutions are customized to meet the unique energy requirements of local communities, businesses, or remote areas, delivering flexibility and strengthening the resilience of power systems. The intermittent availability of renewable energy sources presents a challenge to maintaining a consistent energy supply. To overcome this, advanced energy storage technologies like batteries, supercapacitors, and thermal storage systems are being implemented. These storage solutions enable energy to be captured during peak production times and utilized when renewable sources are not available, ensuring a reliable power supply. Furthermore, the integration of smart energy management systems optimizes energy distribution, significantly improving the efficiency of microgrids. In rural or remote regions with limited access to the central grid, microgrids powered by small-scale renewable energy sources provide a self-sufficient and cost-effective solution. These systems decrease reliance on fossil fuels, reduce electricity costs, and promote the decentralization of energy generation. As renewable energy technologies and storage solutions continue to advance, microgrids are poised to play a crucial role in the future of sustainable energy infrastructure, offering a more resilient and environmentally friendly energy landscape.

Keywords: Microgrids; renewable energy; energy storage; distributed generation; sustainability; grid integration

6.1 Introduction

The global energy landscape is experiencing a significant transformation driven by the increasing demand for cleaner, more sustainable, and decentralized energy

[1]Department of Physical and Mathematical Sciences, Baba Farid College, India

systems. In response to this change, microgrids are emerging as localized energy networks that can function autonomously or in coordination with the main grid (Lasseter, 2002). Microgrids incorporate distributed energy resources (DERs), energy storage systems (ESS), and advanced control technologies to improve energy reliability, resilience, and sustainability (Lund *et al.*, 2015). In light of the negative environmental effects of fossil fuels and the fluctuations in global energy markets, microgrids that utilize small-scale renewable energy (RE) sources offer a crucial solution for a decentralized, efficient, and cleaner energy supply.

RE sources, especially on a smaller scale, are essential for sustainably powering microgrids. Technologies like solar photovoltaic (PV) panels, small wind turbines, and biomass generators enable localized energy production, thereby decreasing reliance on centralized fossil-fuel-based power plants (Zhang and Zhang, 2018). Small-scale solar and wind systems are especially well suited for microgrids because of their modularity, scalability, and adaptability to various geographical locations. For example, solar PV technology has experienced significant improvements in efficiency and cost-effectiveness, making it accessible for residential, commercial, and community microgrids (Saeed *et al.*, 2021).

Alongside solar and wind energy, other renewable technologies like micro-hydropower and bioenergy systems are being increasingly incorporated into microgrids. While micro-hydropower may be restricted by geographical limitations, it can offer a consistent and dependable power source in rural regions (Bhattacharyya, 2014). Biomass-based energy, which transforms organic waste into usable electricity, is also becoming more popular, especially in agricultural areas where biomass resources are plentiful (Kumar *et al.*, 2015). The integration of these small-scale renewable resources allows microgrids to meet various energy demands while lowering greenhouse gas emissions and enhancing energy access.

Incorporating RE into microgrids necessitates efficient ESS, which are crucial for managing the intermittent nature of renewable sources (Li and Wang, 2022). Battery storage solutions allow for the capture and distribution of energy generated from renewables as needed, helping to balance supply and demand fluctuations while enhancing grid stability. Additional storage technologies, including flywheels and supercapacitors, are employed in microgrids to facilitate short-term power management and stabilize fluctuations in voltage and frequency (Singh and Kumar, 2021). Innovative storage solutions, including hydrogen fuel cells, pumped hydro storage, and compressed air energy storage, are being investigated for microgrid applications. For example, hydrogen fuel cells provide the capability for large-scale energy storage and can be used for electricity generation as well as transportation applications within a microgrid (Kwon *et al.*, 2024). In a similar vein, pumped hydro storage while dependent on specific geographical conditions proves effective for long-term energy storage and has been successfully deployed in various regions around the globe. These diverse storage options not only enhance the stability of renewable-powered microgrids but also improve their responsiveness to dynamic fluctuations in energy demand and supply. RE-powered microgrids provide a variety of environmental, economic, and social advantages. From an environmental perspective, they significantly lower greenhouse gas emissions by substituting fossil-fuel-based generation with clean energy sources. Economically, renewable-powered microgrids decrease reliance on imported fuels and shield communities from fluctuating energy prices. Additionally, they generate local employment opportunities in the installation, maintenance, and operation of RE systems (El-Khattam and Salama, 2004).

Despite these advantages, renewable-powered microgrids encounter various challenges, especially regarding reliability, financial viability, and regulatory support (Zhang *et al.*, 2021). The intermittent nature of RE sources necessitates advanced control systems and energy storage solutions to maintain stability and reliability, which can raise the overall costs of microgrid implementation. Furthermore, regulatory policies and incentives differ significantly across regions, potentially hindering or promoting the adoption of renewable-powered microgrids. Addressing these barriers necessitates collaborative efforts among governments, industries, and communities to establish a supportive environment for the integration of RE in microgrids. This chapter investigates the different facets of small-scale RE and storage solutions for microgrids, analyzing their potential advantages, challenges, and future opportunities.

6.2 Renewable energy technologies for microgrids

Integrating RE technologies into microgrids provides multiple benefits, including a lower environmental impact, enhanced energy security, and improved resilience against grid failures. Due to the scalability and adaptability of renewable sources, microgrids can incorporate various technologies to capture clean energy for local consumption. This section delves into the primary RE sources utilized in microgrids, highlighting their distinct characteristics, challenges, and how they synergistically enhance one another within microgrid systems as shown in Figure 6.1.

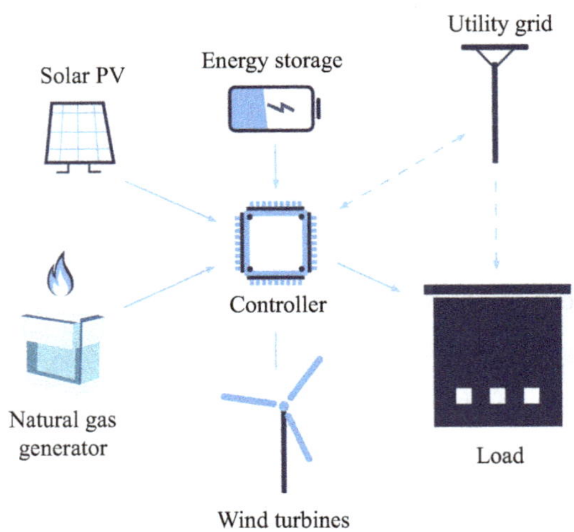

Figure 6.1 Key roles of microgrid controllers in managing the integration, coordination, and optimization of energy sources, loads, and storage systems within a microgrid

6.2.1 Solar power

Solar energy stands out as one of the most readily available renewable sources for microgrid applications, largely because of the simplicity, modularity, and swift deployment of PV systems. Small-scale PV systems can be conveniently installed on rooftops or in designated community areas, making them particularly well suited for microgrids that cater to remote or rural regions.

6.2.1.1 Overview of photovoltaic systems

PV systems convert sunlight directly into electricity through semiconductor materials, usually silicon, which produce electric currents when exposed to sunlight. These systems can be categorized into grid-tied and off-grid systems, with microgrids frequently utilizing off-grid or hybrid configurations to ensure reliable power supply. Improvements in PV technology have resulted in higher efficiency and lower costs, establishing solar energy as a feasible choice for both large-scale and small-scale energy production (Green *et al.*, 2022). Additionally, the versatility of PV systems enable their integration with energy storage devices, such as batteries, to store excess energy for later use, which is especially beneficial for off-grid microgrids.

6.2.1.2 Solar energy potential for microgrids

The potential for solar energy is influenced by geographical location, climate, and solar irradiance. Areas with high sunlight exposure offer greater opportunities for PV installations. Microgrids situated in these regions can take advantage of reliable energy generation during the day, which can be stored in batteries for use at night. For example, regions such as sub-Saharan Africa and parts of Asia, characterized by high solar irradiance, find solar PV technology to be an outstanding choice for microgrid applications (Rosa *et al.*, 2020). Conversely, in areas with lower sunlight or frequent cloud cover, incorporating solar power necessitates the use of additional renewable sources or energy storage solutions to ensure a reliable power supply. In summary, the low operational costs and minimal environmental impact of PV systems position solar energy as one of the most practical choices for microgrids as shown in Figure 6.2.

6.2.2 Wind energy

Wind energy is a significant renewable resource for microgrids, particularly in areas with adequate wind resources. Small-scale wind turbines, specifically designed for microgrid applications, capture kinetic energy from the wind to generate electricity.

6.2.2.1 Overview of small-scale wind turbines

Small-scale wind turbines are distinct from their larger counterparts mainly in terms of size and power output, which generally ranges from a few kilowatts to approximately 100 kW. These turbines are well suited for distributed generation within microgrids and can be installed in diverse settings, including rural areas and small islands. Contemporary small-scale turbines are engineered to function

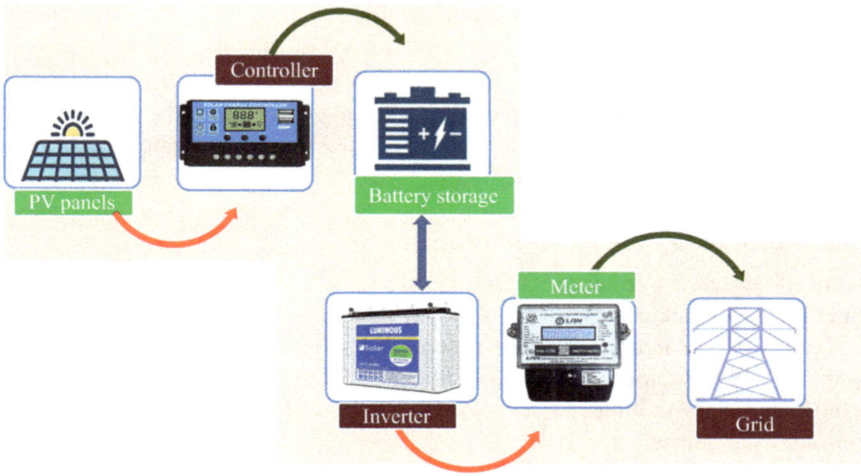

Figure 6.2 Solar energy potential for microgrids from photovoltaic systems

efficiently at lower wind speeds, making them appropriate for regions with consistent, albeit moderate, winds (Chen *et al.*, 2021). However, similar to solar PV systems, the output of wind energy is influenced by local weather conditions, highlighting the importance of analyzing wind patterns during the planning phase.

6.2.2.2 Challenges and benefits of integrating wind energy into microgrids

The intermittent nature of wind energy poses a significant challenge for its integration into microgrids, as changes in wind speed result in fluctuations in power output. To mitigate this issue, wind-based microgrids frequently utilize energy storage solutions or hybrid systems to maintain a reliable power supply (Bui *et al.*, 2020). Moreover, small-scale wind turbines need to be appropriately sited to maximize efficiency and reduce environmental and noise impacts. Despite these challenges, the advantages of wind energy encompass zero fuel costs, low greenhouse gas emissions, and decreased dependence on external energy sources, all of which enhance the sustainability and resilience of microgrids.

6.2.3 Synergy of wind with other renewables

The complementary characteristics of wind and solar energy facilitate a synergistic approach in microgrids, as solar power generally peaks during the day while wind energy may reach its maximum at night. By combining both sources, microgrids can establish a balanced and continuous energy supply, minimizing the requirement for extensive energy storage and enhancing the overall stability of the grid (Rao *et al.*, 2019). This integration is especially advantageous for rural microgrids, where reliable energy access can significantly improve community life and support vital services such as healthcare and education.

6.3 Biomass and other renewable sources

Biomass energy is another crucial renewable resource for microgrids, particularly in rural and agricultural regions where organic waste materials are abundant. Biomass-based microgrids transform organic materials into electricity, heat, or fuel, offering versatility and sustainability.

6.3.1 Bioenergy for rural microgrids

Bioenergy generated from agricultural waste, wood, and other organic materials can ensure a reliable energy supply for rural microgrids. In these settings, biomass gasification and anaerobic digestion processes are frequently employed to convert organic waste into biogas or bioelectricity (Mishra and Chakraborty, 2020). This method provides multiple advantages: it lowers waste disposal costs, reduces environmental impact, and bolsters local economies by generating energy from local resources. Bioenergy systems are especially beneficial in regions where solar or wind resources may be scarce, enabling microgrids to deliver reliable power and enhance resilience.

6.3.2 Geothermal and hydropower potential

Geothermal and small-scale hydropower systems provide additional renewable options for microgrids, particularly in areas with appropriate natural resources. Geothermal energy taps into heat from beneath the Earth's surface, delivering a stable and dependable energy source that can function continuously. In contrast, small-scale hydropower systems harness flowing water in rivers or streams to produce electricity. Both of these technologies have a minimal environmental impact and can provide microgrid operations with a consistent energy output (Anderson and Malakar, 2019). However, these technologies are site-specific, restricting their application to areas with geothermal resources or flowing water bodies. In suitable locations, they offer substantial advantages for microgrids by decreasing reliance on intermittent sources such as solar and wind.

6.3.3 Hybrid renewable systems in microgrids

Hybrid renewable systems integrate multiple RE sources within a single microgrid, enhancing energy reliability and supply stability. For example, a hybrid system could incorporate solar, wind, and biomass, with each source contributing to a balanced power supply. This strategy not only decreases reliance on a single energy source but also enables more efficient resource utilization by leveraging the strengths of each technology. For instance, in a hybrid solar-wind-biomass system, solar and wind can meet daytime and nighttime power demands, while biomass serves as backup during periods of low renewable generation (Garcia and Prakash, 2021).

Hybrid systems improve the resilience of microgrids, especially in remote or off-grid areas, by providing a reliable energy supply despite fluctuations in individual sources. Microgrids are small-scale power grids engineered to function independently or in conjunction with the main grid. These systems play a crucial role in utilizing RE sources like solar, wind, and hydro while ensuring reliable power delivery. Energy storage technologies are essential for improving microgrid performance by stabilizing intermittent energy sources, maintaining power quality, and ensuring a reliable energy supply. This section examines various storage technologies suitable for microgrids, including batteries, flywheels, super-capacitors, pumped hydro storage, and hydrogen storage, highlighting their types, applications, and contributions to creating cost-effective, durable, and sustainable microgrid solutions.

6.4 Battery storage systems

Battery storage systems are the most prevalent form of energy storage utilized in microgrids, offering the flexibility necessary to store and distribute energy as needed. Batteries provide energy stability and support for intermittent renewable sources such as solar and wind, making them essential for the effective operation of microgrids. Table 6.1 provides a comparison of different battery types commonly used in micro-grid applications, highlighting factors such as energy density, cycle life, cost, and efficiency. These characteristics are essential for selecting the most suitable battery technology based on the specific needs and goals of a microgrid system.

6.4.1 Types of batteries

There are various types of batteries suitable for microgrids, each providing distinct advantages.

Lithium-ion (Li-ion) batteries: Li-ion batteries are commonly employed because of their high energy density, lightweight design, and long cycle life. They are especially effective in microgrid applications where space is constrained and energy demands are significant. However, they also have drawbacks, including higher costs and the potential risk of thermal runaway.

Flow batteries: Flow batteries, such as vanadium redox flow batteries, offer longer discharge durations and the capability to endure repeated cycles without significant capacity loss. They are better suited for larger microgrid systems with high energy demands over extended periods. In contrast to Li-ion batteries, flow batteries have a lower energy density but excel in scalability and longevity, making them ideal for microgrids that need large-scale, long-duration storage (Wang *et al.*, 2022).

Table 6.1 Comparison of battery types for microgrid applications

Battery	**Energy density**	**Cost ($/kWh)**	**Safety**	**Cycle life**
Lithium-ion	High	Moderate	Moderate	5,000+
Flow (vanadium redox)	Medium	High	High	12,000+
Lead-acid	Low	Low	Low	500–1,500
Sodium-ion	Medium	Low	High	3,000+

Lead-acid batteries: Traditional lead-acid batteries continue to be a cost-effective option for microgrids, particularly in developing areas. However, their lower energy density and shorter lifespan compared to Li-ion and flow batteries render them less suitable for high-demand applications.

6.4.2 Role in stabilizing intermittent renewable energy

Batteries are crucial for stabilizing the variable output of RE sources. In microgrids, they store excess energy generated during peak production periods and release it when production is low or demand is high. For example, on sunny days, a solar-powered microgrids can store surplus energy in batteries for use during nighttime or cloudy days, thereby ensuring a continuous energy supply and enhancing grid stability (Kandari *et al.*, 2023). By smoothing out fluctuations in power supply, batteries help maintain voltage and frequency levels, thereby ensuring power quality and reliability. This stabilization is especially crucial in remote or off-grid areas, where energy interruptions can be costly or hazardous, highlighting the role of batteries in enhancing energy resilience and autonomy within microgrids.

6.4.3 Cost and lifespan considerations for microgrids

While batteries offer essential energy storage capabilities, they also come with challenges related to cost and lifespan. The initial investment for battery storage can be significant, particularly for Li-ion batteries, which are among the pricier options. However, their long lifespan and minimal maintenance requirements often justify the initial expense over time. Flow batteries, though more cost-effective in terms of cycle life, demand more space, and their high installation costs may restrict their application in smaller microgrids (Smith and Lee, 2024). Lifecycle costs for batteries encompass maintenance, replacement, and disposal expenses. Effective battery management systems (BMS) are vital for extending battery life and enhancing return on investment for microgrid operators. A well-functioning BMS not only monitors battery health but also optimizes charging and discharging cycles, which is essential for maintaining microgrid efficiency and reducing overall operational costs.

6.4.4 Flywheels, supercapacitors, and pumped hydro storage

Alongside batteries, other advanced energy storage technologies like flywheels, supercapacitors, and pumped hydro storage are emerging as viable alternatives for microgrids.

6.4.4.1 Emerging technologies in energy storage

Flywheels: Flywheel energy storage systems (FESS) store kinetic energy in rotating masses, delivering instantaneous power for short durations. Flywheels are advantageous for applications that demand high power density and rapid

Table 6.2 Performance comparison of flywheels, supercapacitors, and pumped hydro storage

Technology	Energy density	Cycle life	Response time	Applications
Flywheels	Low	100,000+	Milliseconds	Frequency regulation, short-term storage
Supercapacitors	Very low	1 million+	Instantaneous	Power smoothing, bridging power
Pumped hydro	High	50,000+	Minutes	Bulk energy storage, peak shaving

charge–discharge cycles, making them ideal for stabilizing frequency and voltage in microgrids (Sutikno *et al.*, 2022).

Supercapacitors: Supercapacitors, known for their high-power density and rapid charge–discharge capabilities, are well suited for applications that require short bursts of energy. Although they do not possess the energy density of traditional batteries, supercapacitors are effective in balancing power quality in microgrids and supporting peak demand periods by discharging energy almost instantaneously (Salam *et al.*, 2024).

Pumped hydro storage: Pumped hydro storage, which involves pumping water to a higher elevation and then releasing it to generate electricity, is the most established large-scale energy storage technology. While it has typically been used in centralized grids, small-scale pumped hydro systems can also benefit microgrids, especially in areas with appropriate topography. This storage method is highly sustainable, boasting long lifespans and minimal environmental impact (Liu *et al.*, 2023). Table 6.2 presents a performance comparison of flywheels, supercapacitors, and pumped hydro storage, focusing on key parameters such as energy storage capacity, efficiency, response time, and lifecycle. This comparison aids in understanding the strengths and limitations of each technology for different energy storage applications, particularly in balancing power supply and demand in energy systems.

6.4.4.2 Applications in microgrid scenarios

Flywheels and supercapacitors are increasingly being utilized in microgrids for applications that demand rapid response times and high-power surges such as load balancing and frequency regulation. In microgrids with significant RE integration, these technologies play a crucial role in maintaining grid stability by compensating for the variability of solar and wind power. For example, flywheels can reduce frequency fluctuations by swiftly absorbing or releasing energy, ensuring a smooth power supply (Ahmed *et al.*, 2023).

Pumped hydro, conversely, is better suited for larger microgrids or those with extended energy storage requirements. In islanded microgrid systems, pumped hydro can offer long-duration storage, assisting in energy management during prolonged periods of low renewable generation.

6.4.5 Hydrogen storage

Hydrogen storage is emerging as a promising solution for long-term energy storage in microgrids. Hydrogen can be generated through water electrolysis powered by RE, stored, and subsequently converted back to electricity using fuel cells, offering a clean and sustainable storage option.

6.4.5.1 Hydrogen as a long-term storage option

Hydrogen storage provides distinct advantages in microgrid applications due to its high energy density and scalability. Unlike batteries, which are best suited for short-term storage, hydrogen can retain large amounts of energy for extended periods. This feature makes hydrogen especially valuable in microgrids that depend on seasonal or intermittent renewable resources, as it can store surplus energy generated during peak production times for use during periods of low production. Hydrogen's function as a long-term storage solution complements the shorter-term storage capabilities of batteries, allowing microgrids to maintain a balanced energy supply across various time frames. For example, surplus solar or wind energy generated during the summer months can be stored as hydrogen and utilized in winter, ensuring energy security and independence.

6.4.5.2 Integration with renewable energy systems in microgrids

Integrating hydrogen storage with RE systems in microgrids is a complex undertaking that requires specialized infrastructure and technology. To optimize the conversion, storage, and retrieval of energy in the form of hydrogen, electrolyzers, fuel cells, and storage tanks must be meticulously managed. In microgrids with substantial RE generation, hydrogen production is strategically carried out during peak production periods, helping to alleviate grid congestion and store energy that might otherwise be curtailed (Thompson and Pescaroli, 2024).

Hydrogen storage in microgrids is an evolving field, with ongoing research aimed at improving the efficiency of hydrogen production, lowering costs, and enhancing storage technology. Furthermore, advancements in solid-state hydrogen storage and fuel cell efficiency hold the potential to make hydrogen storage a more viable and cost-effective options for microgrids in the future.

6.5 Design and optimization of microgrids

In recent years, microgrids have become increasingly recognized as sustainable energy solutions that promote decentralized energy production and storage, particularly in areas where traditional grid expansion is neither practical nor cost-effective (Gonzalez *et al.*, 2020). The incorporation of small-scale RE sources within microgrids has facilitated a reduction in carbon emissions while improving the reliability of energy supply in both off-grid and grid-connected systems. This section examines the key aspects of microgrid design and optimization, addressing topics such as balancing energy generation and demand, microgrid architecture, energy management systems (EMS), and the use of optimization techniques.

6.5.1 Balancing energy generation and demand

A key challenge in microgrid design is achieving alignment between energy generation and demand. This balance is essential, especially since microgrids heavily depend on intermittent renewable sources such as solar and wind energy. Strategies like demand response (DR) programs can greatly assist in maintaining this equilibrium by adjusting demand according to generation availability (Shirazi and Gomez, 2021). Moreover, ESS are vital, as they store excess energy produced during peak generation and release it during times of high demand (Kumar and Singh, 2019).

Demand forecasting through machine learning (ML) models has become crucial for estimating future energy requirements and facilitating effective planning. These predictive models assist microgrid operators in maintaining balance by allowing them to make informed decisions based on anticipated generation and consumption trends. In areas with limited energy storage capacity, these models are vital for reducing energy waste and managing generation levels as necessary.

6.5.2 Microgrid architecture and components

The architecture of a microgrid includes multiple components such as RE sources (like solar panels and wind turbines), ESS (including batteries and flywheels), inverters, and control units (Zhu and Chen, 2023). Depending on operational needs, a microgrid can be configured to function in either islanded or grid-connected mode, each requiring distinct architectural designs and control mechanisms. A microgrid generally consists of three key components: the generation unit, the storage unit, and the load. The generation unit includes both renewable and non-RE sources, typically selected based on local availability. The storage unit, which can feature battery banks or other ESS, acts as a buffer to provide a stable power supply by offsetting variability in RE generation. Finally, the load segment encompasses all energy-consuming devices linked to the microgrid. A critical aspect of microgrid design is the power electronics interface, which regulates the power flow among the generation, storage, and load units. DC/AC inverters are typically employed to convert the DC output from RE sources into AC, ensuring compatibility with the majority of electrical devices (Ramirez *et al.*, 2023). Additionally, contemporary microgrid designs incorporate sophisticated communication networks that monitor real-time performance, facilitating swift fault detection and recovery.

6.5.3 Role of energy management systems

EMS are essential for the operation of microgrids, providing centralized oversight and control over generation, storage, and consumption (Zeng and Li, 2021). An EMS generally consists of a user interface, a data acquisition system, and control algorithms that collaborate to optimize energy flows within the microgrid. The main role of an EMS is to schedule and manage the deployment of renewable and storage assets in response to variations in demand and generation. This management typically relies on optimization algorithms that take into account factors such as cost, emissions, and grid reliability. EMS also supports demand-side

Table 6.3 Key functionalities offered by EMS in microgrids

Functionality	Description
Monitoring	Tracks energy flows, system status, and component performance
Forecasting	Predicts demand and generation to adjust supply dynamically
Optimization	Determines optimal operational settings to minimize costs and maximize efficiency
Control	Executes control commands to manage energy generation, storage, and load balancing

management, enabling consumers to decrease or adjust their energy usage based on price signals or grid conditions (Thakur and Srinivasan, 2022). Furthermore, EMS can enable energy trading between multiple microgrids or between a microgrid and the main grid, enhancing overall energy efficiency. Table 6.3 outlines the key functionalities provided by the EMS in microgrids, including real-time monitoring, optimization of energy generation and storage, DR management, and fault detection. These functionalities are crucial for enhancing the efficiency, reliability, and sustainability of microgrid operations.

Recent developments in EMS have integrated blockchain technology to facilitate secure and transparent energy transactions (Patil and Shah, 2023). This incorporation enables a decentralized EMS, decreasing reliance on a single control point and enhancing resilience and scalability within microgrid networks.

6.5.4 Optimization techniques in managing renewable and storage assets

Optimization in microgrids has evolved significantly, with techniques such as artificial intelligence (AI) and ML improving decision-making accuracy and operational efficiency (Khan and Rahman, 2020). The main objective of these optimization strategies is to lower operational costs, minimize energy losses, and improve grid stability while maximizing the use of RE resources.

Artificial neural networks: Artificial neural networks (ANNs) are used to predict energy demand and generation in microgrids, making it easier for EMS to plan energy distribution. ANNs are especially beneficial in forecasting solar irradiance and wind speeds, aiding in the accurate prediction of RE production (Ahmed and Zhao, 2021).

Genetic algorithms): Genetic algorithms (GA) is a popular optimization technique for scheduling energy resources in a microgrid, particularly in mixed-integer nonlinear programming problems. GA is useful in solving complex optimization problems by mimicking natural selection, helping to optimize dispatch schedules and minimize energy costs (Zhu and Chen, 2023).

Reinforcement learning: Reinforcement learning (RL) has found significant applications in energy management by training EMS to adapt to dynamic operating conditions. RL-based systems learn from the environment, improving decision-making over time and enhancing the system's resilience to unforeseen fluctuations in demand or generation.

In addition to these techniques, hybrid optimization models that combine AI, ML, and statistical methods offer a promising approach to microgrid management. By using hybrid models, microgrids can simultaneously account for weather forecasts, demand patterns, and generation costs, leading to more comprehensive optimization strategies (Thakur and Srinivasan, 2022).

6.6 Challenges of small-scale renewable energy and storage for microgrids

Localized energy systems known as microgrids can function both separately and in tandem with the wider grid. Small-scale RE sources including solar, wind, and biomass have emerged as essential parts of microgrids as the global energy sector moves toward resilience (Dawn *et al.*, 2024). Notwithstanding their benefits, integrating these RE sources into microgrids poses several difficulties that need to be resolved to create dependable and effective energy systems. The main technical, financial, legal, and social issues of small-scale RE and storage facilities in microgrids are examined in this chapter (Rossi *et al.*, 2024).

6.6.1 Technical challenges

6.6.1.1 Intermittency and reliability

The intrinsic intermittency of small-scale RE is one of the main technological obstacles. Variability in climate and the precise moment of day can affect RE sources like wind and solar, which can cause variations in electricity production. Microgrid stability may be hampered by this intermittency, necessitating strong management techniques to guarantee a steady supply of electricity (Kumar *et al.*, 2024). DR tactics and sophisticated forecasting technologies must be used to handle these swings. Reliability can also be improved by incorporating hybrid systems, which integrate several renewable sources. For instance, combining solar energy with wind or biomass can result in a more consistent energy output because these sources frequently enhance one another on a daily and seasonal basis.

6.6.1.2 Energy storage systems

ESS are essential for reducing the erratic nature of RE. Deploying efficient storage solutions still comes with a few difficulties. Despite its efficiency, current battery technologies like lithium-ion have drawbacks in terms of price, longevity, and environmental effects (Patwary *et al.*, 2024). Furthermore, development and research are currently ongoing to determine whether these storage technologies can be scaled to suit the needs of microgrids. Supercapacitors, pumped hydro storage, flow batteries, and other cutting-edge storage technologies are being investigated. Interoperability with different RE sources and the requirement for sophisticated control mechanisms to maximize storage use are just two of the extra difficulties brought on by the intricacy of integrating these systems into current microgrid infrastructures.

6.6.1.3 Grid integration

Microgrid integration of small-scale renewables need sophisticated control systems capable of handling DERs. Microgrids distributed framework makes energy flow synchronization more difficult, requiring advanced algorithms and communication systems. Additionally, for the microgrid to function smoothly, compatibility between various systems and technologies is crucial. By offering the data required for effective decisions, smart grid technologies such as sensors and real-time monitoring can improve the integration of small-scale renewables (Güven and Yücel, 2025). Fortunately, putting these technologies into practice frequently necessitates a large expense for infrastructure and staff training.

6.7 Economic challenges

6.7.1 Capital costs

One major obstacle to the development of microgrids is the initial capital costs involved in setting up small-scale RE systems and the systems for storing energy that go along with them. Despite recent price reductions, installing, maintaining, and operating renewable technologies still come with a hefty cost, particularly for towns with tight finances (Quispe *et al.*, 2024). Innovative finance techniques, like public–private collaborations, grassroots efforts funding models, and government incentives, can be used to lessen these financial difficulties. These strategies can promote expenditures in renewable microgrids and aid in distributing the upfront expenses.

6.7.2 Operational costs

Microgrid administration, monitoring, and preservation can come with significant operating costs in addition to capital expenditures. To maintain their optimum efficiency, small-scale RE systems need constant maintenance, which can put a strain on carriers' finances, particularly in isolated or underserved populations (Gao *et al.*, 2024). Operational costs can be decreased by utilizing preventive care technologies and creating cost-effective preservation strategies. The sustainability and effectiveness of microgrid operations can also be increased by offering local staff training and initiatives to improve capacity.

6.7.3 Market dynamics

Market conditions, such as energy pricing, statutes and regulations, and rivalry against conventional energy sources, also affect the economic feasibility of small-scale RE. Fossil fuels continue to get subsidies in many areas, which makes it challenging for RE sources to compete fairly. Policy measures like carbon pricing, RE certificates, and feed-in tariffs can be crucial to creating a fairer market for RE. These processes have the potential to improve the economic climate for microgrid adoption of small-scale RE sources (Wright *et al.*, 2024).

6.7.4 Regulatory challenges

6.7.4.1 Policy framework

Distinct countries have quite distinct regulatory environments when it comes to microgrids and small-scale RE. The development of renewable microgrids might be hampered by inconsistent policies that provide uncertainty for algorithms and financiers. Fostering a favorable climate for microgrid development requires the establishment of transparent and encouraging regulatory frameworks that enable the integration of RE, expedite the approval process, and facilitate innovative business strategies (Cochran, 2024).

6.7.4.2 Grid interconnection standards

Operational issues, such as adherence to grid interconnect rules and protocols, arise when small-scale RE installations are connected to the current grid. The safe and dependable operation of microgrids depends on these requirements (Rodriguez-Gil *et al.*, 2024). However, the development of renewable systems may be deterred by excessively strict or ambiguous interrelationships regulations. Policymakers must strike a balance between the need to support the expansion of microgrids and RE technologies and the requirements for safety and dependability.

6.7.5 Social challenges

6.7.5.1 Community acceptance

Recognition by neighbors is frequently a determining factor in the successful deployment of small-scale RE in microgrids. Residents in the area could be worried about the potential risks, land use, or aesthetic impact of RE infrastructure. Acceptance can be increased by involving communities early in the design phase, giving them clear information, and showcasing the advantages of renewable microgrids (Zaman *et al.*, 2024). Community support can be further increased by educational initiatives that highlight the benefits of RE such as sustainability and energy independence.

6.7.5.2 Equity and access

One major social concern is ensuring that everyone has equal access to the advantages of small-scale RE. It is possible that marginalized groups lack the knowledge or resources necessary to take part in or profit from renewable microgrid projects.

6.8 Future directions of small-scale RE and storage for microgrids

Small-scale RE and their incorporation into microgrids are becoming more and more important in the shift to a sustainable energy future (Kiasari *et al.*, 2024). Microgrids, which provide granular control and resiliency, are a revolutionary approach to energy generation, delivery, and consumption. This chapter examines the future possibilities of small-scale RE and storage techniques for microgrids considering the quickening pace of technological breakthroughs and the escalating worries about reliability of electricity and sustainability (Elalfy *et al.*, 2024).

6.8.1 The rise of small-scale RE

Due to its capacity to deliver clean energy at a decentralized level, small-scale RE systems such as solar PV, wind turbines, and biomass are becoming more and more popular (Figaj, 2024).

6.8.1.1 Solar energy

Developments in solar PV technology: Perovskite solar cells and bifacial panels are two new technologies that are pushing the limits of power conversion rates, and the performance of solar cells has increased dramatically. These developments increase the appeal of solar energy for microgrid applications, particularly in isolated and off-grid locations. Interaction with building-integrated photovoltaics (BIPV): upcoming advancements are probably going to concentrate on incorporating solar PV systems into construction products, which will allow energy to be produced through structures without the need for extra space.

6.8.1.2 Wind energy

Small wind turbines: It is anticipated that more small-scale wind turbines will be installed, especially in rural and semi-urban areas. To promote wider acceptability, future designs will place a higher priority on efficiency, noise reduction, and aesthetics. Hybrid systems: By integrating solar and wind power, microgrids can become more reliable and less reliant on fossil fuels. It will be essential to conduct research on hybrid systems that maximize energy output according to local conditions.

6.8.1.3 Biomass and biogas

Sustainable biomass utilization: Going forward, sustainable methods for producing biomass energy including using trash and agricultural residues will be the main focus. To maximize energy yield, technologies for effective conversion processes will be essential.

Digesters for biogas production: Anaerobic digesters can be used in microgrids to convert biological waste into biogas, solving the problem of waste disposal and producing renewable power.

6.8.2 Energy storage solutions

Effective energy storage solutions are necessary to maintain stability and dependability in microgrids due to the intermittent nature of RE sources (Jafarizadeh *et al.*, 2024; Rashid, 2024).

6.8.2.1 Battery technologies

Lithium-ion batteries: As lithium-ion technology continues to progress, it will increase energy density, lower costs, and improve lifecycle performance. Additional solutions might be available through research into alternate chemistries including solid-state and sodium-ion batteries.

Flow batteries: Flow battery systems, which store energy using liquid electrolytes, are becoming more and more popular due to their long discharge times and scalability, which makes them appropriate for larger-scale microgrid applications.

6.8.2.2 Systems of mechanical storage

Although pumped hydro storage has historically been used on a large scale, advancements in this technology can be modified for smaller uses by making use of pre-existing water reservoirs in microgrid environments.

Flywheel energy storage: Flywheels are perfect for applications needing brief power spikes because they release energy quickly. They are appropriate for microgrid integration due to their low maintenance and robustness.

6.8.2.3 New developments in storage technologies

Hydrogen storage: Green hydrogen's application as a storage medium for energy will grow as its generation via electrolysis develops more feasible. Hydrogen, which can store and transform back into power when needed, might be produced by microgrids using surplus RE.

Thermal energy storage: Materials with phase change are one example of a thermal storage invention that might store unused electricity as heat. This method can be very helpful when combining heating systems and RE sources.

6.8.3 Smart microgrid technologies

The future of small-scale RE and storage in microgrids is closely tied to the development of smart technologies (Nivolianiti *et al.*, 2024).

6.8.3.1 Advanced control systems

Decentralized control algorithms: Future microgrids will leverage decentralized control systems, allowing local energy management based on real-time data. This will enhance the adaptability and resilience of microgrids against fluctuations in energy supply and demand.

Predictive analytics: Utilizing ML and AI to forecast energy production and consumption patterns will optimize the dispatch of renewable resources and storage systems.

6.8.3.2 Blockchain and distributed ledger technologies

Peer-to-peer energy trading: Blockchain technology can facilitate peer-to-peer energy trading among microgrid participants, allowing consumers to buy and sell excess energy directly, thereby incentivizing local generation and consumption.

Enhanced transparency: Distributed ledger technologies can provide transparent tracking of energy transactions and production, fostering trust among stakeholders in microgrid operations.

6.8.3.3 Internet of things (IoT) integration

Smart meters and sensors: The integration of IoT devices will enable real-time monitoring and management of energy resources, enhancing operational efficiency and facilitating DR strategies.

Consumer engagement: IoT applications will empower consumers by providing insights into energy usage patterns, encouraging energy conservation and enabling active participation in DR programs.

6.8.4 *Policy and regulatory frameworks*

The future of small-scale RE and microgrids will also be shaped by supportive policy and regulatory environments.

6.8.4.1 Incentives for renewable energy adoption

Tax credits and subsidies: Governments can incentivize the adoption of small-scale renewables and energy storage through tax credits, grants, and subsidies, reducing the financial barriers for individuals and communities.

Feed-in tariffs and power purchase agreements: Long-term contracts for RE generation can provide financial stability for microgrid operators, encouraging investment in renewable technologies.

6.8.4.2 Standardization and interoperability

Regulatory standards: Establishing standards for the interconnection of small-scale renewables and storage systems with the grid is essential for ensuring safety and reliability. Regulatory bodies should promote interoperability among different technologies to facilitate seamless integration.

Community engagement in policy development: Involving local stakeholders in the policymaking process can lead to more effective regulations that reflect community needs and promote local RE initiatives.

6.9 Impact of small-scale renewable microgrids on global sustainability efforts

Global sustainability initiatives are greatly aided by small-scale renewable microgrids, especially as the globe struggles with the urgent issues of social justice, reliable electricity, and worldwide warming (Islam *et al.*, 2024). Through the utilization of nearby RE sources and encouraging community involvement, these systems support a comprehensive strategy for longevity. Small-scale renewable microgrids have the following significant effects on global sustainability:

6.9.1 *Reduction of greenhouse gas emissions*

The capacity of small-scale renewable microgrids to lower greenhouse gas emissions is one of their primary accomplishments. Microgrids reduce reliance on fossil fuels, which are a significant source of greenhouse gas emissions, by using RE sources including sun, wind, and biomass. Areas' carbon footprints are lessened by the local production of RE sources, which is in line with international emissions reduction goals (Tatar and Aydin, 2024).

6.9.2 *Enhancing energy access and equity*

For impoverished and isolated towns without dependable links to conventional grids, microgrids are very useful for supplying energy. Microgrids can provide economical and sustainable energy solutions by implementing small-scale renewable technologies, raising living standards and fostering economic growth

(Dadjiogou *et al.*, 2024). By strengthening communities and lowering energy poverty which disproportionately harms vulnerable groups equitable access to energy promotes a sustainable society.

6.9.3 Resilience and adaptation to climate change

As climate change, small-scale renewable microgrids' resilience is essential. During outages, these systems can function independently of the main grid, supplying vital electricity during crises like natural catastrophes. Because they are decentralized, societies are better equipped to adapt to the effects of climate change and maintain resources like crisis management and healthcare (Salam *et al.*, 2024). Additionally, microgrids lessen the risks brought on by centralized energy systems and lengthy logistical chains by encouraging decentralized energy generation.

6.9.4 Fostering local economies and job creation

By generating jobs in the installation, upkeep, and operation of RE systems, investments in small-scale renewable microgrids boost communities. In addition to promoting financial resilience, this local employment creation fosters innovation and improving skills in the nearby regions. The economic activity that results from more areas using renewable technologies has the potential to create a sustainable circular economy, where resources are repurposed and recovered, thus increasing the advantages to the environment (Vetrivel *et al.*, 2025).

6.9.5 Encouraging sustainable practices and community engagement

Microgrids frequently encourage community involvement and involvement with managing electricity, so cultivating a sustainable culture. Communities are more committed to environmentally friendly initiatives like preserving resources and energy conservation when they take charge of local energy systems (Wazirali *et al.*, 2023). A wider cultural shift favoring sustainability may result from this beginnings engagement, encouraging other communities to follow suit.

6.9.6 Facilitating technological innovation

Small-scale renewable microgrid deployment spurs energy technology innovation, ranging from smart grid applications to energy storage options. As these systems develop, they can incorporate cutting-edge technology like cryptocurrencies and AI, potentially improving grid dependability and energy management (Quizhpe *et al.*, 2024). In addition to increasing microgrid productivity, this invention establishes a standard for internationally applicable scalable solutions, which will have an impact on more comprehensive energy regimes and legislation.

6.10 Conclusion

In conclusion, the incorporation of microgrids with small-scale RE sources signifies a revolutionary change in contemporary energy systems, permitting localized energy production, improved resilience, and a decreased dependence on conventional grid infrastructures. Particularly in isolated and underdeveloped regions, the various renewable technologies from solar PV and small-scale wind turbines to biomass and hybrid systems offer significant promise for supplying energy needs. As this chapter has made clear, energy storage technologies such as flywheels, batteries, and hydrogen systems are essential for reducing the erratic nature of RE sources and guaranteeing a steady and dependable supply of energy. The future of microgrids is bright, notwithstanding the difficulties presented by conventional electricity networks such as grid stability and regulatory barriers. Cutting-edge optimization methods that make use of AI and ML can greatly improve operational effectiveness and energy management, opening the door for creative answers to new energy-related problems. Additionally, continuous advancement and research of novel materials and methods will keep lowering costs and enhancing the functionality of storage and RE systems. Small-scale renewable microgrids have the potential to have an influence that goes beyond neighborhood electricity security; by lowering emission of greenhouse gases and encouraging energy independence, they greatly aid global sustainability initiatives. Ensuring a sustainable and equitable energy environment will depend on the smart deployment of microgrids with integrated renewable sources as we transition to a more decentralized and resilient energy future.

References

Ahmed, F., Al Kez, D., McLoone, S., Best, R. J., Cameron, C., and Foley, A. (2023). Dynamic grid stability in low carbon power systems with minimum inertia. *Renewable Energy*, 210, 486–506.

Ahmed, M., and Zhao, Y. (2021). Forecasting renewable energy generation using artificial neural networks in microgrid applications. *Energy Systems Research Journal*, 8(2), 120–138.

Anderson, B., and Malakar, P. (2019). Distributed energy and microgrid solutions for grid reliability. *IEEE Transactions on Power Systems*, 34(2), 303–312.

Bhattacharyya, S. C. (2014). *Rural Electrification through Decentralized Off-grid Systems in Developing Countries*. Berlin: Springer.

Bui, V., Kim, E., and Kwon, J. (2020). Challenges in integrating renewable energy into traditional grids. *Energy Policy*, 129, 231–245.

Chen, R., Xiao, T., and Zhang, M. (2021). Microgrid resilience in modern energy systems. *Applied Energy*, 278, 115678.

Cochran, C. (2024). Fortifying the grid in the Southeastern US through microgrids: Navigating an uncertain regulatory environment. *Environmental Progress & Sustainable Energy*, 43(1), e14337.

Dadjiogou, K. Z., Ajavon, A. S. A., and Bokovi, Y. (2024). Enhancing energy access in rural areas: Intelligent microgrid management for universal telecommunications and electricity. *Cleaner Energy Systems*, 9, 100136.

Dawn, S., Ramakrishna, A., Ramesh, M., *et al.* (2024). Integration of renewable energy in microgrids and smart grids in deregulated power systems: A comparative exploration. *Advanced Energy and Sustainability Research*, 5(10), 2400088.

Elalfy, D. A., Gouda, E., Kotb, M. F., Bureš, V. and Sedhom, B. E. (2024). Comprehensive review of energy storage systems technologies, objectives, challenges, and future trends. *Energy Strategy Reviews*, 54, 101482.

El-Khattam, W., and Salama, M. M. A. (2004). Distributed generation technologies, definitions, and benefits. *Electric Power Systems Research*, 71(2), 119–128.

Figaj, R. (2024). Energy and economic sustainability of a small-scale hybrid renewable energy system powered by biogas, solar energy, and wind. *Energies*, 17(3), 706.

Gao, J., Maalla, A., Li, X., Zhou, X., and Lian, K. (2024). Comprehensive model for efficient microgrid operation: Addressing uncertainties and economic considerations. *Energy*, 306, 132407.

Garcia, F., and Prakash, R. (2021). Environmental benefits of microgrids in decentralized energy systems. *Environmental Research*, 198, 112563.

Gonzalez, J., Lee, S., and Kim, D. (2020). Microgrid integration for sustainable energy supply. *Journal of Renewable Energy Research*, 15(1), 30–45.

Green, M., Hisham, D., and Leung, K. (2022). Advances in photovoltaic technology for decentralized applications. *Solar Energy Journal*, 56, 111–122.

Güven, A. F., and Yücel, E. (2025). Sustainable energy integration and optimization in microgrids: Enhancing efficiency with electric vehicle charging solutions. *Electrical Engineering*, 107, 1541–1573.

Islam, M. A., Ali, M. N., Mollick, T., *et al.* (2024). Assessing the feasibility and quality performance of a renewable energy-based hybrid microgrid for electrification of remote communities. *Energy Conversion and Management: X*, 23, 100674.

Jafarizadeh, H., Yamini, E., Zolfaghari, S. M., Esmaeilion, F., Assad, M. E. H. and Soltani, M. (2024). Navigating challenges in large-scale renewable energy storage: Barriers, solutions, and innovations. *Energy Reports*, 12, 2179–2192.

Kandari, R., Gupta, P., and Kumar, A. (2023). Battery state of charge based improved adaptive droop control for power management of a microgrid having large scale renewable generation. *Sustainable Energy Technologies and Assessments*, 57, 103146.

Khan, A., and Rahman, S. (2020). Optimization techniques in microgrid energy management systems. *Energy Optimization Journal*, 12(4), 45–58.

Kiasari, M., Ghaffari, M., and Aly, H. H. (2024). A comprehensive review of the current status of smart grid technologies for renewable energies integration and future trends: The role of machine learning and energy storage systems. *Energies*, 17(16), 4128.

Kumar, R., Pal, P., and Ranjan, S. (2015). Bioenergy and biomass utilization in microgrids. *Energy Procedia*, 76, 118–127.

Kumar, R., and Singh, P. (2019). Role of energy storage in demand response and load balancing in microgrids. *Smart Grid and Microgrid Journal*, 6(3), 255–270.

Kumar, A., Kumar, S., Abdelaziz, A. Y., Saket, R. K. and Kothari, D. P. (2024). Reliability aspects in the design and development of microgrids. In Saket, R. K. and Sanjeevikumar, P. (eds.) *Reliability Analysis of Modern Power Systems*. Piscataway, NJ: Wiley-IEEE Press, pp. 493–506.

Kwon, K., Lee, H.-B., Kim, N., Park, S., and Joshua, S. R. (2024). Integrated battery and hydrogen energy storage for enhanced grid power savings and green hydrogen utilization. *Applied Sciences*, 14(17), 7631.

Lasseter, R. H. (2002). Microgrids and distributed generation. *Journal of Energy*, 26 (1), 90–95.

Li, Y., and Wang, H. (2022). Energy storage technologies for renewable integration in microgrids. *Sustainable Energy Reviews*, 42, 17–31.

Liu, W., Deines, E., and Müller, T. O. (2024). Sustainable urban transformations based on integrated microgrid designs. *Nature Sustainability*, 7(8), 1067–1079.

Liu, X., Zhao, T., Wang, P., Liu, J., and Blaabjerg, F. (2023). Microgrid energy management with energy storage systems: A review. *CSEE Journal of Power and Energy Systems*, 9(2), 483–504.

Lund, P. D., Lindgren, J., Mikkola, J., and Salpakari, J. (2015). Review of energy system flexibility measures to enable high levels of variable renewable electricity. *Renewable and Sustainable Energy Reviews*, 45, 785–807

Mishra, P., and Chakraborty, S. (2020). Biomass utilization in distributed energy systems. *BioEnergy Research*, 25, 215–229.

Nivolianiti, E., Karnavas, Y. L. and Charpentier, J. F. (2024). Energy management of shipboard microgrids integrating energy storage systems: A review. *Renewable and Sustainable Energy Reviews*, 189, 114012.

Patil, V., and Shah, K. (2023). Blockchain in microgrid energy management: A decentralized approach. *Journal of Emerging Technologies in Power Systems*, 14(2), 112–129.

Patwary, A. K., Sayem, M. A., Hossain, M. A. and Halim, M. A. (2024). A review of energy storage systems (ESS) for integrating renewable energies in microgrids. *Control Systems and Optimization Letters*, 2(1), 103–112.

Quispe, J. C., Obispo, A. E., and Alcantara, F. J. (2024). Economic feasibility assessment of microgrids with renewable energy sources in Peruvian rural areas. *Clean Technologies and Environmental Policy*, 26(5), 1415–1438.

Quizhpe, K., Arévalo, P., Ochoa-Correa, D., and Villa-Ávila, E. (2024). Optimizing microgrid planning for renewable integration in power systems: A comprehensive review. *Electronics*, 13(18), 3620.

Ramirez, F., Ochoa, H., and Morales, L. (2023). Power electronics in modern microgrid architecture. *Electric Power Systems Review*, 10(1), 33–50.

Rao, N., Patel, S., and Bhattacharya, A. (2019). Wind energy applications for microgrids. *Wind Engineering*, 43(5), 519–532.

Rashid, S. M. (2024). Employing advanced control, energy storage, and renewable technologies to enhance power system stability. *Energy Reports*, 11, 3202–3223.

Rodriguez-Gil, J. A., Mojica-Nava, E., Vargas-Medina, D., *et al.* (2024). Energy management system in networked microgrids: An overview. *Energy Systems*, 1–32. https://doi.org/10.1007/s12667-024-00676-6.

Rosa, J., Delgado, F., and Cruz, M. (2020). Solar energy applications in rural microgrids. *Energy Policy Review*, 52, 215–229.

Rossi, M., Jin, L., Monforti Ferrario, A., *et al.* (2024). Energy hub and micro-energy hub architecture in integrated local energy communities: Enabling technologies and energy planning tools. *Energies*, 17(19), 4813.

Ruth, M., *et al.* (2015). Climate change and microgrid resilience. *Environmental Research Letters*, 10(10), 104009.

Saeed, M. H., Fangzong, W., Kalwar, B. A., and Iqbal, S. (2021). A review on microgrids' challenges & perspectives. *IEEE Access*, 9, 166502–166517.

Salam, S. S. A., Raj, V., Petra, M. I., Azad, A. K., Mathew, S., and Sulthan, S. M. (2024). Charge scheduling optimization of electric vehicles: A comprehensive review of essentiality, perspectives, techniques, and security. *IEEE Access*, 12, 121010–121034.

Salam, I. U., Yousif, M., Numan, M., and Billah, M. (2024). Addressing the challenge of climate change: The role of microgrids in fostering a sustainable future – A comprehensive review. *Renewable Energy Focus*, 48, 100538.

Shirazi, Z., and Gomez, M. (2021). Balancing renewable energy generation and consumption in microgrids. *Microgrid Journal*, 5(4), 198–210.

Singh, P., and Kumar, N. (2021). Advanced storage solutions for microgrid stability. *Renewable Energy Science*, 49, 85–92.

Smith, D., and Lee, J. (2024). *Lifecycle Costs of Battery Energy Storage Systems for Microgrids*. Sustainable Microgrid Technologies.

Sutikno, T., Arsadiando, W., Wangsupphaphol, A., Yudhana, A., and Facta, M. (2022). A review of recent advances on hybrid energy storage system for solar photovoltaics power generation. *IEEE Access* 10, 42346–42364.

Tatar, S. M., and Aydin, E. (2024). Design and operation of renewable energy microgrids under uncertainty towards green deal and minimum carbon emissions. *Sustainable Energy, Grids and Networks*, 37, 101233.

Thakur, M., and Srinivasan, P. (2022). Energy management systems in microgrid design: Optimization and demand response. *Journal of Renewable and Sustainable Energy*, 9(3), 66–80.

Thompson, D., and Pescaroli, G. (2024). Financing electricity resilience in local communities: A review of the literature. *Environment Systems and Decisions*, 44, 740–762.

Vetrivel, S. C., Arun, V. P., Saravanan, T. P., Maheswari, R., and Mohanasundari, M. (2025). Socio-economic impacts of digital innovations in renewable energy adoption. In Jaheer Mukthar, K. P., Jain, V., Tsai, S., and Wu, C. (eds.). *Digital Innovations for Renewable Energy and Conservation*. Hershey, PA: IGI Global, pp. 277–312.

Wang, Y., Sun, Y., Zhang, Y., *et al.* (2022). Optimal modeling and analysis of microgrid lithium iron phosphate battery energy storage system under different power supply states. *Journal of Power Sources*, 521, 230931.

Wazirali, R., Yaghoubi, E., Abujazar, M. S. S., Ahmad, R., and Vakili, A. H. (2023). State-of-the-art review on energy and load forecasting in microgrids using artificial neural networks, machine learning, and deep learning techniques. *Electric Power Systems Research*, 225, 109792.

Wright, S., Frost, M., Wong, A., and Parton, K. (2024). Australian microgrids: Navigating complexity in the regional energy transition. *Energy Research & Social Science*, 113, 103540.

Zaman, M. R., Sarker, S., Halim, M. A., Ibrahim, S., and Haque, A. (2024). A thorough analysis of the opportunities and challenges of community microgrid system based on renewable energy in Bangladesh. *Control Systems and Optimization Letters*, 2(1), 28–35.

Zhang, H., Xiang, W., Lin, W., and wen, J. (2021). Grid forming converters in renewable energy sources dominated power grid: Control strategy, stability, application, and challenges. *Journal of Modern Power Systems and Clean Energy*, 9(6), 1239–1256.

Zhang, Y., and Zhang, Z. (2018). Small-scale renewable energy systems for microgrids. *Journal of Renewable Energy*, 13(4), 43–57.

Zeng, H., and Li, R. (2021). Centralized and decentralized control strategies in microgrid energy management. *International Journal of Microgrid Studies*, 8(1), 95–109.

Zhu, Q., and Chen, Y. (2023). Renewable energy integration in microgrid architecture: Components and design principles. *Renewable Energy Journal*, 21(3), 120–137.

Chapter 7

Advancing sustainable energy: integrating small-scale renewables and storage in microgrids

Abdulhalim Musa Abubakar[1], Ehime Irene Itamah[2], Kiran Batool[3], Cemre Avsar[4], Muhammad Asif[5], Volodymyr Volodymyrovich Kutarov[6], Baudilio Coto[7] and Kashif Abbas[8]

Abstract

Integration of small-scale renewable energy sources and storage systems into microgrids represent a pivotal advancement in sustainable energy management. Harnessing wind, photovoltaic (PV), and battery storage technologies creates resilient, efficient, and eco-friendly microgrids. Exploring the latest developments in renewable energy technologies, storage solutions, and energy management systems provides a comprehensive overview of the design, implementation, and optimization of microgrids. Case studies and real-world applications illustrate the economic and environmental benefits of deploying small-scale renewable energy systems. Challenges and opportunities in policy, grid integration, and technological advancements are discussed, underscoring the importance of innovation and collaboration in achieving global energy sustainability. Replacement of macrogrids with microgrids is a political issue, because there is a possibility that local utilities might not be aware of the benefits of microgrid technology. This chapter aims to equip readers with the knowledge and tools necessary to contribute to the future of clean energy through the effective management of small-scale renewable energy

[1]Department of Chemical Engineering, Faculty of Engineering, Modibbo Adama University, Nigeria
[2]Department of Electrical and Electronics Engineering, Faculty of Engineering, Federal Polytechnic Daura, Nigeria
[3]Beijing Key Laboratory of New Energy and Low Carbon Development, School of Economics and Management, North China Electric Power University, China
[4]Department of Chemical Engineering, Ankara University, Turkey
[5]Department of Energy and Environment, Faculty of Agricultural Engineering (FAE), Sindh Agriculture University, Pakistan
[6]Physics Research Institute, Odesa I. I. Mechnikov National University, Ukraine
[7]Chemical, Energy and Mechanical Technology Department, School of Experimental Sciences and Technology (ESCET), Rey Juan Carlos University, Spain
[8]School of Mechanical Engineering, Faculty of Engineering, Xian Jiaotong University, China

and storage in microgrids. However, there are still some developments required to achieve more efficient and available energy and to introduce this microgrid technology as a commercial output.

Keywords: Microgrids; energy storage technologies; renewable energy; battery storage; grid integration

7.1 Introduction

The integration of small-scale renewable energy sources/resources (RER) into microgrids mark a significant advancement in sustainable energy management [1]. As the global demand for cleaner, more reliable, and decentralized power grows, microgrids offer a solution that harnesses local renewable resources such as wind, solar, hydropower, and biomass [2,3]. Uddin *et al.* [4] described microgrids as new emerging distributed energy resources (DERs), divided into renewable or conventional sources. Microgrids are not only essential for reducing dependency on centralized power systems but also for enhancing energy security and resilience, especially in remote or underserved regions. It encompasses various DERs, storage units, electrical loads, and power electronic converters [5–8]. With the world facing increasing climate challenges [9], there is an urgent need to transition to renewable energy systems that contribute to a lower carbon footprint while ensuring energy reliability. Microgrids, by design, provide flexibility to operate either in conjunction with the main power grid or independently, allowing for greater control over energy production and consumption [10,11]. Ahmed *et al.* [11] gave the typical control type of microgrid mode – namely, droop control, PQ control, current control, and voltage/frequency control. Anvaru-Moghaddam *et al.* [12] and Muhtadi *et al.* [13] classified the aforementioned control types under primary, secondary, or tertiary controls. Previously, market for microgrid grew to $18.9 billion in 2020 [14]. The capability is particularly important in areas prone to grid instability or natural disasters, where maintaining energy supply is critical. Also, the shift toward small-scale renewable energies is driven by advancements in technologies such as battery storage, energy management systems (EMS) [15], and smart grids [16], all of which enhance the efficiency and effectiveness of these systems.

Numerous studies have examined the role of microgrids in modern energy systems, highlighting their potential to not only provide energy independence but also reduce operational costs and emissions [17]. Previous research has demonstrated the effectiveness of renewable energy integration in microgrids, with examples of successful implementations around the world. China is currently planning to launch the largest microgrid in the world [18]. Other successful implementations are in Canada [19–21], Greece [22–24], the USA [10,25,26], the Netherlands [27,28], Germany [9,29,30], Italy [31,32], and the UK [33,34]. Already, numerous campus microgrids have been installed globally [35], including a laboratory microgrid using the proposed gateway and energy analyzer emulator by Gomes *et al.* [36] and testbeds mentioned by Hossain *et al.* [37]. Often, institutions have been relied upon for innovations and implementation of this technology [38]. Notably, Centre for Energy Systems and

Control at Howard University previously entered an agreement with Pareto Energy to build microgrid in its campus [14]. The same author reported that the University of California had already combined steam turbine, gas turbines, and solar cell installation plus some storage. Those systems have proven their ability to reduce reliance on fossil fuels and mitigate the environmental impacts associated with traditional energy generation methods. But Bellido *et al.* [39] list some bottlenecks that must be removed to enable its use in Brazil, based on a case study of Federal University of Rio de Janeiro. Hence, this study aims to explore the technological foundations, design considerations, and optimization strategies for small-scale renewable energy systems in microgrids. Case studies and real-world applications analysis will provide insights into the economic and environmental benefits of microgrids, while also addressing the challenges and opportunities in policy, regulation, and technological advancement. Some of the identified challenges have been formerly pinpointed by Huayllas *et al.* [40]. Through this discussion, the study seeks to equip readers with the knowledge necessary to contribute to the future of sustainable energy solutions. Because Boche *et al.* [41] realized that without a clear solution to a sustainable microgrid development, many identified challenges may remained unaddressed.

7.2 Technological foundations

7.2.1 Wind energy systems

Wind energy systems are essential components of small-scale renewable energy setups, especially in microgrids designed to power remote or decentralized communities. The systems convert kinetic energy from the wind into electricity through wind turbines [42]. For microgrids, small- to medium-sized wind turbines are typically used, offering a clean and sustainable source of energy that can complement other renewable sources like solar and hydropower. Wind turbines are highly effective in areas with consistent and strong wind patterns, making them a reliable solution in many regions. Specific advantages of wind energy systems are their ability to generate electricity without relying on fuel, thus producing no direct greenhouse gas emissions. Alhassan [43] reported a successful installation of a wind power farm in Katsina, Nigeria in 2021, as shown in Figure 7.1 – banking on

Figure 7.1 Commissioned 10 MW wind farm in Katsina, Nigeria. Source: Author's screenshot.

Table 7.1 Wind energy power generation in small communities – case study

Location	Country	Purpose	Reference
Gotland Island	Sweden	Provide clean and renewable energy to a remote community and reduce reliance on fossil fuels.	[49,50]
King Island	Australia	Integrate wind power with other renewable sources for a hybrid energy system.	[51]
Samso Island	Denmark	Make the island 100% renewable through a combination of wind and other renewable energy sources.	[52]
El Hierro	Spain	To achieve energy self-sufficiency through wind power in combination with a pumped hydro storage system.	[53,54]
Kodiak Island	USA	Transition the island's energy supply from diesel to renewable wind energy.	[55]

earlier potential revealed by Balarabe *et al.* [44], Jonathan *et al.* [45], and Salisu and Garba [46]. It thus contributes significantly to reducing the carbon footprint of microgrids and supporting global efforts toward sustainability. Wind turbines can also operate in tandem with other energy systems (Table 7.1), such as solar panels, to create a hybrid setup that enhances the reliability and consistency of energy supply, particularly when solar energy is less available [47,48].

However, wind energy systems also come with certain challenges, especially in the context of small-scale applications. Because Argungu and Dabai's [56] reliability analysis of the wind farm in Katsina show that the turbine-site matching is very low, as the plant will only be operating between 1 and 2.3 MW instead of the 10 MW design capacity. Jigawa, Potiskum, Kaduna, Gusau, Kano, and Bauchi are other locations in Nigeria with proven wind-to-electricity potential [57–59]. The initial investment for installing wind turbines can be high [60], and maintenance can be demanding due to mechanical wear and tear. Moreover, wind energy is inherently variable, as it depends on weather conditions. To address this variability, energy storage systems (ESS), such as batteries, are often integrated into microgrids [61], allowing excess energy generated during peak wind periods to be stored and used during times of low wind availability. In spite of these challenges, advances in technology are making wind energy systems more accessible and efficient for microgrid applications. Innovations in turbine design, improved energy storage options, and better forecasting tools for wind patterns are helping to optimize the use of wind energy, making it a viable and increasingly popular option for small-scale renewable energy systems.

7.2.2 Photovoltaic systems

Photovoltaic (PV) systems, commonly known as solar power systems, are a cornerstone of small-scale renewable energy setups, particularly in microgrids. Solar panels are used by these systems to convert sunlight directly into electricity through

Figure 7.2 Large-scale solar power to rural areas of Guinea-Bissau. Source: Author's coupling.

the PV effect, making them an incredibly sustainable and eco-friendly energy source. In microgrids, PV systems are highly valued for their scalability and flexibility, allowing for small installations on individual homes or larger arrays to power entire communities. Since solar energy is abundant in most regions, especially those with high sun exposure, PV systems provide a reliable and renewable energy option for decentralized energy production. Among the primary advantages of PV systems is their ability to generate electricity without emitting greenhouse gases, contributing to a significant reduction in environmental impact [62]. Furthermore, solar panels require minimal maintenance once installed and have a long lifespan, making them a cost-effective solution over time. In small-scale applications [63], PV systems can be easily integrated into existing infrastructure, such as rooftops, which maximizes land use and simplifies installation [64].

However, like all renewable energy sources, PV systems face certain limitations. Solar energy is intermittent according to Abou El-Ela *et al.* [65], meaning electricity production depends on sunlight availability, which can fluctuate throughout the day and across seasons. To mitigate this, energy storage solutions, such as batteries, are often paired with PV systems in microgrids. Typically, Obara *et al.* [66] discovered that PV account for 20% of power generation based on data obtained from Japan's Teuri and Yagishiri islands. In Africa, PV solar electrification have been successfully applied in several locations due to easy access of the microgrid technology [67]. Specifically, in the community of Bigene, Guinea-Bissau (Figure 7.2), TRNSYS software was used to analyze the performance of the PV microgrid [68]. PV storage systems capture excess electricity generated during peak sunlight hours and store it for use when sunlight is insufficient, ensuring a steady energy supply. Advances in solar technology are continually improving the efficiency and affordability of PV systems. Innovations like more efficient solar cells, better energy storage solutions, and smart grid integration are making PV systems increasingly viable for small-scale microgrid applications, offering communities a clean, reliable, and renewable energy source [69].

7.2.3 Battery storage technologies

Battery storage technologies play a crucial role in small-scale renewable energy systems, particularly in microgrids, where the reliability and consistency of energy supply are essential. Examples are lithium-ion, lead-acid, flow, sodium–sulfur, nickel–cadmium, solid-state batteries, compressed air energy storage (CAES), and flywheel energy storage shown in Figure 7.3. Here, excess electricity generated by renewable sources like solar and wind is being stored by the system, making the energy available when production is low or demand is high. For microgrids, batteries are vital in balancing the variability of renewable energy, ensuring that power is available even during periods of low sunlight or weak winds. The most commonly used battery technologies in small-scale microgrids are lithium-ion batteries [70]. They are favored for their high energy density, long lifespan, and relatively fast charging and discharging capabilities. Lithium-ion batteries are well suited to store energy from intermittent renewable sources, providing backup power during peak demand or when the renewable generation is not sufficient [71,72]. In remote areas (Table 7.2), battery storage ensures energy independence [60], allowing communities to operate without relying on grid access or fossil fuel generators [34].

Battery storage technologies face certain challenges, despite their advantages. After their review, Boche *et al.* [41] found that one of the main issues is the cost,

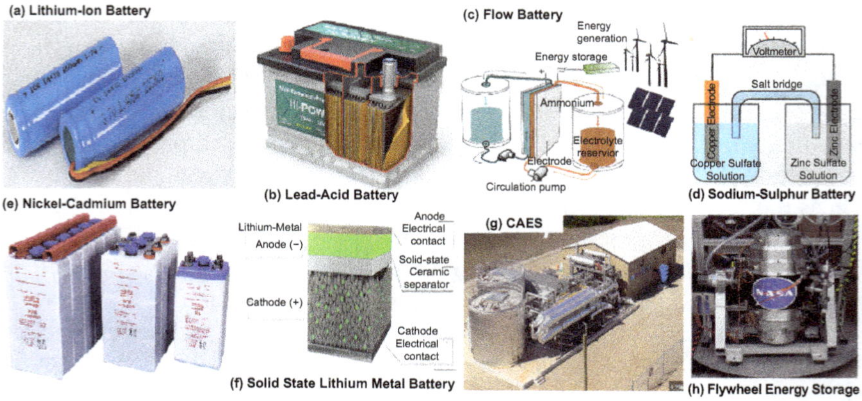

Figure 7.3 Types of battery storage systems. (a) Lithium-ion battery, (b) lead-acid battery, (c) flow battery, (d) sodium–sulfur battery, (e) nickel–cadmium battery, (f) solid-state lithium metal battery, (g) CAES, and (h) flywheel energy storage. Source: Author's merger.

Table 7.2 List of successful implementation of energy storage system

Battery/storage type	Capacity	Community and country	Project	Reference
Lithium-ion battery	13.6 MWh	Hornsdale Power Reserve, Australia	Tesla Report	[73,74]
Flow battery	200 kWh	Isle of Eigg, Scotland	Eigg Electric	[75,76]
Lead-acid battery	1 MWh	El Hierro, Canary Islands, Spain	Gorona del Viento	[77–79]
Sodium–sulfur battery	34 MWh	Fujian Province, China	NGK Insulators Ltd	[80,81]
Pumped hydro storage	10 MW	King Island, Australia	Hydro Tasmania	[82,83]

which can be significant, especially for large storage capacities. Additionally, while batteries are improving in terms of efficiency and longevity, they still degrade over time, leading to a need for eventual replacement [48,62]. Environmental concerns related to the mining of materials for batteries, such as lithium and cobalt, also raise sustainability questions [84]. Significant advances in battery recycling and the development of alternative storage technologies like flow batteries or solid-state batteries are helping to address some of these challenges. Inclusively, battery storage technologies are essential in making renewable energy systems more reliable and flexible. As technology evolves, battery storage is becoming more affordable and efficient, making it a cornerstone of small-scale renewable energy solutions for microgrids. Specific examples include vanadium redox flow batteries and reversible solid oxide cells given by Baldinelli *et al.* [85]; CAES system discussed by Shahinzadeh *et al.* [86]; supercapacitor and super magnetic ESS explained by Fathima and Palanisamy [87, 88] and; flywheels, pumped hydro system (PHS), electrochemical and thermal energy storage described by Sadati [89]. The singular act of storing surplus energy in batteries facilitates a stable and continuous power supply, fostering a transition toward cleaner and more resilient energy systems [90].

7.3 Design and implementation of microgrids

The design and implementation of microgrids are critical steps in creating decentralized energy systems that can efficiently manage power generation, distribution [91], and storage [92]. Microgrids are designed to operate either independently or in connection with the main power grid, offering flexibility and resilience. The success of a microgrid largely depends on careful planning, the integration of renewable energy sources, and the selection of effective storage and energy management solutions. Kamal *et al.* [93] reported the designing and operation of a rural standalone microgrid with electrical loads modeled for the electrification energy scarce village of Uttarakhand, India. In his Masters' Thesis, Sabhlok [14] divulges several issues associated with microgrid design and implementation.

7.3.1 Planning and design considerations

Planning and designing a microgrid require an in-depth analysis of the specific energy needs of the community or facility it will serve, sometimes using a multi-criteria energy planning procedures [94]. The size [95], location [96], and purpose of the microgrid play a major role in determining its overall design [18,97]. Stadler and Naslé [98] gave the steps and methods involved in creating bankable microgrids by focusing on a unified microgrid planning software. Key factors to consider include the local availability of RER, such as sunlight and wind, and the expected energy demand throughout the day and across different seasons. Additionally, the microgrid's ability to seamlessly switch between grid-connected and islanded (off-grid) modes, depending on grid conditions or emergencies, is an important feature that enhances its reliability [70,99,100]. Zheng *et al.* [91] divulge some examples of islanding detection mode of microgrid that include local loads, a doubly fed induction generator type wind generator, natural gas-fired generator, and PV generator. Another critical consideration in microgrid design is the regulatory and economic framework. Planners must navigate local laws, grid interconnection standards, and economic incentives for renewable energy [7,101]. Cost analysis is essential to balance upfront investments with long-term benefits, ensuring the microgrid is financially viable while meeting sustainability goals. Notably, Chebabhi *et al.* [102] presented a case study of the optimal design and sizing of renewable energies in microgrids based on financial considerations for Biskra, Egypt. Akter *et al.* [64] does the same in a case study of its applicability in St. Martin's Island, Bangladesh. Scalability is also important, as microgrids should be able to grow with future energy needs or technological advancements.

7.3.2 Integration of renewable energy sources

RER, such as solar, wind, and biomass, are often central to the design of microgrids due to their sustainability and reduced environmental impact, as reviewed by Patwary *et al.* [2]. The integration of these energy sources requires careful consideration of local climate and geography [103]. For example, areas with abundant sunlight may rely heavily on PV systems [104], while windy regions may prioritize wind turbines [105]. Simple combination of multiple renewable sources can enhance the microgrid's overall reliability by reducing dependence on any single energy source, ensuring a more consistent power supply [106,107]. When integrating renewable energy, it's also important to design for variability, in accordance with Abou El-Ela *et al.*'s [65] study. Since solar and wind energy are intermittent, microgrids must include strategies to balance energy production with consumption. Hybrid systems, which combine renewable sources with traditional generators or other renewables, provide flexibility, as discussed by Kamal *et al.* [108] and Homa *et al.* [109]. For instance, solar panels can be used during the day, while wind turbines generate power during the night or cloudy periods, creating a complementary system that maximizes resource availability [110]. In accordance with SDG7, Kumar *et al.* [111] proposed a hybrid renewable energy microgrid coupled with sensitivity analysis. To meet the energy shortage outlined by the Paris Agreement, a case study was performed by Prabatha *et al.* [20] with the developed framework for a small off-grid community in British Columbia, Canada. Not just micro-grids, macro- and mini-grids are typically coupled with distributed energy generation systems, especially with RER [112–114]. Table 7.3 differentiates between these three grid systems.

However, there is a fourth grid system called nanogrid that offers connectivity between micro, macrogrid, and adjacent nanogrid, as exemplified by Dahiru *et al.* [112].

7.3.3 Storage solutions and energy management

ESS is a critical component of microgrid design, as it allows for the capture and later use of excess electricity generated by renewable sources [110]. Batteries, particularly lithium-ions, are the most common storage solution, enabling energy to be stored during periods of high production and used when demand exceeds generation [115]. Other technologies, such as flow batteries or thermal storage, may

Table 7.3 Differences between mini-, micro- and macro-grid system

Feature	Microgrid	Mini-grid	Macro-grid
Definition	Small-scale power system, often localized	Intermediate-scale system, often serving a community or cluster	Large-scale power system, typically covering multiple regions or countries.
Size	Usually less than 1 MW	Ranges from 1 MW to 10 MW	Generally over 10 MW
Connectivity	Can operate independently or connected to the main grid.	Usually operates independently but can connect to the main grid.	Primarily interconnected across vast areas.
Energy sources	Renewable sources (solar, wind), diesel	A mix of renewables and conventional sources.	Primarily conventional sources (coal, gas) with some renewables.
Control system	Localized control with advanced automation	Semi-centralized control	Centralized control with extensive infrastructure
Application	Remote areas, military bases, hospitals	Rural electrification, community power	National or regional electricity supply
Flexibility	Highly flexible and adaptable	Moderate flexibility	Lower flexibility, relies on large infrastructure
Reliability	High reliability with local resources	Good reliability with backups	Varies; can be affected by transmission issues
Cost	High upfront costs but low operational	Moderate costs	High investment but economies of scale
Environmental impact	Low, especially with renewables	Moderate, depending on the energy mix	Variable; can be high due to fossil fuels

Source: Author's view.

also be considered depending on the specific needs and resources of the microgrid. Abid *et al.* [92] and Soshinskaya *et al.* [116] collectively gave an example of an electric vehicle charging station. Sadati [89] and Al-Ghussain *et al.* [95] looked at the implementation of PHS in reservoirs of the northern part of Cyprus to pump water from low to higher reservoirs. The choice of storage technology depends on factors like energy capacity, discharge rate, and longevity. ESS also deliver ancillary services to the grid, like peak shaving, energy arbitrage, and frequency regulation [117]. Effective EMS are essential for optimizing the performance of microgrids. Energy production, consumption, and storage levels in real time are being monitored by these systems, making smart decisions about when to store energy or draw from reserves. An EMS can also prioritize energy usage based on critical loads, ensuring that essential systems remain powered during grid outages or times of low energy production. Following an intelligent way of managing resources, EMS help maximize the efficiency and reliability of microgrids, and make them more adaptable to changing energy needs and conditions.

7.4 Optimization techniques

Optimizing the performance of small-scale renewable energy systems and microgrids is crucial for maximizing energy efficiency, reliability, and sustainability. To achieve these goals, a range of optimization techniques can be employed [112]. Liu *et al.* [118] rightly state that the optimal design, control, and scheduling of such complex systems are of great importance toward their successful practical realization in actual application studies. These include strategies to enhance energy efficiency, effective load management, and the integration of smart grid technologies [108,119]. In their study, Planas *et al.* [90] gave the common objective of microgrid optimal control, which is to minimize the operating cost.

7.4.1 Energy efficiency strategies

Energy efficiency is a key factor in optimizing the operation of microgrids and ensuring that renewable energy sources are utilized effectively [94]. In Kiptoo *et al.* [120], it is mentioned that a common strategy is the deployment of demand-side management programs that focus on reducing energy consumption by encouraging consumers to adjust their usage patterns. It was, respectively, found in Roman-Barri *et al.* [121] and Junior and Freitas [122] that intelligent electronic devices (IED) and power electronics will improve energy efficiency. For example, energy-efficient appliances, smart thermostats, and LED lighting can significantly reduce electricity demand, enabling the microgrid to operate more efficiently with the available RER. Another approach involves optimizing the efficiency of renewable energy systems themselves. For instance, maximum power point tracking (MPPT) technology in solar panels ensures that PV systems generate the maximum possible energy from sunlight. Optimization as described by Kamal *et al.* [93] utilizes gray wolf algorithm, producing an energy cost of 0.313 \$/kWh. Optimization basically justifies the cost of microgrid implementation, according to Fathima and Palanisamy [87]. Several optimization techniques including multi-agent-based method and meta-heuristics algorithms, which is said to outperform conventional techniques in terms of efficacy, have been systematically reviewed by Thirunavukkarasu *et al.* [3]. Shahinzadeh *et al.* [86] developed a teaching-learning-based optimization (TLBO) algorithm for teachers and student demonstration. Similarly, regular maintenance and upgrades to wind turbines can help increase their energy output. In combination, these measures reduce energy losses and enhance overall system performance, allowing microgrids to meet local energy needs while minimizing waste.

7.4.2 Load management and balancing

Hossain *et al.* [6] mentioned that load management and balancing are essential for ensuring that the supply of energy within a microgrid aligns with demand at all times. One critical optimization technique is the use of load forecasting [35], which predicts energy consumption patterns based on historical data, weather forecasts, and real-time monitoring. MATLAB and PSCAD software can be used to simulate and analyze different load models [15,64]. More so, reconfiguration is a control action for load/generation shedding investigated using genetic algorithm by Shariatzadeh *et al.* [123] and Shariatzadeh *et al.* [124]. In the UK, Abu-Sharkh *et al.* [33] found that the optimum amalgamation of microgrid generators (having 1.4 kWp PV array), if supplemented by energy storage of 2.7 kWh per household, will be beneficial to 45% of households. By anticipating demand spikes, microgrid operators can proactively adjust energy generation and storage levels, preventing shortages or surpluses. Demand response (DR) programs are another key component of load management [120]. DR programs encourage consumers to reduce their energy usage during peak periods in exchange for financial incentives [86]. By smoothing out demand fluctuations, DR helps prevent grid congestion and reduces the need for costly backup power generation [36,125]. Additionally, effective energy storage integration plays a vital role in balancing loads – especially in urban settings [115,126]. Battery storage systems can absorb excess energy during periods of low demand and release it when demand is high. This not only ensures a reliable energy supply but also maximizes the use of renewable energy by preventing curtailment during times of overproduction. Using a linear programming algorithm, a novel optimization model involving load demand, historical weather and installation prices data – i.e., the Bayesian regularized artificial neural networks (BRANN) model in MATLAB, was applied in Corvo, Portugal [127]. Ultimately, a hydrogen-based microgrid model was developed and applied at the Renewable Energy Facility of Sardegna Ricerche, Italy, to analyze the anticipated performance of the microgrid bearing in mind dissimilar load profiles and equipment sizes [104].

7.4.3 Smart grid technologies

Smart grid technologies offer a range of tools to optimize the performance of microgrids through enhanced communication [69,100], automation, and data analytics. Smart microgrid can aid the transition from centralized to decentralized power systems [128]. Advanced metering infrastructure (AMI), for example, allows for real-time monitoring of energy consumption and production, providing both consumers and grid operators with accurate data to inform decision-making [129]. With AMI or smart metering, microgrids can automatically adjust energy flows in response to changes in demand or generation, improving efficiency and reliability [99]. Purposeful integration of artificial intelligence (AI) [16] and machine learning (ML) further enhances optimization by enabling predictive analytics and automated decision-making. Both technologies can optimize energy distribution in real time by analyzing large datasets to identify patterns and forecast energy needs. AI-powered algorithms can also optimize battery storage management, ensuring that stored energy is utilized at the most cost-effective and efficient times. Lastly, smart inverters help manage the integration of distributed renewable energy sources into the grid [130]. Smart inverter devices can automatically adjust

Table 7.4 Developed model for optimizing energy outputs from microgrid

S. No.	Optimization objective	Model name	Reference
1.	Minimize operational costs	Cost minimization model	[133]
2.	Maximize renewable energy utilization	Renewable optimization model	[134]
3.	Enhance reliability and resilience	Resilience enhancement model	[135,136]
4.	Optimize energy distribution	Distribution optimization model	[137]
5.	Reduce carbon emissions	Emission reduction model	[138]

the flow of electricity from renewable systems like solar panels or wind turbines, stabilizing voltage and frequency levels within the microgrid [131]. By ensuring that energy flows are balanced and consistent, smart inverters improve the reliability of microgrid operations, even as renewable energy generation fluctuates [132]. Most times, the reasons for optimizing microgrid systems can be found in Table 7.4.

Table 7.4 serves as a summary of different approaches taken in optimizing energy outputs from microgrids, showcasing the diversity of objectives and methodologies in the field.

7.5 Case studies

Case studies from diverse locations showcase the adaptability of these systems, illustrating both the successes and challenges encountered in implementation. Several successful microgrid implementations around the world demonstrate the viability of small-scale renewable energy systems. One notable example is its implementation in California, USA [139], especially the Stone Edge Farm Microgrid. This microgrid integrates solar power, fuel cells, and battery storage to provide sustainable energy to a vineyard, ensuring energy reliability and autonomy from the grid. It showcases how a combination of renewable energy sources can supply power to both residential and agricultural operations, even in regions prone to grid instability. According to García *et al.* [140] and as supported by Ramos *et al.* [141], farms in Southern Spain can make use of hybrid battery storage and renewable energies. Elsewhere in the same country (Barcelona), microgrid implementation adapts to different sorts of storage devices, load, and energy sources [121]. In another case, the Isle of Eigg in Scotland implemented a community-owned microgrid powered by a mix of wind, solar, and hydropower. As desired, the project transformed the island from complete reliance on diesel generators to 98% renewable energy, improving the quality of life for residents and providing a model for off-grid communities globally [55,142]. The system has also proven resilient in harsh weather conditions, reinforcing the concept of microgrids as reliable alternatives to conventional power networks. Sometimes, a low-load diesel integration may be applied, because it allows the system to operate to its full capacity [143].

Uluski *et al.* [144] wrote the implementation of a microgrid at the Navy Yard in Philadelphia, Pennsylvania, which began operation in 2017.

Hirsch *et al.* [17] and Chandak and Rout [145] ascribe DER as being able to meet energy needs from New York to rural India. The economic and environmental impacts of successful microgrid implementations are substantial, offering both cost savings and significant environmental benefits [88]. Example is the DC-solar microgrids developed for villages or rural settings [72,146]. In terms of economic impact, the Brooklyn Microgrid in New York City is a pioneering example of a peer-to-peer (P2P) energy trading system [147]. It allows residents to trade excess solar energy with each other using blockchain technology, reducing reliance on the central grid and lowering electricity costs. Eales *et al.*'s [67] case study revealed that solar microgrids are cost competitive with diesel microgrids in all locations in Malawi. This decentralized approach demonstrates how microgrids can not only provide energy security but also create new economic opportunities through innovative trading models [148]. Another P2P microgrid development is the home solar electrification going on in Bangladesh [149]. From an environmental perspective, the Bronx Zoo in New York installed a microgrid that utilizes solar panels and battery storage to provide clean energy for its operations. This project reduced the zoo's dependence on fossil fuels, cutting both its energy costs and carbon footprint. Over time, the zoo's transition to renewable energy has contributed to a marked reduction in greenhouse gas emissions, helping the organization meet its sustainability goals. In view of the foregone, these case studies underscore the dual economic and environmental advantages of microgrids, including reduced operating costs, enhanced energy independence, and a lowered carbon footprint, contributing to global climate mitigation efforts [103].

7.6 Challenges and opportunities

One of the major challenges facing the adoption of small-scale renewable energy and microgrids is the policy and regulatory environment [101]. Social, technical, economic, environmental, and policy (STEEP) model was applied in remote communities in Nigeria by Akinyele *et al.* [150] to analyze the failure factors associated with microgrid. In many regions, outdated regulations designed for centralized energy grids can hinder the deployment of decentralized systems like microgrids. Issues such as grid interconnection standards, permitting, and energy tariffs often make it difficult for microgrid operators to integrate seamlessly with larger utility grids [40]. Furthermore, policies may not always provide adequate financial incentives or subsidies for renewable energy projects, limiting investment and slowing adoption [151]. However, there are significant opportunities for growth as governments increasingly recognize the importance of renewable energy in combating climate change. Emerging policies, such as feed-in tariffs, tax incentives, and grants for renewable energy projects, are encouraging the deployment of small-scale energy systems [152]. Moreover, energy tariff structure and governmental support were found to be the most imperative in the implementation of a feasible microenergy grid model for workable and low-carbon cities development in China [153]. Additionally, regulations that support net metering, where excess energy generated by microgrids is sold back to the grid, create economic benefits for microgrid operators. Policy reforms focused on grid modernization and flexibility, along with international climate agreements, are paving the way for wider adoption and innovation in this space [38].

From a technological perspective, the implementation of small-scale renewable energy and storage systems in microgrids faces several challenges [4]. Among the key barriers is the intermittency of renewable energy sources like solar and wind, which require reliable storage solutions to ensure a stable energy supply. A transition to decentralized system (microgrid) in both the UK and the USA, majorly operating a centralized system (relying on nuclear and fossil fuel) at the moment, is not completely feasible [25,33]. Current battery technologies, while advancing, still pose limitations in terms of cost, efficiency, and lifespan, particularly for large-scale storage needs. Based on a techno-economic analysis of a microgrid in residential community in Beijing, China, He *et al.* [126] found that a system with modernized batteries are more effective than those without. Additionally, the integration of multiple renewable sources into microgrids can be complex [16], requiring advanced EMS to optimize generation, storage, and consumption in real time. Bellido *et al.* [39] in their study highlighted some barriers or constraints affecting the implementation of DER in the Brazilian Power Sector. Despite these challenges, technological innovations present significant opportunities. Warneryd and Karltorp [154] mentioned that, to increase sustainability for the future, communities should be engaged as system-active participants. Breakthroughs in battery storage, such as the development of solid-state batteries and flow batteries [155], are improving energy storage capacity and durability. Advances in smart grid technology and EMS are making it easier to manage energy flows in microgrids, allowing for better integration of variable renewable sources. Furthermore, innovations in renewable generation, such as more efficient solar panels and smaller, more effective wind turbines, are expanding the potential for microgrids to provide reliable power even in remote or challenging environments.

7.6.1 Government incentives and support

Government incentives and support programs play a crucial role in driving the deployment of renewable energy systems and microgrids [25]. Moreover, Hossain *et al.* [6] negates the implementation of flywheel as very expensive compared to other RER. Provision of incentives, which may include tax breaks, grants, subsidies, and low-interest loans, is essential for offsetting the high upfront costs associated with renewable energy technologies. Thus, governments can encourage more individuals, businesses, and communities to invest in small-scale renewable energy projects by reducing financial barriers [148]. A key opportunity for growth is the increasing number of countries recognizing the importance of transitioning to clean energy. Many governments have introduced feed-in tariffs that allow microgrid operators to sell excess energy back to the grid at favorable rates, creating a financial return for renewable energy generation [152,153]. Renewable energy credits, green funding [39], and carbon pricing mechanisms are also encouraging investment in cleaner technologies by providing additional revenue streams for those generating or consuming renewable energy. However, the availability and effectiveness of government incentives can vary widely between regions. In Nigeria, a Feed-in Tariff for Renewable Energy Sourced Electricity has been established by the government in 2015 [150]. A national R&D project called Korea Micro Energy Grid has been funded by South Korea since 2010 [156]. Some areas may lack sufficient policy frameworks to incentivize the adoption of renewable energy, while in others, bureaucratic processes or unstable policy environments may hinder long-term investment [42]. Those locations may rely on donations; typical of Bronsbergen microgrid storage system in the Netherlands [39]. Consistent and transparent support from governments is essential to overcome these challenges and create a stable market for small-scale renewable energy solutions, based on the study by Abdul Muqeet *et al.* [35].

7.6.2 Standards and compliance

Standards and compliance regulations are critical in ensuring the safety, efficiency, and interoperability of small-scale renewable energy systems and microgrids. Regulatory standards help maintain consistent quality across energy technologies, ensuring that installations meet necessary safety protocols and performance benchmarks. Normally, standards cover a wide range of aspects, including the design, installation, and maintenance of energy systems, as well as their integration with larger grids [157]. Lack of uniformity across different regions is one of the major challenges with standards and compliance. Varying national and local standards can create confusion and slow down the deployment of renewable energy systems. For instance, microgrids that operate independently or in conjunction with the central grid must adhere to specific grid interconnection standards, which may differ significantly between jurisdictions. This can complicate the design and implementation process, especially for multinational projects or those located in border regions. Notwithstanding these challenges, there are significant opportunities in the area of standards and compliance. The development of international standards, such as those set by the International Electrotechnical Commission (IEC) [158] or the Institute of Electrical and Electronics Engineers (IEEE) [159], can help harmonize regulations, making it easier for manufacturers and operators to implement microgrids and renewable energy systems worldwide. Still, compliance with strict standards can improve the reliability and efficiency of these systems, increasing consumer confidence and boosting market adoption [160]. In the long term, advances in smart grid technology and better regulatory frameworks can help streamline compliance, further promoting the integration of renewable energy into microgrids.

7.7 Economic and environmental benefits

7.7.1 Cost-benefit analysis

Economic viability of small-scale renewable energy systems and microgrids can be assessed through a detailed cost-benefit analysis. While the upfront capital costs for renewable energy systems like solar panels, wind turbines, and battery storage can be high, the long-term savings are substantial [63]. Microgrids allow communities and businesses to reduce their dependence on conventional energy sources, thereby saving on electricity bills and reducing exposure to fluctuating energy prices [145]. Additionally, the cost of renewable technologies has steadily decreased over the past decade, making the transition to small-scale systems more financially attractive. On the benefits side, small-scale renewable energy systems often require less maintenance compared to traditional power plants, reducing operational expenses over time. South Africa was found to boost of high potential implementation capacity, after an assessment of rural areas in the country by Patel and Chowdhury [161] and Xu *et al.* [162]. Furthermore, financial incentives like tax credits, subsidies, and grants for renewable energy installations can significantly offset initial costs, enhancing the economic case for microgrid adoption. Beyond direct financial gains, these systems offer societal benefits by fostering energy independence and local job creation, especially in installation and maintenance roles.

7.7.2 Reduction in carbon footprint

One of the most significant environmental benefits of small-scale renewable energy systems is the reduction in carbon footprint. Adefarati *et al.* [157] stated that, by utilizing clean, renewable energy sources such as solar, wind, and biomass, microgrids eliminate the need for fossil fuel-based power generation, which is responsible for a large share of global greenhouse gas emissions. This transition directly contributes to mitigating climate change and improving air quality. For businesses and communities, adopting renewable energy systems provides an opportunity to meet environmental regulations and sustainability goals, often required by corporate social responsibility initiatives. Moreover, as governments implement stricter carbon policies and emissions reduction targets, microgrids offer a proactive way to comply with these regulations. The ability to generate and use renewable energy locally reduces transportation-related emissions and avoids the inefficiencies and losses associated with long-distance power transmission [1].

7.7.3 Enhancing energy security and reliability

Energy security and reliability are critical issues in today's energy landscape, especially as extreme weather events and cyber threats increase the vulnerability of centralized power grids [158]. Small-scale renewable energy systems, combined with storage solutions, offer a way to enhance energy security by creating decentralized power systems that can operate independently from the main grid during outages or emergencies. Microgrids can switch to "island mode" in the event of a grid failure [14,132], ensuring a continuous power supply for critical infrastructure such as hospitals, schools, and emergency services. Also, renewable energy sources like solar and wind, when combined with advanced EMS, provide reliable energy production throughout the day and night. The incorporation of battery storage ensures that energy generated during periods of excess production can be stored and used during peak demand or when renewable resources are unavailable (a backup power), further enhancing system reliability [163]. Likely, such resilience helps protect against energy price volatility, fuel shortages, and external disruptions, offering both economic and energy stability.

7.8 Emerging research area

Looking ahead, there are several promising trends and research directions in the field of small-scale renewable energy and microgrids. One major trend is the increasing focus on decentralization and community-based energy solutions that empower local regions to generate and manage their own energy. Relative to this, Justo *et al.* [164] distinguished between the application of AC or DC microgrids. DC-based microgrids have now been reported to be a better and more feasible solution for meeting local consumer level of reliable, controllable, and stable power demand, according to Modu *et al.* [8], Zubieta [165], and Hossain *et al.* [37]. It is due to the fact that AC microgrid systems present numerous problems for microgrids deployment, as mentioned by Lopes *et al.* [131], even though they can operate both in an island or grid-tied mode [159]. As more communities seek energy independence and resilience, especially in the face of climate change and grid instability, the demand for microgrids is expected to grow significantly – as demonstrated in swarm electrification in the Global South and Germany [9,149]. Aberilla *et al.* [70] affirmed that stand-alone home systems or community microgrids known as small-scale off-grid renewable energy systems are progressively being used for rural electrification. Research in energy storage and renewable energy generation is also a key focus. Ongoing studies into next-generation storage technologies, such as hydrogen-based storage and supercapacitors, are aiming to solve the challenges of energy intermittency. Additionally, the development of more efficient solar cells, such as perovskite and bifacial panels, could dramatically increase energy generation from PV systems. Another emerging research area is the use of AI and ML to optimize microgrid performance, enabling smarter energy distribution and consumption patterns. The use of a supervisory control and data acquisition (SCADA) software makes configuration and supervision of microgrids easier, said Roman-Barri *et al.* [121]. Salam *et al.* [160] previously pinpointed some future research areas, which includes full-scale development, experimental performance evaluation, microgrid transformation into intelligent system and field demonstration. Nonetheless, the majority of the aforementioned recommendations have been addressed, as stated in this study.

7.9 Conclusion

The integration of small-scale renewable energy systems into microgrids represent a transformative approach to addressing modern energy challenges. As the global energy landscape shifts toward sustainability, microgrids offer a practical and efficient solution for decentralized energy generation, ensuring greater energy independence, security, and resilience. Utilizing local renewable resources like wind, solar, and battery storage decreases microgrids' dependence on fossil fuels and plays a key role in lowering carbon emissions, supporting global initiatives to address climate change. Throughout this review, the technological foundations, design considerations, and optimization techniques that are critical to the successful implementation of small-scale renewable energies in microgrids are explored. Case studies have demonstrated the viability and benefits of microgrids in various contexts, from improving energy access in remote communities to reducing operational costs and enhancing environmental sustainability. While the potential of microgrids is undeniable, challenges remain, particularly in areas such as policy, regulatory frameworks, and technological barriers. However, advancements in battery storage technologies, smart grid systems, and renewable energy integration are paving the way for more efficient and reliable microgrid solutions. Government incentives and regulatory reforms will continue to play a crucial role in accelerating the adoption of these systems.

References

[1] T. Adefarati and R. C. Bansal, "Application of renewable energy resources in a microgrid power system," in The 7th International Conference on Renewable Power Generation (RPG 2018), 2019, pp. 5308–5313. doi:10.1049/joe.2018.9261.

[2] A. K. Patwary, A. Sayem, A. Hossain, and A. Halim, "A review of energy storage Systems (ESS) for integrating renewable energies in microgrids," *Control Syst. Optim. Lett.*, vol. 2, no. 1, pp. 103–112, 2024, doi:10.59247/csol.v2i1.68.

[3] G. S. Thirunavukkarasu, M. Seyedmahmoudian, E. Jamei, B. Horan, S. Mekhilef, and A. Stojcevski, "Role of optimization techniques in microgrid energy management systems—A review," *Energy Strateg. Rev.*, vol. 43, no. 100899, pp. 1–22, 2022, doi:10.1016/j.esr.2022.100899.

[4] M. Uddin, H. Mo, D. Dong, S. Elsawah, J. Zhu, and J. M. Guerrero, "Microgrids: A review, outstanding issues and future trends," *Energy Strateg. Rev.*, vol. 49, no. 101127, pp. 1–16, 2023, doi:10.1016/j.esr.2023.101127.

[5] A. M. R. Lede, M. G. Molina, M. Martinez, and P. E. Mercado, "Microgrid architectures for distributed generation: A brief review," in 2017 IEEE PES Innovative Smart Grid Technologies Conference - Latin America (ISGT Latin America). Quito, Ecuador, 2017, pp. 1–6. doi:10.1109/ISGT-LA.2017.8126746.

[6] E. Hossain, E. Kabalci, R. Bayindir, and R. Perez, "A comprehensive study on microgrid technology," *Int. J. Renew. Energy Res.*, vol. 4, no. 4, pp. 1094–1107, 2014, doi:10.20508/ijrer.v4i4.2005.g6455.

[7] M. Wolsink, "Distributed energy systems as common goods: Socio-political acceptance of renewables in intelligent microgrids," *Renew. Sustain. Energy Rev.*, vol. 127, no. 109841, pp. 1–14, 2020, doi:10.1016/j.rser.2020.109841.

[8] B. Modu, P. Abdullah, M. A. Sanusi, and M. F. Hamza, "DC-based microgrid: Topologies, control schemes, and implementations," *Alexandria Eng. J.*, vol. 70, pp. 61–92, 2023, doi:10.1016/j.aej.2023.02.021.

[9] H. Kirchhoff, N. Kebir, K. Neumann, P. W. Heller, and K. Strunz, "Developing mutual success factors and their application to swarm electrification: Microgrids with 100% renewable energies in the Global South and Germany," *J. Clean. Prod.*, vol. 128, pp. 1–23, 2016, doi:10.1016/j.jclepro.2016.03.080.

[10] W. Feng, M. Jin, X. Liu, *et al.*, "A review of microgrid development in the United States–A decade of progress on policies, demonstrations, controls, and software tools," *Appl. Energy*, vol. 228, pp. 1656–1668, 2018, doi:10.1016/j.apenergy.2018.06.096.

[11] M. N. Ahmed, M. Hojabri, A. M. Humada, H. Bin Daniyal, and H. F. Frayyeh, "An overview on microgrid control strategies," *Int. J. Eng. Adv. Technol.*, vol. 4, no. 5, pp. 93–98, 2015, [Online]. Available: https://portal.arid.my/Publications/fbedd8d3-ea1b-48.pdf.

[12] A. Anvaru-Moghaddam, H. Abdi, B. Mohammadi-Ivatloo, and N. Hatziargyriou, *Microgrids: Advances in Operation, Control, and Protection*. Springer, Cham, 2021. doi:10.1007/978-3-030-59750-4.

[13] A. Muhtadi, D. Pandit, N. Nguyen, and J. Mitra, "Distributed energy resources based microgrid: Review of architecture, control, and reliability," *IEEE Trans. Ind. Appl.*, vol. 57, no. 3, pp. 2223–2235, 2021, doi:10.1109/TIA.2021.3065329.

[14] V. P. Sabhlok, "Dynamics and challenges of microgrids implementation," Degree of Master of Science in Management Studies, Massachusetts Institute of Technology (MIT) Sloan School of Management, 2013. [Online]. Available: http://hdl.handle.net/1721.1/80685.

[15] A. Singh and B. S. Surjan, "Microgrid: A review," *Int. J. Res. Eng. Technol.*, vol. 3, no. 2, pp. 185–198, 2014, [Online]. Available: http://www.ijret.org.

[16] V. N. Coelho, M. W. Cohen, I. M. Coelho, N. Liu, and F. G. Guimarães, "Multi-agent systems applied for energy systems integration: State-of- the-art applications and trends in microgrids," *Appl. Energy*, vol. 187, pp. 820–832, 2017, doi:10.1016/j.apenergy.2016.10.056.

[17] A. Hirsch, Y. Parag, and J. Guerrero, "Microgrids: A review of technologies, key drivers, and outstanding issues," *Renew. Sustain. Energy Rev.*, vol. 90, pp. 402–411, 2018, doi:10.1016/j.rser.2018.03.040.

[18] J. Romankiewicz, C. Marnay, N. Zhou, and M. Qu, "Lessons from international experience for China's microgrid demonstration program," *Energy Policy*, vol. 67, pp. 198–208, 2014, doi:10.1016/j.enpol.2013.11.059.

[19] E. Nasr-Azadani, P. Su, W. Zheng, *et al.*, "Canadian Renewable Energy Laboratory (CANREL): A testbed for microgrids," *IEEE Electrif. Mag.*, vol. 8, no. 1, pp. 49–60, 2020, doi:10.1109/MELE.2019.2962889.

[20] T. Prabatha, J. Hager, B. Carneiro, K. Hewage, and R. Sadiq, "Analyzing energy options for small-scale off-grid communities: A Canadian case study," *J. Clean. Prod.*, vol. 249, p. 119320, 2020, doi:10.1016/j.jclepro.2019.119320.

[21] T. Stringer and M. Joanis, "Decarbonizing Canada's remote microgrids," *Energy*, vol. 264, no. 126287, 2023, doi:10.1016/j.energy.2022.126287.

[22] X. Li, "Sustainable island energy systems: a case study of Tilos Island, Greece," in S. Löbbe, F. Sioshansi, and D. Robinson (eds.) *Energy Communities Customer-Centered, Market Driven, Welfare-Enhancing?* AP Academic Press, 2022, ch. 25, pp. 435–448. doi:10.1016/B978-0-323-91135-1.00023-7.

[23] N. Hatziargyriou, A. Dimeas, N. Vasilakis, D. Lagos, and A. Kontou, "The Kythnos microgrid: A 20-year history," *IEEE Electrification Magazine*, vol. 8, no. 4, pp. 46–54, Dec. 2020. doi:10.1109/MELE.2020.3026439.

[24] B. Kroposki, R. Lasseter, T. Ise, S. Morozumi, S. Papathanassiou, and N. D. Hatziargyriou, "Making microgrids work," *IEEE Power and Energy Magazine*, vol. 6, no. 3, pp. 40–53, 2008. doi:10.1109/MPE.2008.918718.

[25] W. Ajaz and D. Bernell, "Microgrids and the transition toward decentralized energy systems in the United States: A multi-level perspective," *Energy Policy*, vol. 149, no. 112094, pp. 1–11, 2021, doi:10.1016/j.enpol.2020.112094.

[26] D. T. Ton and M. A. Smith, "The U.S. Department of Energy's microgrid initiative," *Electr. J.*, vol. 25, no. 8, pp. 84–94, 2012, doi:10.1016/j.tej.2012.09.013.

[27] F. Norouzi, T. Hoppe, L. M. Kamp, C. Manktelow, and P. Bauer, "Diagnosis of the implementation of smart grid innovation in The Netherlands and corrective actions*," *Renew. Sustain. Energy Rev.*, vol. 175, no. 113185, pp. 1–20, 2023, doi:10.1016/j.rser.2023.113185.

[28] T. L. KULeuven, "The first micro grid in The Netherlands: Bronsbergen," 2009. [Online]. Available: http://www.leonardo-energy.org/.

[29] R. Kohlmann and K. Mittelbach, "Smart grids in Germany-Fields of action for distribution system operators on the way to Smart Grids," 2011, *BDEW-Verteilnetzstudie 2011*. [Online]. Available: https://www.bdew.de/media/documents/Pub_20120601_Brochure_Smart-Grids-Germany.pdf.

[30] A. Dautzenberg, M. Kaiser, M. Weibelzahl, and J. Weissflog, "Industrial multi-energy communities as grid-connected microgrids: Understanding the role of asymmetric grid-charge regulation," *J. Clean. Prod.*, vol. 466, no. 142738, pp. 1–16, 2024, doi:10.1016/j.jclepro.2024.142738.

[31] M. Caliano, F. Delfino, M. Di Somma, *et al.* "An energy management system for microgrids including costs, exergy, and stress indexes," *Sustain. Energy, Grids Networks*, vol. 32, p. 100915, 2022, doi:10.1016/j.segan.2022.100915.

[32] F. Pasimeni, "Adoption and diffusion of micro-grids in Italy. An analysis of regional factors using agent-based modelling," SWPS 2017-09 (June), 2017. doi:10.2139/ssrn.2986198.

[33] S. Abu-Sharkh, R. J. Arnold, J. Kohler, *et al.*, "Can microgrids make a major contribution to UK energy supply?," *Renew. Sustain. Energy Rev.*, vol. 10, no. 2, pp. 78–127, 2006, doi:10.1016/j.rser.2004.09.013.

[34] V. K. Sood and H. Abdelgawad, "Microgrids architectures," in R. K. Chauhan and K. Chauhan (eds.) *Distributed Energy Resources in Microgrids*. Elsevier Inc., 2019, ch. 1, pp. 1–31. doi:10.1016/B978-0-12-817774-7.00001-6.

[35] H. Abdul Muqeet, H. M. Munir, H. Javed, M. Shahzad, M. Jamil, and J. M. Guerrero, "An energy management system of campus microgrids: State-of-the-art and future challenges," *Energies*, vol. 14, no. 6525, pp. 1–34, 2021, doi:10.3390/en14206525.

[36] L. Gomes, P. Faria, Z. Vale, and J. Silva, "Energy analyzer emulator for microgrid implementation and demonstration and respective gateway," *IEEE Trans. Ind. Appl.*, vol. 99, pp. 1–10, 2018, doi:10.1109/TIA.2018.2866451.

[37] E. Hossain, E. Kabalci, R. Bayindir, and R. Perez, "Microgrid testbeds around the world: State of art," *Energy Convers. Manag.*, vol. 86, pp. 132–153, 2014, doi:10.1016/j.enconman.2014.05.012.

[38] M. Warneryd, M. Håkansson, and K. Karltorp, "Unpacking the complexity of community microgrids: A review of institutions' roles for development of microgrids," *Renew. Sustain. Energy Rev.*, vol. 121, no. 109690, pp. 1–16, 2020, doi:10.1016/j.rser.2019.109690.

[39] M. H. Bellido, L. P. Rosa, A. O. Pereira, D. M. Falcão, and S. K. Ribeiro, "Barriers, challenges and opportunities for microgrid implementation: The case of Federal University of Rio de Janeiro," *J. Clean. Prod.*, vol. 188, pp. 1–33, 2018, doi:10.1016/j.jclepro.2018.03.012.

[40] T. E. D. C. Huayllas, S. Ramos, and R. L. Vasquez-Arnez, "Microgrid systems: Current status and challenges," in *IEEE/PES Transmission and Distribution Conference and Exposition: Latin America (T&D-LA), Sao Paulo, Brazil*, 2010, pp. 7–12. doi:10.1109/TDC-LA.2010.5762853.

[41] A. Boche, C. Foucher, and L. F. L. Villa, "Understanding microgrid sustainability: A systemic and comprehensive review," *Energies*, vol. 15, no. 2906, pp. 1–29, 2022, doi:10.3390/en15082906.

[42] L. Mariam, M. Basu, and M. F. Conlon, "Microgrid: Architecture, policy and future trends," *Renew. Sustain. Energy Rev.*, vol. 64, pp. 477–489, 2016, doi:10.1016/j.rser.2016.06.037.

[43] R. Alhassan, "FG to commission 10 megawatt Katsina wind farm in December," *Daily Nigerian*, Nov. 24, 2021. [Online]. Available: https://dailynigerian.com/commission-megawatt-katsina/.

[44] M. Balarabe, K. Abdullah, M. Nawawi, and F. Tan, "Accessement of wind power potential in Katsina State from meteorological data," *Mod. Appl. Sci.*, vol. 8, no. 6, pp. 53–60, 2014, doi:10.5539/mas.v8n6p53.

[45] P. O. Jonathan, O. Ojo, and I. Davidson, "Design considerations of the Katsina Wind Farm in Nigeria," in 2016 IEEE PES Power Africa Conference at Livingstone, Zambia, 2016. doi:10.1109/PowerAfrica.2016.7556611.

[46] L. Salisu and I. Garba, "Electricity generation using wind in Katsina State, Nigeria," *Int. J. Eng. Res. Technol.*, vol. 2, no. 2, pp. 1–13, 2013, [Online]. Available: https://www.ijert.org/research/electricity-generation-using-wind-in-katsina-state-nigeria-IJERTV2IS2611.pdf.

[47] Q. Hassan, S. Algburi, A. Z. Sameem, H. M. Salman, and M. Jaszczur, "A review of hybrid renewable energy systems: Solar and wind-powered solutions: Challenges, opportunities, and policy implications," *Results Eng.*, vol. 20, no. 101621, pp. 1–25, 2023, doi:10.1016/j.rineng.2023.101621.

[48] T. Adefarati and R. C. Bansal, "Reliability, economic and environmental analysis of a microgrid system in the presence of renewable energy resources," *Appl. Energy*, vol. 236, pp. 1089–1114, 2019, doi:10.1016/j.apenergy.2018.12.050.

[49] S. Skaros, "Investigation on the energy consumption in the built environment of Gotland," Degree Project in the Built Environment, Second Cycle: Kth Royal Institute of Technology, School of Architecture and the Built Environment, 2020. [Online]. Available: https://www.diva-portal.org/smash/get/diva2:1504077/FULLTEXT01.pdf.

[50] F. M. Mundbjerg and M. B. Thomsen, "(Em)powering community energy on Gotland-An exploration of energy justice through a case study of 'an island in an island,'" Lund University International Master's Programme in Environmental Studies and Sustainability Science-Lund University Centre for Sustainable Studies (LUCSUS), 2023. [Online]. Available: https://lup.lub.lu.se/luur/download?func=downloadFile&recordOId=9120318&fileOId=9120326.

[51] J. Hamilton, M. Negnevitsky, X. Wang, and S. Lyden, "High penetration renewable generation within Australian isolated and remote power systems," *Energy*, vol. 168, pp. 684–692, 2019, doi:10.1016/j.energy.2018.11.118.

[52] J. Jantzen, M. Kristensen, and T. H. Christensen, "Sociotechnical transition to smart energy: The case of Samso 1997–2030," *Energy*, vol. 162, pp. 20–34, 2018, doi:10.1016/j.energy.2018.07.174.

[53] F. Briongos, C. A. P. Gaona, and J. A. Sánchez-Fernández, "Increasing demand covered by renewable energy in El Hierro Island," *Sustainability*, vol. 15, no. 16185, pp. 1–14, 2023, doi:10.3390/su152316185.

[54] G. Jastrzebska, "El Hierro renewable energy hybrid system: A tough compromise," *Energies*, vol. 11, no. 2812, pp. 1–20, 2018, doi:10.3390/en11102812.

[55] K. Bunker, S. Doig, K. Hawley, and J. Morris, *Renewable microgrids: Profiles from Islands and remote communities across the Globe*, vol. 11. Rocky Mountain Institute and Carbon War Room, 2015. [Online]. Available: http://www.rmi.org/islands-renewable-microgrids.

[56] G. M. Argungu and K. A. Dabai, "The reliability analysis of the ongoing 10MW electricity generation from wind in Katsina State, Nigeria," *Int. J. Adv. Sci. Res. Eng.*, vol. 3, no. 11, pp. 55–64, 2017, doi:10.7324/IJASRE.2017.32552.

[57] O. Eseosa and U. Okechukwu, "Harnessing and modelling wind power plant with permanent magnet syncrhonous generator for power generation in Nigeria," *Res. J. Nanosci. Eng.*, vol. 4, no. 2, pp. 1–15, 2020, [Online]. Available: https://sryahwapublications.com/article/download/2637-5591.0402001.

[58] O. S. Ohunakin, "Wind resource evaluation in six selected high altitude locations in Nigeria," *Renew. Energy*, vol. 36, no. 12, pp. 3273–3281, 2011, doi:10.1016/j.renene.2011.04.026.

[59] A. F. Akpaneno and A. B. Jamila, "Nonlinear dynamical characterization of wind speed and wind power in Kano, Jigawa and Katsina States of Nigeria," *Niger. J. Phys.*, vol. 31, no. 2, pp. 215–223, 2022, [Online]. Available: https://njp.nipngr.org/index.php/njp/article/view/78.

[60] F. Li, K. Xie, and J. Yang, "Optimization and analysis of a hybrid energy storage system in a small-scale standalone microgrid for Remote Area Power

Supply (RAPS)," *Energies*, vol. 8, pp. 4802–4826, 2015, doi:10.3390/en8064802.

[61] R. Hemmati, "Technical and economic analysis of home energy management system incorporating small-scale wind turbine and battery energy storage system," *J. Clean. Prod.*, vol. 159, pp. 106–118, 2017, doi:10.1016/j.jclepro.2017.04.174.

[62] J. Jurasz, B. Ceran, and A. Orłowska, "Component degradation in small-scale off-grid PV-battery systems operation in terms of reliability, environmental impact and economic performance," *Sustain. Energy Technol. Assessments*, vol. 38, no. 100647, pp. 1–11, 2020, doi:10.1016/j.seta.2020.100647.

[63] Y. Astriani, G. M. Shafiullah, M. Anda, and H. Hilal, "Techno-economic evaluation of utilizing a heating small-scale microgrid," in *10th International Conference on Applied Energy (ICAE 2018), 22-25 August 2018, Hong Kong, China*, Elsevier B.V., 2019, pp. 3131–3137. doi:10.1016/j.egypro.2019.01.1013.

[64] H. Akter, H. O. R. Howlader, A. Nakadomari, R. Islam, A. Y. Saber, and T. Senjyu, "A short assessment of renewable energy for optimal sizing of 100% renewable energy based microgrids in remote Islands of developing countries: A case study in Bangladesh," *Energies*, vol. 15, no. 1084, pp. 1–30, 2022, doi:10.3390/en15031084.

[65] A. A. Abou El-Ela, R. A. El-Sehiemy, S. M. Allam, A. M. Shaheen, N. A. Nagem, and A. M. Sharaf, "Renewable energy micro-grid interfacing: Economic and environmental issues," *Electronics*, vol. 11, no. 815, pp. 1–27, 2022, doi:10.3390/electronics11050815.

[66] S. Obara, S. Fujimoto, K. Sato, and Y. Utsugi, "Planning renewable energy introduction for a microgrid without battery storage," *Energy*, vol. 215, no. 119176, pp. 1–41, 2020, doi:10.1016/j.energy.2020.119176.

[67] A. Eales, A. Alsop, D. Frame, S. Strachan, and S. Galloway, "Assessing the market for solar photovoltaic (PV) microgrids in malawi," *J. Sustain. Res.*, vol. 2, no. 1, pp. 1–44, 2020, doi:10.20900/jsr20200008.

[68] J. A. Aguilar-Jiménez, L. Hernández-Callejo, J. A. Suástegui-Macías, *et al.*, "Energy and economic analysis of renewable energy-based isolated microgrids with AGM and lithium battery energy storage: Case study Bigene, Guinea-Bissau," *Urban Sci.*, vol. 7, no. 2, p. 6, 2023, doi:10.3390/urbansci7020066.

[69] C. Marnay, S. Chatzivasileiadis, C. Abbey, *et al.*, "Microgrid evolution roadmap," *International Symposium on Smart Electric Distribution Systems and Technologies (EDST)*, pp. 139–144, 2015, [Online]. Available: http://microgrid-symposiums.org.

[70] J. M. Aberilla, A. Gallego-Schmid, L. Stamford, and A. Azapagic, "Design and environmental sustainability assessment of small-scale off-grid energy systems for remote rural communities," *Appl. Energy*, vol. 258, no. 114004, pp. 1–16, 2020, doi:10.1016/j.apenergy.2019.114004.

[71] E. Bergstedt and S. Nyström, "Possibilities of batteries in the distribution system as substitute for cable reinforcements while maintaining reliable electricity distribution," Master's thesis in Electric Power Engineering, Department of Electrical Engineering, Chalmers University of Technology, 2018. [Online]. Available: https://odr.chalmers.se/server/api/core/bitstreams/c8345cb4-6d42-4654-8d8d-6c9235b0913b/content.

[72] D. Magdefrau, T. Taufik, M. Poshtan, and M. Muscarella, "Analysis and review of DC microgrid implementations," in *2016 International Seminar on Application for Technology and Communication*, 2016, pp. 241–246. doi:10.1109/ISEMANTIC.2016.7873845.

[73] Y. Sun, "Research on integrated multi-energy system to improve energy efficiency and enhance technological progress of renewable energy in the APEC region," Asia-Pacific Economic Cooperation (APEC), 2019. [Online]. Available: https://www.apec.org/docs/default-source/Publications/2019/12.

[74] F. Keck, M. Lenzen, A. M. Vassallo, and L. I. Mengyu, "The impact of battery energy storage for renewable energy power grids in Australia," *Energy*, vol. 173, pp. 647–657, 2019, doi:10.1016/j.energy.2019.02.053.

[75] D. Infield and J. Hill, "Literature review: Electrical energy storage for Scotland," 2014. [Online]. Available: https://www.climatexchange.org.uk/wp-content/uploads/2023/09/electrical_energy_storage_in_scotland.pdf.

[76] L. Breen, "Modelling, optimisation and the lessons learned of a renewable based electrical network–The Isle of Eigg," Master of Science Degree, Sustainable Engineering: Renewable Energy Systems and the Environment-Department of Mechanical and Aerospace Engineering, University of Strathclyde, 2015. [Online]. Available: https://www.esru.strath.ac.uk/Documents/MSc_2015/Breen.pdf.

[77] A. M. Quevedo, E. J. M. Domínguez, J. M. de León Izquier, *et al.*, "Gorona del Viento wind-hyrdo power plant-Results, improvement actuations, and next steps," in *3rd Interntional Hybrid Power System Workshop, Tenerife, Spain, 8-9 May, 2018*, 2018, pp. 1–9. [Online]. Available: https://hybridpowersystems.org/wp-content/uploads/sites/9/2018/05/1_2_TENE18_036_paper_Marrero_Agustin.pdf.

[78] M. Uche-Soria and C. Rodríguez-Monroy, "Special regulation of isolated power systems: The Canary Islands, Spain," *Sustainability*, vol. 10, no. 2572, pp. 1–20, 2018, doi:10.3390/su10072572.

[79] O. Novykh, J. A. Mendez, B. Gonzalez-Diaz, and I. Sviridenko, "Performance analysis of hybrid hydroelectric Gorona del Viento and the basic directions of its perfection," *Renew. Energy Power Qual. J.*, vol. 17, pp. 489–494, 2019, doi:10.24084/repqj17.353.

[80] M. Kamibayashi and K. Tanaka, "Recent sodium sulfur battery applications," in *Transmission and Distribution Conference and Exposition [2001 IEEE/PES]*, 2001. doi:10.1109/TDC.2001.971425.

[81] Z. Wen, J. Cao, Z. Gu, X. Xu, F. Zhang, and Z. Lin, "Research on sodium sulfur battery for energy storage," *Solid State Ionics*, vol. 179, no. 27–32, pp. 1697–1701, 2008, doi:10.1016/j.ssi.2008.01.070.

[82] P. Hearps, R. Dargaville, D. McConnel, M. Sandiford, T. Forcey, and P. Seligman, "Opportunities for pumped hydro energy storage in Australia," 2014. [Online]. Available: https://energy.unimelb.edu.au/__data/assets/pdf_file/0007/1526587/Opps-for-pumped-hydro-in-Australia.pdf.

[83] A. Blakers, M. Stocks, B. Lu, K. Anderson, and A. Nadolny, "An atlas of pumped hydro energy storage," 2017. [Online]. Available: http://re100.eng.anu.edu.au/research/phes/.

[84] A. Paneri, K. Moore, A. Beylot, S. Muller, M. Braun, and X. Yan, "Renewable energy can make small-scale mining in Europe more feasible," *Resour. Conserv. Recycl.*, vol. 172, no. 105674, pp. 1–12, 2021, doi:10.1016/j.resconrec.2021.105674.

[85] A. Baldinelli, L. Barelli, G. Bidini, and G. Discepoli, "Economics of innovative high capacity-to-power energy storage technologies pointing at 100% renewable micro-grids," *J. Energy Storage*, vol. 28, no. 101198, pp. 1–13, 2020, doi:10.1016/j.est.2020.101198.

[86] H. Shahinzadeh, H. Shayeghi, and G. B. Gharehpetian, "A short-term energy management of microgrids considering renewable energy resources, micro-compressed air energy storage and DRPs," *Int. J. Renew. Energy Res.*, vol. 9, no. 4, pp. 1712–1723, 2019, [Online]. Available: https://www.researchgate.net/publication/338375686.

[87] A. H. Fathima and K. Palanisamy, "Optimization in microgrids with hybrid energy systems – A review," *Renew. Sustain. Energy Rev.*, vol. 45, pp. 431–446, 2015, doi:10.1016/j.rser.2015.01.059.

[88] P. Kokkotis, C. S. Psomopoulos, G. C. Ioannidis, and S. D. Kaminaris, "Small scale energy storage systems. A short review in their potential environmental impact," *Fresenius Environ. Bull.*, vol. 26, no. 9, pp. 5658–5665, 2017, [Online]. Available: https://www.researchgate.net/publication/319755301.

[89] S. M. S. Sadati, "Assessment of renewable energy based micro-grids for small communities," The Degree of Master of Science in Sustainable Environment and Energy Systems Program, Middle East Technical University, Northern Cyprus Campus, 2016. doi:10.13140/RG.2.2.28631.93608.

[90] E. Planas, J. Andreu, J. I. Gárate, I. M. De Alegría, and E. Ibarra, "AC and DC technology in microgrids: A review," *Renew. Sustain. Energy Rev.*, vol. 43, pp. 726–749, 2015, doi:10.1016/j.rser.2014.11.067.

[91] T. Zheng, H. Yang, R. Zhao, Y. C. Kang, and V. Terzija, "Design, evaluation and implementation of an Islanding Detection Method for a micro-grid," *Energies*, vol. 11, no. 323, pp. 1–24, 2018, doi:10.3390/en11020323.

[92] S. Abid, R. Ahshan, R. Al Abri, A. Al-Badi, and M. Albadi, "Techno-economic and environmental assessment of renewable energy sources, virtual synchronous generators, and electric vehicle charging stations in microgrids," *Appl. Energy*, vol. 353, no. 122028, 2024, doi:10.1016/j.apenergy.2023.122028.

[93] M. Kamal, I. Ashraf, and E. Fernandez, "Sustainable electrification planning of rural microgrid using renewable resources and its environmental impact assessment," *Environ. Sci. Pollut. Res.*, vol. 29, pp. 86376–86399, 2022, doi:10.1007/s11356-022-19949-7.

[94] A. Kumar, B. Sah, A. R. Singh, *et al.*, "Multicriteria decision-making methodologies and their applications in sustainable energy system/microgrids," in S. H. E. Abdel Aleem, A. Y. Abdelaziz, A. F. Zobaa, and R. Bansal (eds.) *Decision Making Applications in Modern Power Systems*. Elsevier Inc., 2020, ch. 1, pp. 1–40. doi:10.1016/B978-0-12-816445-7.00001-3.

[95] L. Al-Ghussain, R. Samu, O. Taylan, and M. Fahrioglu, "Sizing renewable energy systems with energy storage systems in microgrids for maximum cost-efficient utilization of renewable energy resources," *Sustain. Cities Soc.*, vol. 55, no. 102059, pp. 1–12, 2020, doi:10.1016/j.scs.2020.102059.

[96] P. Liu, Y. Wu, Y. Su, and B. Duan, "Fault detection and location of microgrid based on distributed decision," in *43rd Annual Conference of the IEEE Industrial Electronics Society [IECON 2017]*, 2017. doi:10.1109/IECON.2017.8216873.

[97] K. Zitelman, K. Verclas, S. Cramer, D. S. Byrnett, and D. Liberatore, "User objectives and design approaches for microgrids: Options for delivering reliability and resilience, clean energy, energy savings, and other priorities," 2021. [Online]. Available: https://pubs.naruc.org/pub/E1F332D4-155D-0A36-31CB-889ABED753D5.

[98] M. Stadler and A. Naslé, "Planning and implementation of bankable microgrids," *Electr. J.*, vol. 32, pp. 24–29, 2020, doi:10.1016/j.tej.2019.05.004.

[99] K. I. Ibekwe, P. E. Ohenhen, O. Chidolue, *et al.*, "Microgrid systems in U. S. energy infrastructure: A comprehensive review: Exploring decentralized energy solutions, their benefits, and challenges in regional implementation," *World J. Adv. Res. Rev.*, vol. 21, no. 1, pp. 973–987, 2024, doi:10.30574/wjarr.2024.21.1.0112.

[100] J.-S. Kim, S. M. So, J.-T. Kim, *et al.*, "Microgrids platform: A design and implementation of common platform for seamless microgrids operation," *Electr. Power Syst. Res.*, vol. 167, pp. 21–38, 2019, doi:10.1016/j.epsr.2018.10.019.

[101] R. Wallsgrove, J. Woo, J.-H. Lee, and L. Akiba, "The emerging potential of microgrids in the transition to 100% renewable energy systems," *Energies*, vol. 14, no. 1687, pp. 1–28, 2021, doi:10.3390/en14061687.

[102] A. Chebabhi, I. Tegani, A. D. Benhamadouche, and O. Kraa, "Optimal design and sizing of renewable energies in microgrids based on financial considerations a case study of Biskra, Algeria," *Energy Convers. Manag.*, vol. 291, no. 117270, 2023, doi:10.1016/j.enconman.2023.117270.

[103] T. S. Ustun, C. Ozansoy, and A. Zayegh, "Recent developments in microgrids and example cases around the world—A review," *Renew. Sustain. Energy Rev.*, vol. 15, no. 8, pp. 4030–4041, 2011, doi:10.1016/j.rser.2011.07.033.

[104] F. Serra, M. Lucariello, M. Petrollese, and G. Cau, "Optimal integration of hydrogen-based energy storage systems in photovoltaic microgrids: A techno-economic assessment," *Energies*, vol. 13, no. 4149, pp. 1–18, 2020, doi:10.3390/en13164149.

[105] G. Mendes and N. Nigmatulina, "Grid-connected microgrids: From research to sustainable implementation," in W. Leal Filho, A. Azul, L. Brandli, A. Lange Salvia, and T. Wall (eds.) *Affordable and Clean Energy.*

Encyclopedia of the UN Sustainable Development Goals. Springer, Cham, 2020, pp. 1–15. doi:10.1007/978-3-319-71057-0_113-1.

[106] J. M. Schwidtal, *Comprehensive evaluation of electricity market design elements for the strategic integration of renewable energies*. Department of Industrial Engineering, Thermo-Mechanical Energy Engineering, Università degli Studi di Padova, PhD Course in Industrial Engineering, 2022. [Online]. Available: https://www.research.unipd.it/bitstream/11577/3454107/2/Tesi_JanMarc_Schwidtal_3.pdf.

[107] X. Liu and B. Su, "Microgrids-An integration of renewable energy technologies," in *Technical Session 3 Protection, Control, Communication and Automation of Distribution Network [CICED 2008]*, 2008, pp. 1–7. doi:10.1109/CICED.2008.5211651.

[108] M. Kamal, A. Mohammad, I. Ashraf, and E. Fernandez, "Rural electrification using renewable energy resources and its environmental impact assessment," *Environ. Sci. Pollut. Res.*, vol. 29, pp. 86562–86579, 2022, doi:10.1007/s11356-022-22001-3.

[109] M. Homa, A. Pałac, M. Zoładek, and R. Figaj, "Small-scale hybrid and polygeneration renewable energy systems: Energy generation and storage technologies, applications, and analysis methodology," *Energies*, vol. 15, no. 9152, pp. 1–52, 2022, doi:10.3390/en15239152.

[110] A. A. K. Arani, G. B. Gharehpetian, and M. Abedi, "Review on energy storage systems control methods in microgrids," *Electr. Power Energy Syst.*, vol. 107, pp. 745–757, 2019, doi:10.1016/j.ijepes.2018.12.040.

[111] N. M. Kumar, S. S. Chopra, A. A. Chand, R. M. Elavarasan, and G. M. Shafiulla, "Hybrid renewable energy microgrid for a residential community: A techno-economic and environmental perspective in the context of the SDG7," *Sustainability*, vol. 12, no. 3944, pp. 1–30, 2020, doi:10.3390/su12103944.

[112] A. T. Dahiru, D. Daud, C. W. Tan, Z. T. Jagun, S. Samsudin, and M. D. Abdulhakeem, "A comprehensive review of demand side management in distributed grids based on real estate perspectives," *Environ. Sci. Pollut. Res.*, vol. 30, pp. 81984–82013, 2023, doi:10.1007/s11356-023-25146-x.

[113] P. J. Levi, S. D. Kurland, M. Carbajales-Dale, J. P. Weyant, A. R. Brandt, and S. M. Benson, "Macro-energy systems: Toward a new discipline," *Joule*, vol. 3, no. 10, pp. 2282–2286, 2019, doi:10.1016/j.joule.2019.07.017.

[114] J. Yan, Y. Zhai, P. Wijayatunga, A. M. Mohamed, and P. E. Campana, "Editorial-Renewable energy integration with mini/micro-grids," *Appl. Energy*, vol. 201, pp. 241–244, 2017, doi:10.1016/j.apenergy.2017.05.160.

[115] A. Kapur, "Evaluating energy storage systems for renewable energy integrated urban community microgrids," *Electrochem. Soc. Trans.*, vol. 107, no. 1, pp. 1981–2001, 2022, doi:10.1149/10701.1981ecst.

[116] M. Soshinskaya, W. H. J. Crijns-Graus, J. M. Guerrero, and J. C. Vasquez, "Microgrids: Experiences, barriers and success factors," *Renew. Sustain. Energy Rev.*, vol. 40, pp. 659–672, 2014, doi:10.1016/j.rser.2014.07.198.

[117] D. Q. Oliveira, O. R. Saavedra, K. Santos-Pereira, *et al.*, "A critical review of energy storage technologies for microgrids," *Energy Syst.*, pp. 1–30, 2021, doi:10.1007/s12667-021-00464-6.

[118] P. Liu, M. C. Georgiadis, and E. N. Pistikopoulos, "An energy systems engineering approach for the design and operation of microgrids in residential applications," *Chem. Eng. Res. Des.*, vol. 91, no. 10, pp. 2054–2069, 2013, doi:10.1016/j.cherd.2013.08.016.

[119] O. R. Saavedra, B. Venkatesh, M. Resener, and P. M. Pardalos, "Energy systems special issue on 'Smart Microgrids,'" *Energy Syst.*, pp. 1–3, 2021, doi:10.1007/s12667-021-00487-z.

[120] M. K. Kiptoo, M. E. Lotfy, O. B. Adewuyi, A. Conteh, A. M. Howlader, and T. Senjyu, "Integrated approach for optimal techno-economic planning for high renewable energy-based isolated microgrid considering cost of energy storage and demand response strategies," *Energy Convers. Manag.*, vol. 215, no. 112917, pp. 1–16, 2020, doi:10.1016/j.enconman.2020.112917.

[121] M. Roman-Barri, I. Cairo-Molins, A. Sumper, and A. Sudria-Andreu, "Experience on the implementation of a microgrid project in Barcelona," *IEEE Xplore*, pp. 1–7, 2011, doi:10.1109/ISGTEUROPE.2010.5638996.

[122] M. E. T. S. Junior and L. C. G. Freitas, "Power electronics for modern sustainable power systems: Distributed generation, microgrids and smart grids—A review," *Sustainability*, vol. 14, no. 3597, pp. 1–22, 2022, doi:10.3390/su14063597.

[123] F. Shariatzadeh, R. Zamora, and A. K. Srivastava, "Real time implementation of microgrid reconfiguration," in *North American Power Symposium*, 2011, pp. 1–6. doi:10.1109/NAPS.2011.6025181.

[124] F. Shariatzadeh, C. B. Vellaithurai, S. S. Biswas, R. Zamora, and A. K. Srivastava, "Real-time implementation of intelligent reconfiguration algorithm for microgrid," *IEEE Trans. Sustain. Energy*, vol. 5, no. 2, pp. 598–607, 2014, doi:10.1109/TSTE.2013.2289864.

[125] F. C. Robert, G. S. Sisodia, and S. Gopalan, "A critical review on the utilization of storage and demand response for the implementation of renewable energy microgrids," *Sustain. Cities Soc.*, vol. 40, pp. 1–32, 2018, doi:10.1016/j.scs.2018.04.008.

[126] L. He, S. Zhang, Y. Chen, L. Ren, and J. Li, "Techno-economic potential of a renewable energy-based microgrid system for a sustainable large-scale residential community in Beijing, China," *Renew. Sustain. Energy Rev.*, vol. 93, pp. 631–641, 2018, doi:10.1016/j.rser.2018.05.053.

[127] J. G. Gomes, H. J. Xu, Q. Yang, and C. Y. Zhao, "An optimization study on a typical renewable microgrid energy system with energy storage," *Energy*, vol. 234, no. 121210, pp. 1–15, 2021, doi:10.1016/j.energy.2021.121210.

[128] F. Norouzi, T. Hoppe, L. R. Elizondo, and P. Bauer, "A review of socio-technical barriers to smart microgrid development*," *Renew. Sustain. Energy Rev.*, vol. 167, no. 112674, pp. 1–17, 2022, doi:10.1016/j.rser.2022.112674.

[129] E. Kabalcı, Y. Kabalcı, and P. Siano, "Design and implementation of a smart metering infrastructure for low voltage microgrids," *Int. J. Electr.*

Power Energy Syst., vol. 134, no. 107375, pp. 1–13, 2022, doi:10.1016/j.ijepes.2021.107375.

[130] G. Shahgholian, "A brief review on microgrids: Operation, applications, modeling, and control," *Int. Trans. Electr. Energy Syst.*, vol. 31, no. e12885, pp. 1–28, 2021, doi:10.1002/2050-7038.12885.

[131] J. A. P. Lopes, A. G. Madureira, and C. Moreira, "A view of microgrids," in P. D. Lund, J. A. Byrne, R. Haas, and D. Flynn (eds.) *Advances in Energy Systems: The Large-scale Renewable Energy Integration Challenge*, 1st ed. John Wiley & Sons Ltd., 2019, ch. 9, pp. 149–166. [Online]. Available: https://dokumen.pub/download/advances-in-energy-systems-the-large-scale-renewable-energy-integration-challenge-9781119508281-1119508282.html.

[132] G. Platt, A. Berry, and D. Cornforth, "What role for microgrids?," in F. P. Sioshansi (ed.) *Smart Grid-Part II Smart Supply: Integrating Renewable & Distributed Generation*. Elsevier Inc., 2012, ch. 8, pp. 185–207. doi:10.1016/B978-0-12-386452-9.00008-5.

[133] A. Shufian and N. Mohammad, "Modeling and analysis of cost-effective energy management for integrated microgrids," *Clean. Eng. Technol.*, vol. 8, no. 100508, pp. 1–19, 2022, doi:10.1016/j.clet.2022.100508.

[134] S. K. Bhoi, K. Kasturi, and M. R. Nayak, "Optimisation of the operation of a microgrid with renewable energy sources and battery storage," *e-Prime - Adv. Electr. Eng. Electron. Energy*, vol. 4, no. 100159, pp. 1–9, 2023, doi:10.1016/j.prime.2023.100159.

[135] H. Qin and T. Liu, "Resilience enhancement of multi-microgrid system of systems based on distributed energy scheduling and network reconfiguration," *J. Electr. Eng. Technol.*, vol. 19, pp. 2135–2147, 2024, doi:10.1007/s42835-023-01724-4.

[136] Y. Wang, A. O. Rousis, and G. Strbac, "A resilience enhancement strategy for networked microgrids incorporating electricity and transport and utilizing a stochastic hierarchical control approach," *Sustain. Energy, Grids Networks*, vol. 26, no. 100464, pp. 1–12, 2021, doi:10.1016/j.segan.2021.100464.

[137] K. Zhou, S. Yang, Z. Chen, and S. Ding, "Optimal load distribution model of microgrid in the smart grid environment," *Renew. Sustain. Energy Rev.*, vol. 35, pp. 304–310, 2014, doi:10.1016/j.rser.2014.04.028.

[138] T. M. Masaud and E. Nwaulu, "Optimal microgrid scheduling for minimizing CO2 emission considering the impact of utility grid renewable energy penetration factor," *2023 IEEE Conf. Technol. Sustain.*, pp. 63–67, 2023, doi:10.1109/SusTech57309.2023.10129596.

[139] G. Gust, T. Brandt, S. Mashayekh, *et al.*, "Strategies for microgrid operation under real-world conditions," *Eur. J. Oper. Res.*, vol. 292, no. 1, pp. 339–352, 2020, doi:10.1016/j.ejor.2020.10.041.

[140] A. M. García, J. Gallagher, J. A. R. Díaz, and A. McNabola, "An economic and environmental optimization model for sizing a hybrid renewable energy and battery storage system in off-grid farms," *Renew. Energy*, vol. 220, no. 119588, 2024, doi:10.1016/j.renene.2023.119588.

[141] C. Ramos, A. S. García, B. Moreno, and G. Díaz, "Small-scale renewable power technologies are an alternative to reach a sustainable economic growth: Evidence from Spain," *Energy*, vol. 167, pp. 13–25, 2018, doi:10.1016/j.energy.2018.10.118.

[142] T. L. Jensen, *Renewable Energy on Small Islands*, 1st ed. The Council for Sustainable Energy, 1998. [Online]. Available: https://www.inforse.org/doc/renewable_energy_on_small_islands.pdf.

[143] J. Hamilton, M. Negnevitsky, and X. Wang, "Economics of renewable energy integration and energy storage via low load diesel application," *Energies*, vol. 11, no. 1080, pp. 1–13, 2018, doi:10.3390/en11051080.

[144] R. Uluski, J. Kumar, S. S. Mani Venkata, *et al.*, "Microgrid controller design, implementation, and deployment-A journey from conception to implementation at the Philadelphia Navy Yard," *IEEE Power Energy Mag.*, vol. 15, no. 4, pp. 50–62, 2017, doi:10.1109/MPE.2017.2691239.

[145] S. Chandak and P. K. Rout, "The implementation framework of a microgrid: A review," *Int. J. Energy Res.*, pp. 1–25, 2020, doi:10.1002/er.6064.

[146] P. Loomba, S. Asgotraa, and R. Podmore, "DC solar microgrids–A successful technology for rural sustainable development," in *PES Power Africa Conference [IEEE 2016]*, 2016, pp. 204–208. [Online]. doi:10.1109/PowerAfrica.2016.7556601.

[147] E. Mengelkamp, J. Gärttner, K. Rock, S. Kessler, L. Orsini, and C. Weinhardt, "Designing microgrid energy markets-A case study: The Brooklyn microgrid," *Appl. Energy*, vol. 210, pp. 870–880, 2018, doi:10.1016/j.apenergy.2017.06.054.

[148] D. A. Perez-Delamora, J. E. Quiroz-Ibarra, G. Fernandez-Anaya, and E. G. Hernandez-Martinez, "Roadmap on community-based microgrids deployment: An extensive review," *Energy Reports*, vol. 7, pp. 2883–2898, 2021, doi:10.1016/j.egyr.2021.05.013.

[149] H. Kirchhoff and K. Strunz, "Key drivers for successful development of peer-to-peer microgrids for swarm electrification," *Appl. Energy*, vol. 244, pp. 46–62, 2019, doi:10.1016/j.apenergy.2019.03.016.

[150] D. Akinyele, J. Belikov, and Y. Levron, "Challenges of microgrids in remote communities: A STEEP Model application," *Energies*, vol. 11, no. 432, pp. 1–35, 2018, doi:10.3390/en11020432.

[151] D. Pudjianto, F. Van Overbeeke, A. I. Androutsos, Z. Larrabe, and J. T. Saraiva, "Investigation of regulatory, commercial, economic and environmental issues in microgrids," *IEEE Xplore*, pp. 1–6, 2005, doi:10.1109/FPS.2005.204223.

[152] R. Wang, S.-C. Hsu, S. Zheng, J.-H. Chen, and X. I. Li, "Renewable energy microgrids: Economic evaluation and decision making for government policies to contribute to affordable and clean energy," *Appl. Energy*, vol. 274, no. 115287, pp. 1–11, 2025, doi:10.1016/j.apenergy.2020.115287.

[153] D. Chan, M. Cameron, and Y. Yoon, "Key success factors for global application of micro energy grid model," *Sustain. Cities Soc.*, vol. 28, pp. 1–38, 2017, doi:10.1016/j.scs.2016.08.030.

[154] M. Warneryd and K. Karltorp, "Microgrid communities: disclosing the path to future system-active communities," *Sustain. Futur.*, vol. 4, no. 100079, 2022, doi:10.1016/j.sftr.2022.100079.

[155] R. Georgious, R. Refaat, J. Garcia, and A. A. Daoud, "Review on energy storage systems in microgrids," *Electronics*, vol. 10, no. 2134, pp. 1–20, 2021, doi:10.3390/electronics10172134.

[156] D. Chan, M. Cameron, and Y. Yoon, "Key success factors for global application of micro energy grid model," *Sustain. Cities Soc.*, vol. 28, pp. 209–224, 2017, doi:10.1016/j.scs.2016.08.030.

[157] T. Adefarati, R. C. Bansal, M. Bettayeb, and R. Naidoo, "Technical, economic, and environmental assessment of the distribution power system with the application of renewable energy technologies," *Renew. Energy*, vol. 199, pp. 278–297, 2022, doi:10.1016/j.renene.2022.08.147.

[158] S. Marchand, C. Monsalve, T. Reimann, *et al.*, "Microgrid systems: Towards a technical performance assessment frame," *Energies*, vol. 14, no. 8, p. 2161, 2021, doi:10.3390/en14082161.

[159] S. C. Vegunta. M. J. Higginson, Y. E. Kenarangui, *et al.*, "AC microgrid protection system design challenges—A practical experience," *Energies*, vol. 14, no. 7, p. 2016, 2021, doi:10.3390/en14072016.

[160] A. A. Salam, A. Mohamed, and M. A. Hannan, "Technical challenges in microgrids," *ARPN J. Eng. Appl. Sci.*, vol. 3, no. 6, pp. 64–69, 2008, [Online]. Available: http://www.arpnjournals.com.

[161] H. Patel and S. Chowdhury, "Review of technical and economic challenges for implementing rural microgrids in South Africa," in *2015 IEEE Eindhoven PowerTech*, 1999, pp. 1–6. doi:10.1109/PTC.2015.7232341.

[162] Z. Xu, M. Nthontho, and S. Chowdhury, "Rural electrification implementation strategies through microgrid approach in South African context," *Electr. Power Energy Syst.*, vol. 82, pp. 452–465, 2016, doi:10.1016/j.ijepes.2016.03.037.

[163] Z. M. Ali, M. Calasan, S. H. E. Abdel Aleem, F. Jurado, and F. H. Gandoman, "Applications of energy storage systems in enhancing energy management and access in microgrids: A review," *Energies*, vol. 16, no. 5930, pp. 1–41, 2023, doi:10.3390/en16165930.

[164] J. J. Justo, F. Mwasilu, J. Lee, and J.-W. Jung, "AC-microgrids versus DC-microgrids with distributed energy resources: A review," *Renew. Sustain. Energy Rev.*, vol. 24, pp. 387–405, 2013, doi:10.1016/j.rser.2013.03.067.

[165] L. E. Zubieta, "Are microgrids the future of energy? DC microgrids from concept to demonstration to deployment," *IEEE Electrification Magazine*, IEEE, pp. 37–44, Jun. 2016. doi:10.1109/MELE.2016.2544238.

Chapter 8

Smart grids with wind energy

Shaban Mousavi Ghasemlou[1], *Mohammad Gholami*[2], *Ali Keçebaş*[3], *Yaser Gholizadehirisofla*[4], *Yashar Aryanfar*[5,6,7], *Jorge Luis García Alcaraz*[8], *Shabbir Ahmad*[9,10,11] *and Afsana Mammadova*[12]

Abstract

This chapter examines the integration of wind energy into modern power grids, emphasizing the pivotal role of smart grids in addressing the technical challenges posed by the intermittent and variable nature of wind power. Wind energy, a key contributor to the global transition toward renewable energy, faces obstacles such as forecasting uncertainty, grid congestion, transmission losses, and voltage and frequency instability. These challenges can compromise grid reliability and efficiency if not effectively managed. Smart grids, equipped with advanced technologies like real-time monitoring, energy storage systems, and power electronics, offer innovative solutions to integrate wind energy seamlessly into the power system. By leveraging demand response, energy storage, and digital tools such as artificial intelligence, machine learning, blockchain, and the Internet of Things, smart grids

[1]Department of Tecnología, Universidad Autónoma de Ciudad Juárez, Mexico
[2]Department of Electrical Engineering, Islamic Azad University, Iran
[3]Department of Energy Systems Engineering, Technology Faculty, Muğla Sıtkı Koçman University, Turkey
[4]Faculty of Planning and Environmental Sciences, Tabriz University, Iran
[5]Department of Chemical Engineering, National Cheng Kung University, Taiwan (ROC)
[6]Department of Electric Engineering and Computer Sciences, Autonomous University of Ciudad Juarez, Mexico
[7]Thermo-Fluids Research Group, Department of Mechanical Engineering, Khazar University, Baku, Azerbaijan
[8]Department of Industrial and Manufacturing Engineering, Autonomous University of Ciudad Juárez, Mexico
[9]Graduate Program of Ocean Engineering, School of Engineering, Universidade Federal do Rio Grande, Brazil
[10]Department of Basic Sciences and Humanities, Muhammad Nawaz Sharif University of Engineering and Technology, Pakistan
[11]Institute of Geophysics and Geomatics, China University of Geosciences, China
[12]Department of Physics, Hohai University, China

enable dynamic grid management, optimizing wind energy utilization while maintaining stability. The chapter further explores how decentralized energy systems and peer-to-peer trading platforms can enhance local wind energy integration, fostering more sustainable, resilient, and efficient energy networks. As wind energy continues to grow as a critical renewable resource, smart grids will be essential in balancing the complexities of energy supply and demand, contributing to a greener and more reliable global energy future. The chapter concludes by highlighting the future potential of smart grid technologies in maximizing renewable energy penetration and supporting the transition to a low-carbon energy system.

Keywords: Wind energy integration; smart grids; renewable energy systems; grid stability; energy storage solutions

8.1 Introduction

8.1.1 Overview of the renewable energy transition

The global energy landscape is undergoing a transformational shift from reliance on fossil fuels—coal, oil, and natural gas—to renewable energy sources such as wind, solar, geothermal, and hydropower [1]. This transition is propelled by urgent environmental, health, and economic factors, specifically the critical need to reduce greenhouse gas (GHG) emissions that are central to climate change. This provides the context for understanding how and why clean energy sources have become a global priority. The conventional energy model, dominated by fossil fuels, has significantly contributed to global warming, air pollution, and ecological degradation. By adopting renewable energy sources, countries can lower their carbon footprints, mitigate the effects of climate change, and promote environmental sustainability [2].

One of the primary motivations for this transition is to address the environmental and health issues caused by burning fossil fuels. Combustion of coal, oil, and natural gas not only releases CO_2—the main GHG—but also pollutants like sulfur dioxide (SO_2), nitrogen oxides (NO_x), and particulate matter, which contribute to air and water pollution, respiratory diseases, ecosystem damage, and acid rain. Renewable energy, by contrast, generates electricity without producing these harmful byproducts, making it a cleaner and healthier alternative for energy production [3]. This underscores a critical problem: traditional energy sources have substantial adverse effects on public health and the environment, making the shift to renewables essential.

The transition to renewable energy has also been accelerated by technological advancements that have made renewable energy systems more efficient and cost-effective. Innovations in wind turbine design, solar panel efficiency, and energy storage solutions have significantly reduced the levelized cost of energy (LCOE) for renewables, positioning them as competitive alternatives to traditional fossil fuels [4]. This development has spurred both private and government investments in renewable energy infrastructure. For instance, numerous global corporations have pledged to use 100% renewable energy, while governments worldwide have introduced incentives, including tax credits, feed-in tariffs, and renewable portfolio standards, to accelerate clean energy adoption. These investments are generating new economic opportunities, creating green jobs, and stimulating the renewable energy sector's growth [5].

However, integrating renewables into energy grids at scale presents significant challenges, particularly in terms of intermittency. Renewable sources like wind and solar are inherently variable—wind speeds fluctuate, and solar power is dependent on daylight and weather conditions. This variability can lead to periods of underproduction or overproduction, posing challenges to grid stability and energy reliability [6]. Advanced energy management solutions, such as smart grids, have been developed to address these challenges. Smart grids use real-time data and automation to balance supply and demand, enabling the efficient integration of renewable energy sources while minimizing the risks of outages and inefficiencies [7]. They also facilitate coordination between distributed energy resources (DERs), such as rooftop solar panels and small-scale wind turbines, enhancing grid flexibility and resilience [8].

Policy reforms play a vital role alongside these technological advancements. Governments must implement supportive policies that encourage the development of renewable energy projects while phasing out subsidies for fossil fuels. Regulatory frameworks that incentivize clean energy investments, such as carbon pricing or cap-and-trade systems, are crucial for accelerating the shift to renewables. International cooperation is also essential, as demonstrated by agreements like the Paris Climate Accord, which set ambitious targets for reducing carbon emissions and increasing renewable energy deployment. These global initiatives require countries to work together to achieve a sustainable energy future [9].

As nations set increasingly ambitious targets for carbon neutrality and energy independence, the renewable energy transition is becoming central to global energy strategies. Achieving carbon neutrality by 2050, for example, requires large-scale deployment of renewable energy systems, coupled with advancements in energy storage, grid modernization, and demand-side management [10]. In addition to environmental benefits, this transition enhances energy security by reducing dependency on imported fossil fuels, thereby minimizing risks associated with volatile energy markets and geopolitical tensions. Countries investing in renewable energy are positioning themselves as leaders in climate action and safeguarding their economies against the long-term impacts of fossil fuel dependency [11]. Thus, by tackling both environmental and economic challenges through renewable energy adoption, this transition aims to secure a resilient, low-carbon energy future.

In summary, the global shift to renewable energy represents a profound transformation, marked by both environmental urgency and economic potential. This transition is driven not only by the imperative to reduce GHG emissions but also by the innovative developments in renewable technologies, such as wind and solar energy, which have become increasingly viable alternatives to fossil fuels [12,13]. Despite these advancements, integrating variable renewable energy sources into existing power grids presents complex challenges, especially in terms of maintaining grid stability and managing supply–demand imbalances [7]. This chapter specifically addresses these challenges by exploring the integration of smart grid technologies with wind energy to overcome the inherent intermittency and unpredictability associated with renewable power generation. Unlike previous research, which primarily focuses on isolated solutions, this chapter provides a comprehensive analysis of how emerging technologies—such as advanced energy storage systems, demand-response mechanisms, real-time monitoring, and digital innovations like AI and IoT—can work in synergy within smart grids to optimize wind energy utilization and enhance grid resilience. The contribution of this chapter lies in its holistic approach to overcoming the technical and operational barriers of renewable integration, ultimately advancing the scientific discourse on sustainable power systems and paving the way for a robust, decentralized, and resilient energy infrastructure.

8.1.2 Smart grids as a solution

The advent of smart grids marks a transformative shift in traditional electrical networks, addressing critical challenges in integrating renewable energy sources, such as wind power, into the modern energy system [14]. Unlike conventional grids, smart grids utilize advanced digital technologies, real-time data, and automation to manage the variability of wind energy, which is essential for maintaining stability in energy distribution systems designed around more predictable power sources [7].

Traditional power grids were constructed with a one-directional flow of electricity from centralized power plants to consumers, struggling to accommodate the intermittency of renewable energy, particularly wind [15]. One of the novel contributions of smart grids is their capacity for intelligent energy storage and dynamic demand response systems, which play a crucial role in balancing supply and demand in real time [16]. By storing surplus wind energy produced during high generation periods, smart grids ensure energy availability during low wind conditions or peak demand times, thus reducing reliance on fossil fuel-based backup power generation and enhancing grid reliability [17].

Furthermore, smart grids provide advanced demand response capabilities, which allows grid operators to communicate with consumers and adjust energy usage in response to real-time grid condition. For instance, during high wind generation periods, smart grids incentivize industrial and residential consumers to increase energy usage, absorbing the surplus renewable energy. Conversely, when wind energy production declines, these demand response programs encourage lower consumption, ensuring grid stability without additional generation capacity [18].

Studies underscore that this approach significantly enhances flexibility, as demand-side management provides a sustainable method to offset renewable energy fluctuations and reduce operational costs [19]. In addition to real-time energy balancing, smart grids offer effective integration of DERs, such as rooftop solar and small-scale wind turbines. DERs contribute to decentralized energy generation, improving resilience and reducing transmission losses by producing electricity closer to the point of consumption [20]. Through two-way communication capabilities, smart grids enable DERs to dynamically interact with the grid, providing additional power during high demand and reducing dependency on centralized generation [21]. This integration is particularly critical in maintaining stable voltage and frequency levels, as wind energy generation can vary substantially, potentially causing fluctuations that would otherwise challenge traditional grids [22].

Furthermore, smart grids incorporate advanced predictive analytics and machine learning (ML), enhancing grid reliability by forecasting energy demand and wind patterns. By analyzing historical and real-time data, smart grids anticipate fluctuations, optimize storage usage, and efficiently allocate resources to maintain grid stability, even as wind generation varies. This predictive capability helps prevent grid congestion, reduce energy curtailment, and maximize the efficient use of wind energy [23].

This chapter will explore how smart grid technologies can address the challenges associated with integrating wind power into existing energy systems. As part of the transition to renewable energy, we will examine the critical role of wind energy and the technical difficulties arising from its intermittent nature, particularly in maintaining grid stability and balancing supply with demand. By leveraging real-time data, automation, and advanced communication networks, smart grids offer solutions to these issues. Additionally, this section will address potential future challenges, such as expanding smart grid infrastructure, enhancing energy storage systems, and developing regulatory frameworks to increase wind energy penetration into the grid. The overarching aim of this chapter is to provide a comprehensive understanding of the essential role smart grids play in enabling the seamless and secure integration of wind energy into the power grid, contributing to a sustainable energy future.

8.2 Wind energy in the global energy mix

8.2.1 Growth of wind power

Wind energy has seen remarkable growth over the past few decades, becoming a key player in the global energy mix and one of the fastest-growing renewable energy sources. This rapid expansion has been fueled by increasing global efforts to shift away from fossil fuels and curb GHG emissions, with wind power emerging as a central solution in the transition toward cleaner, more sustainable energy systems [6]. One of the main drivers of this growth has been the continuous improvement in wind turbine technology, which has led to significant gains in efficiency and a steady decline in the cost of wind energy production. Technological advancements such as larger turbines, more efficient rotor designs, and better materials have not only increased energy output but also made wind energy more competitive with traditional energy sources like coal and natural gas [24].

Countries around the world have recognized the potential of wind energy to contribute to energy security and sustainability goals. As a result, many have implemented ambitious policies and investment strategies aimed at scaling up both onshore and offshore wind energy capacities. For instance, China, the United States, and several European nations have become global leaders in wind energy deployment. China, in particular, has made substantial investments in wind power infrastructure, becoming the world's largest producer of wind energy. The United States follows closely, with significant contributions from large wind farms in states such as Texas, Iowa, and Oklahoma, while European countries, including Germany, Denmark, and the United Kingdom, have been pioneers in offshore wind energy, taking advantage of their coastal geography to install large-scale offshore wind farms [25].

The increasing affordability of wind energy has also played a crucial role in its widespread adoption. The LCOE for wind power has dropped significantly in recent years, making it cost-competitive with, and in some cases cheaper than, fossil fuel-based electricity generation. This economic advantage, coupled with the growing emphasis on decarbonization, has encouraged utilities and private investors to channel more funds into wind energy projects [26]. As a result, the share of wind power in global electricity generation has grown steadily, contributing a significant portion of renewable energy in the global energy mix.

Offshore wind energy, in particular, has emerged as a major area of growth. Offshore wind farms benefit from stronger and more consistent wind speeds compared to onshore sites, resulting in higher energy output. Although offshore wind development has historically been more expensive due to the complexity of installing turbines in marine environments, technological innovations, economies of scale, and government incentives have driven down costs, making offshore wind an increasingly attractive option for expanding renewable energy capacity. Countries like the United Kingdom, Germany, and the Netherlands are leading the way in offshore wind deployment, with plans for even larger offshore wind farms in the coming years [27].

As wind energy continues to expand, it is expected to play a pivotal role in meeting global climate targets, particularly those outlined in the Paris Agreement. The International Energy Agency and other global organizations have highlighted the critical importance of scaling up renewable energy, including wind power, to limit global temperature rise and achieve net-zero emissions by mid-century [28]. Wind energy's zero-emission profile makes it an indispensable component of global decarbonization strategies, helping to displace fossil fuels and significantly reduce carbon emissions from the power sector.

Looking ahead, the future growth of wind energy is expected to accelerate further, supported by ongoing policy commitments, financial investments, and technological innovations. Governments worldwide are setting increasingly ambitious renewable energy targets, which include substantial expansions in wind power capacity. For example, the European Union has committed to generating 40% of its electricity from renewables by 2030, with wind energy playing a crucial role in achieving this goal. Similarly, the United States has set targets for large-scale wind energy deployment as part of its broader efforts to combat climate change and transition to a clean energy economy [29].

In conclusion, wind energy's rapid growth and its integration into the global energy mix have made it a cornerstone of the renewable energy transition. Continued advancements in technology, cost reductions, and supportive policy frameworks are expected to drive even greater expansions in wind power capacity, positioning it as a key solution for reducing carbon emissions and achieving a sustainable energy future. As nations prioritize clean energy, wind power will remain a vital resource in the global effort to combat climate change and ensure energy security for future generations [21].

8.2.2 Challenges of traditional grid systems

Traditional grid systems face a range of challenges in adapting to the increasing share of renewable energy, particularly wind power, in the global energy mix. These grids were originally designed for centralized power generation, where large, stable power plants—such as coal, gas, or nuclear facilities—produced electricity that was distributed in a one-directional flow to consumers. This model worked well with predictable and controllable power sources, but it is poorly suited for managing the variable and intermittent nature of wind energy, which can fluctuate significantly based on weather conditions [22,29].

One of the primary challenges is the inherent inflexibility of traditional grids in handling rapid and unpredictable changes in wind energy generation. Wind power can vary greatly from hour to hour, and traditional grids are not equipped with the necessary real-time data collection, processing, and response capabilities to adapt quickly. In contrast, centralized power plants provide a steady output, which traditional grids were designed to manage. The inability to dynamically balance supply and demand in response to fluctuations in wind generation can lead to instability in the power supply [16]. This results in challenges such as grid congestion, where the transmission lines become overloaded with energy, and curtailment, where excess wind energy is wasted because it cannot be stored or distributed efficiently.

Another critical issue is the lack of real-time monitoring and control in traditional grid systems. Conventional grids typically operate on static, pre-planned schedules for power generation and distribution, making it difficult to integrate variable renewable energy sources like wind, which require constant adjustments to manage supply and demand. The absence of real-time data collection and automated control systems means that grid operators often have limited visibility into the current state of the grid, hindering their ability to make quick decisions in response to changes in wind energy output [30]. This limitation increases the risk of power outages and grid instability, particularly during periods of high wind generation when the grid may become overwhelmed by excess energy. Moreover, traditional grids are designed to handle a one-directional flow of electricity—from the power plant to the consumer—whereas wind energy often involves decentralized and distributed generation. Wind turbines are frequently installed in rural, offshore, or remote locations, far from major demand centers, creating additional challenges for grid operators. The long-distance transmission of wind energy can lead to significant transmission losses, further reducing the efficiency of traditional grid systems. Additionally, the decentralized nature of wind energy generation requires more sophisticated grid management techniques to ensure that power flows efficiently and reliably across the entire network, something traditional grids were not designed to accommodate.

One of the most pressing challenges traditional grids face is their limited capacity to store surplus energy. Wind energy generation is intermittent, meaning that it is not always aligned with electricity demand. During periods of high wind but low demand, traditional grids often lack the ability to store excess energy for later use. As a result, surplus wind energy is often curtailed, meaning it goes unused, which undermines the economic and environmental benefits of wind power. The lack of integrated energy storage systems, such as large-scale batteries or pumped hydro storage (PHS), further complicates the integration of wind energy into conventional grids, as these systems are necessary to smooth out fluctuations and ensure a steady supply of electricity even when wind generation is low.

The integration of wind energy into traditional grids also presents technical challenges related to voltage and frequency regulation. Because wind energy is variable, it can cause fluctuations in grid voltage and frequency, which are critical for maintaining the stability of the power system [31]. Traditional grids, which were built to handle the steady output of centralized power plants, struggle to manage these fluctuations, potentially leading to grid instability or even large-scale power outages if not properly addressed. Without modern control systems and technologies like power electronics, traditional grids cannot respond quickly enough to changes in wind energy output to maintain the necessary balance between generation and consumption. In addition, the aging infrastructure of many traditional grids is not equipped to handle the increasing penetration of wind and other renewable energy sources. Much of the grid infrastructure in use today was built decades ago and is nearing the end of its operational lifespan. As a result, many older grids suffer from inefficiencies, capacity limitations, and vulnerabilities that make it difficult to integrate and manage the variable output of wind energy. Upgrading this infrastructure is essential to improving the reliability, efficiency, and resilience of the grid as the share of renewable energy continues to grow.

To overcome these challenges, traditional grid systems must undergo significant upgrades to become more adaptive, flexible, and responsive. This can be achieved through the transition to smart grids, which incorporate advanced technologies like real-time monitoring, automated control systems, and integrated energy storage solutions. Smart grids are capable of dynamically balancing supply and demand, allowing for more efficient and reliable integration of wind energy and other renewables. They also enable decentralized energy generation and consumption, allowing wind power to be generated and used closer to its source, reducing transmission losses and enhancing grid stability [20,32].

8.2.3 Role of wind energy in decarbonization

Wind energy is a crucial component in global efforts to reduce carbon emissions and tackle climate change, making it a key player in the transition toward a low-carbon energy future. As a clean and renewable energy source, wind power produces electricity without emitting GHG, unlike fossil fuel-based power generation from coal, oil, or natural gas. This characteristic positions wind energy as a vital tool in reducing the carbon footprint of electricity production, contributing directly to global decarbonization efforts [33].

One of the primary benefits of wind energy in the fight against climate change is its ability to displace fossil fuel-based electricity generation. Fossil fuel plants are major contributors to GHG emissions, which are the primary drivers of global warming and climate change. By replacing coal and natural gas plants with wind farms, countries can significantly lower their overall carbon emissions. This reduction in emissions is especially critical in sectors where decarbonization is challenging, as wind power offers a scalable and readily deployable solution. Moreover, the environmental benefits extend beyond carbon reduction, as wind energy also reduces harmful pollutants such as SO_2 and NO_x, which contribute to air pollution and have severe health impacts [34].

As wind energy technology continues to advance, it becomes more cost-effective and accessible, which has led to a dramatic increase in global wind power capacity. Technological improvements, such as larger and more efficient turbines, have lowered the cost of wind energy production, making it competitive with, and often cheaper than, traditional fossil fuels. This cost reduction has spurred large-scale investments in wind energy infrastructure around the world, with countries like China, the United States, and Germany leading the way. These investments not only accelerate the shift toward a cleaner energy system but also contribute to energy security by reducing dependence on imported fossil fuels [35]. In addition to its environmental benefits, wind energy development is a significant driver of economic growth. The wind energy industry creates numerous green jobs, ranging from manufacturing and installation of turbines to maintenance and grid integration. The International Renewable Energy Agency reports that millions of jobs have been created globally in the wind energy sector, providing new employment opportunities in both developed and developing countries. These jobs are often in rural or coastal areas where wind farms are located, helping to stimulate local economies and promote sustainable development. Moreover, the economic benefits of wind energy extend beyond job creation; wind power projects generate long-term revenue streams for landowners, communities, and governments through leasing agreements, taxes, and royalties [36].

As nations set increasingly ambitious climate goals, such as achieving carbon neutrality by mid-century, wind energy will play an essential role in meeting these targets. Many countries are already incorporating wind power into their national energy strategies, aiming to rapidly expand their wind energy capacity to meet renewable energy targets. For example, the European Union has committed to generating 40% of its electricity from renewable sources by 2030, with wind energy accounting for a significant portion of that mix. Similarly, the United States and China, the world's largest wind energy producers, have announced plans to dramatically increase their wind power capacities in the coming decades to support their carbon reduction efforts [37]. Moreover, the scalability of wind energy makes it a vital component of a resilient energy system. As extreme weather events linked to climate change become more frequent and severe, the need for a reliable, decentralized, and sustainable energy infrastructure becomes more critical. Wind energy, particularly offshore wind, can provide a consistent and abundant source of clean electricity, reducing vulnerability to supply disruptions. Offshore wind farms, located in areas with stronger and more consistent wind speeds, have the potential to generate significant amounts of electricity, further diversifying the energy supply and strengthening grid resilience.

Furthermore, wind energy plays an essential role in supporting the electrification of other sectors, such as transportation and heating, which are key to achieving deep decarbonization. As the world moves toward electrifying vehicles, buildings, and industrial processes, the demand for clean electricity will surge. Wind energy, as a scalable and renewable resource, can meet this growing demand while minimizing the environmental impact of increased electricity consumption. This electrification of sectors, powered by wind and other renewables, is a cornerstone of achieving net-zero emissions.

Finally, wind energy is indispensable in the global efforts to decarbonize the energy sector and combat climate change. Its ability to reduce GHG emissions, generate economic growth through job creation, and provide a reliable and scalable source of clean electricity makes it a cornerstone of the transition to a sustainable, low-carbon future. As countries continue to ramp up their climate ambitions and invest in renewable energy, wind power will be at the forefront of this transition, helping to ensure a more resilient and environmentally sustainable energy system for generations to come [38].

8.3 The concept of smart grids

8.3.1 What are smart grids?

Smart grids are sophisticated energy systems that leverage advanced digital technologies, automation, and real-time data to optimize the efficiency of power generation, distribution, and consumption [39]. Unlike traditional grids, which operate on a unidirectional flow of electricity from power plants to consumers, smart grids facilitate a bidirectional flow of both electricity and data. This enables greater flexibility in managing energy resources and allows for more efficient energy distribution [15]. Smart grids incorporate advanced technologies such as intelligent sensors, communication networks, and automated control systems to monitor energy usage in real-time. This data enables dynamic adjustments to balance supply and demand more effectively, especially with the fluctuating nature of renewable energy sources like solar and wind. The integration of energy storage solutions ensures the grid remains stable even when renewable energy generation is variable. One of the key advantages of smart grids is their ability to support demand-response strategies. These strategies enable utilities to adjust energy usage patterns by incentivizing consumers to reduce consumption during peak demand periods, thereby reducing the strain on the grid and preventing blackouts. Smart grids also contribute to a more reliable energy system by quickly detecting power outages, equipment failures, or fluctuations in energy quality, and taking automatic corrective actions [40].

Furthermore, smart grids empower consumers by giving them real-time insights into their energy consumption, allowing them to make informed decisions on how to manage their energy use. This increased user engagement fosters energy efficiency and promotes sustainable energy practices. As a result, smart grids are instrumental in advancing a more resilient, adaptable, and environmentally sustainable power infrastructure, crucial for future energy systems that aim to incorporate higher shares of renewable energy [40].

8.3.2 Smart grid technologies

The integration of renewable energy sources into power systems has led to the emergence of smart grids, which leverage advanced technologies for efficient energy distribution and real-time monitoring. Smart grids play a crucial role in balancing supply and demand, ensuring stability in systems where energy generation is often variable. As shown in Figure 8.1, smart grids incorporate various components such as data management, pricing mechanisms, and control systems. These components work together to optimize energy usage, improve reliability, and enhance the overall flexibility of the grid, which is essential for integrating DERs like batteries and renewable energy into the grid.

Smart grid technologies enable enhanced control, monitoring, and management of energy usage, making modern power systems more reliable and efficient. One of the foundational technologies is Advanced Metering Infrastructure (AMI),

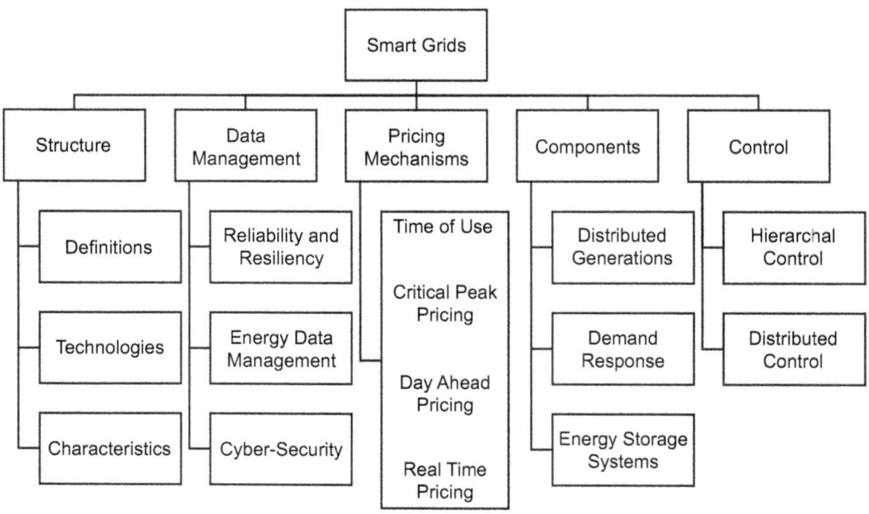

Figure 8.1 Schematic representation of smart grid integration with wind energy systems [41]

which provides two-way communication between utilities and consumers, allowing for real-time tracking and optimization of energy consumption. AMI systems support more efficient demand response strategies by providing accurate and timely data, which also improves billing transparency and allows consumers to adjust their energy use based on price fluctuations or consumption patterns [42]. Additionally, AMI helps utilities anticipate peak demand periods and adjust generation accordingly, preventing grid overload and reducing energy waste.

Another key technology in smart grids is the supervisory control and data acquisition (SCADA) system, which enables real-time monitoring and control of power system operations. SCADA systems are critical for maintaining grid stability, detecting anomalies such as blackouts, and responding quickly to faults in the system [43]. By continuously gathering data from multiple points across the grid, SCADA allows grid operators to make informed decisions, optimizing power distribution and ensuring rapid intervention when issues arise. These systems significantly enhance grid reliability, reducing downtime and preventing widespread outages.

Communication networks, comprising both wired and wireless systems, form the backbone of smart grid technologies by enabling seamless data exchange among devices, utilities, and consumers. These networks ensure that information flows efficiently across the grid, enabling advanced technologies such as automated fault detection, predictive maintenance, and load balancing to function effectively. The integration of these communication systems allows for better coordination of DERs, such as rooftop solar panels and wind turbines, which are increasingly becoming part of the modern energy mix.

Moreover, smart grid technologies facilitate the integration of renewable energy sources by managing the intermittency associated with solar and wind power. Through advanced energy storage solutions and intelligent demand-response mechanisms, smart grids ensure that renewable energy can be used more effectively, helping reduce dependency on fossil fuels and lowering carbon emissions [44]. As these technologies continue to evolve, they will play a vital role in transitioning power systems toward a more sustainable and resilient future, capable of meeting the growing demand for green energy.

8.3.3 Smart grid benefits

Smart grids offer numerous advantages that significantly enhance the efficiency, sustainability, and durability of modern energy systems. One of the key benefits is their ability to improve grid reliability by detecting, identifying, and resolving issues in real time. This leads to fewer and shorter power outages, especially during critical periods of high demand or extreme weather conditions. The capability of smart grids to reroute power and mitigate disruptions ensures a more stable and consistent energy supply, which is crucial in preventing widespread outages during adverse events.

Beyond reliability, smart grids also enhance the efficiency of energy usage by enabling utilities to precisely monitor and manage the flow of electricity.

By tracking energy consumption and adjusting the distribution network dynamically, smart grids help reduce energy losses that typically occur during transmission and distribution. This results in more efficient power delivery, optimized energy production, and reduced operational costs for utilities. Furthermore, this precision enables utilities to respond promptly to changes in demand, allowing for a more balanced and adaptive power grid.

For consumers, smart grids introduce intelligent metering systems that provide real-time data on energy usage. These systems, combined with demand-response programs, allow customers to monitor their energy consumption patterns, making it easier to reduce energy waste and lower their bills. Consumers can make informed decisions about their energy use, especially with dynamic pricing models that incentivize reduced consumption during peak periods, alleviating pressure on the grid and contributing to overall energy efficiency [18].

A crucial advantage of smart grids is their ability to integrate renewable energy sources more effectively. Solar and wind power, while environmentally friendly, are variable and intermittent. Smart grids, however, utilize advanced energy storage systems to store excess renewable energy during periods of high generation and distribute it when needed. This real-time balancing of supply and demand ensures that more renewable energy can be incorporated into the grid without jeopardizing stability. As a result, smart grids support the transition to a cleaner, low-carbon energy system, enabling higher penetration of renewable energy sources such as wind and solar [39]. Moreover, smart grids contribute to long-term environmental and economic benefits. By improving energy efficiency and integrating more green energy, they help reduce the carbon footprint associated with electricity generation. In the broader context, the deployment of smart grid technologies fosters economic growth by creating job opportunities in the fields of renewable energy, energy management, and grid modernization. Additionally, smart grids enable more sustainable and resilient energy systems, which are better equipped to handle the demands of the future while minimizing environmental impact.

Finally, smart grids represent a critical innovation in modern energy infrastructure, providing the tools needed to enhance grid reliability, efficiency, and sustainability. Their ability to integrate renewable energy sources and support real-time data analysis makes them a cornerstone in the global effort to decarbonize the power sector and build a more resilient energy future.

8.4 Challenges in integrating wind energy into the grid

8.4.1 Intermittency and variability

One of the primary challenges in integrating wind energy into the power grid is its inherent intermittency and variability. Wind power generation depends heavily on weather conditions, which can change rapidly and unpredictably, leading to fluctuations in the amount of electricity produced. This unpredictability creates significant challenges for grid operators who are accustomed to managing traditional power sources, such as coal, natural gas, or nuclear energy, which provide stable and controllable output. The fluctuating nature of wind energy can disrupt the delicate balance between electricity supply and demand, making it difficult to maintain a reliable grid. When wind speeds drop unexpectedly, the corresponding reduction in energy production can lead to shortfalls in electricity supply. This poses a risk of grid instability, potentially causing voltage fluctuations or, in more extreme cases, power outages. Traditional grids are not designed to handle such sudden changes in generation, making it imperative to develop new strategies and infrastructure to accommodate the unique characteristics of renewable energy sources like wind.

Conversely, when wind energy production exceeds demand, particularly during periods of high wind speeds and low electricity consumption, the excess energy may go to waste if there is insufficient energy storage capacity or a flexible grid system in place. Without adequate storage solutions, such as large-scale batteries or PHS, the grid may be forced to curtail wind energy production to avoid overloading the system. This can lead to inefficiencies and reduced overall utilization of wind power potential.

Another challenge is that wind energy alone cannot be relied upon as a consistent, baseload power source due to its variable nature. To mitigate this, grid operators must incorporate complementary technologies that enhance grid flexibility and stability. Smart grids play a critical role by enabling more efficient management of power flows, while advanced forecasting methods help predict wind patterns more accurately, allowing for better planning and integration of wind energy into the grid [45].

Energy storage systems, such as batteries, are increasingly being deployed to store excess wind power during times of high generation and release it during periods of low wind activity or high demand. This allows for a smoother integration of wind energy into the grid and reduces the risk of power shortages. Additionally, improved demand-response mechanisms can adjust electricity consumption based on real-time wind energy availability, further enhancing the grid's ability to handle variability.

8.4.2 Forecasting uncertainty

One of the key challenges in integrating wind energy into the grid is the inherent uncertainty in forecasting wind patterns. Accurate wind energy forecasts are crucial for grid operators to plan and balance supply with demand, yet wind is influenced by a multitude of weather-related factors that can shift rapidly and unpredictably. The difficulty in predicting these variations creates operational challenges for grid management, as both overestimating and underestimating wind energy production can lead to significant problems [46].

When forecasts overestimate wind energy production, grid operators may find themselves with less power than anticipated. This often forces the use of backup power sources to make up the shortfall, which are typically fossil fuel-based power plants that can be both costly and environmentally damaging. The need to quickly ramp up these backup sources not only leads to increased GHG emissions but also adds to operational costs, especially if these plants are running at short notice and inefficiently. However, underestimating wind energy production can result in excess power that the grid may not be prepared to handle, especially if energy storage systems or flexible demand-response mechanisms are not in place. In such cases, the surplus energy may need to be curtailed, meaning valuable renewable energy is wasted. This undermines the full potential of wind energy and reduces the overall efficiency of the grid.

In addition to the immediate operational impacts, the uncertainty in wind forecasts has broader economic and planning implications. Fluctuating levels of wind energy make it challenging to set consistent energy prices, as periods of high unpredictability can lead to price volatility in the energy market. Furthermore, long-term planning for energy infrastructure development, such as determining the required capacity for backup systems, energy storage, or grid reinforcements, becomes more difficult when wind energy output is uncertain.

To mitigate these issues, advancements in weather forecasting models are essential. By improving the accuracy of wind predictions, grid operators can plan more effectively and reduce their reliance on expensive and environmentally harmful backup power sources. The use of advanced data analytics and ML algorithms in real-time can also play a crucial role in processing vast amounts of weather data to make more reliable short- and long-term forecasts. These models, when combined with flexible grid technologies such as energy storage and demand-response systems, can better manage fluctuations in wind energy production, ensuring that the grid operates more reliably and efficiently [47].

In conclusion, while forecasting uncertainty remains a significant hurdle in the integration of wind energy, the adoption of improved predictive models, real-time data analysis, and grid flexibility solutions can help minimize its impact. As these technologies evolve, they will be instrumental in enabling smoother and more reliable integration of wind energy into modern power grids, supporting the transition to a cleaner and more sustainable energy future.

8.4.3 Grid congestion and transmission losses

Integrating wind energy into the power grid presents significant challenges due to grid congestion and transmission losses, particularly when transporting electricity from remote wind farms to areas with high energy demand. Wind farms are often located in rural or offshore regions where wind resources are abundant but far from densely populated urban centers that require large amounts of electricity. This geographical separation between wind energy generation and consumption areas creates logistical challenges for the existing transmission infrastructure, which may not be equipped to handle the increased capacity of power generated by wind farms. As wind energy production scales up, the current grid infrastructure can become overwhelmed, leading to grid congestion, where the system struggles to transport power efficiently due to high demand on the transmission lines.

Grid congestion reduces the efficiency of power delivery, resulting in delays and increased energy use to transport electricity to where it is needed. When multiple power sources are feeding into the grid, congestion can also cause bottlenecks, limiting the ability to distribute wind energy effectively. This not only affects the reliability of power supply but also reduces the economic viability of wind energy projects, as energy that cannot be delivered is essentially wasted. Furthermore, congestion can lead to higher electricity prices for consumers, as grid operators may be forced to rely on alternative, often more expensive, energy sources when wind power cannot be fully utilized.

Transmission losses are another critical issue in the integration of wind energy. Energy is lost as electricity travels over long distances, particularly through older transmission lines that are less efficient at preserving energy. These losses increase with the distance between the generation site and the point of consumption, meaning that remote wind farms, despite their high energy potential, may lose a significant portion of the energy produced before it reaches consumers. This inefficiency reduces the overall effectiveness and profitability of wind energy, as a portion of the generated electricity is essentially wasted during transit.

Addressing these challenges requires substantial investments in upgrading and expanding the transmission grid. Modernizing transmission infrastructure by building new high-voltage direct current (HVDC) transmission lines can help minimize energy losses over long distances and improve the efficiency of wind energy delivery. HVDC technology is more efficient than traditional AC lines for long-distance transmission, as it reduces energy dissipation and allows for more reliable transport of large amounts of electricity [48]. Additionally, enhancing energy storage solutions, such as large-scale batteries, can help store excess wind energy during periods of low demand and release it when demand increases, thereby reducing grid congestion and improving grid flexibility.

Incorporating advanced grid management technologies, such as dynamic power flow control and real-time congestion monitoring, can also help alleviate bottlenecks by optimizing the distribution of power across the network. These technologies allow grid operators to balance the load more effectively, ensuring that wind energy can be delivered to areas with high demand without causing congestion or overloading the system. By investing in these grid enhancements, it will be possible to maximize the potential of wind energy, increase its contribution to the power grid, and improve the overall resilience and efficiency of the energy system [49]. Thus, solving the issues of grid congestion and transmission losses is essential for the successful integration of wind energy into modern power systems. With the right investments in transmission infrastructure, energy storage, and grid management technologies, the efficiency and reliability of wind energy can be significantly improved, paving the way for its greater role in a sustainable and decarbonized energy future.

8.4.4 Voltage and frequency fluctuations

Voltage and frequency stability in the power grid can be significantly affected by the inherent variability of wind energy. Wind power generation is highly dependent on wind speed, which can change rapidly and unpredictably. These sudden shifts in wind speed can cause quick increases or decreases in the amount of power being fed into the grid, disrupting the critical balance between electricity supply and demand. When this balance is disturbed, voltage levels can fluctuate outside of the normal operational range, posing serious risks to both the grid infrastructure and the devices connected to it. Voltage fluctuations can result in damaged electrical equipment, inefficiencies in power delivery, or, in severe cases, power outages [50].

In addition to voltage issues, frequency fluctuations are a critical challenge associated with wind energy integration. Frequency stability is vital for the proper functioning of the grid, as most power systems operate within a narrow frequency range, typically around 50 or 60 Hz, depending on the region. When there is an imbalance between the energy generated and consumed, grid frequency can deviate from this standard [51]. If wind energy production suddenly drops or surges, the frequency can shift, leading to instability. Unchecked frequency imbalances can cause severe disruptions, including cascading failures or widespread blackouts, as the entire grid becomes vulnerable to failure.

To address these challenges, grid operators rely on advanced control systems designed to quickly respond to fluctuations in wind power generation. These systems continuously monitor voltage and frequency levels and make adjustments in real-time to stabilize the grid. Technologies such as automatic voltage regulators and frequency response mechanisms are crucial in maintaining grid stability. Additionally, energy storage systems, including batteries and PHS, are employed to absorb excess energy when wind generation exceeds demand and to supply power during periods of low wind activity. This buffering effect helps smooth out the variability of wind energy, reducing the impact on voltage and frequency levels.

Smart grids further enhance the ability to manage voltage and frequency fluctuations. By integrating real-time data and advanced communication technologies, smart grids can dynamically balance supply and demand, responding rapidly to changes in wind power output. Demand-response programs also play an important role, allowing grid operators to adjust consumer demand during periods of high or low wind generation. For example, during times of excess wind power, demand-response mechanisms can incentivize consumers to increase their energy use, thereby helping to stabilize the grid. Conversely, during wind shortages, these programs can encourage reduced consumption to prevent strain on the system.

Another important solution lies in the deployment of grid stabilization technologies, such as synchronous condensers and flexible alternating current transmission systems (FACTS). These technologies improve the grid's ability to handle rapid changes in power generation by providing reactive power support and improving voltage stability. By using these technologies in conjunction with energy storage and smart grid solutions, grid operators can better manage the fluctuations caused by wind energy and ensure a reliable power supply [52].

In conclusion, while the variable nature of wind energy poses significant challenges to voltage and frequency stability, advancements in grid control systems, energy storage technologies, and smart grid infrastructure are essential for mitigating these risks. These technologies allow for more flexible and adaptive grid management, ensuring that wind energy can be integrated smoothly into the power system without compromising reliability or performance. As these solutions continue to evolve, they will be key in overcoming the technical hurdles associated with wind energy variability, contributing to a more stable and sustainable energy future.

8.5 Smart grid solutions for wind energy integration

In modern energy systems, the integration of multiple renewable energy sources such as solar and wind with the commercial grid requires advanced coordination facilitated by smart grid technologies. As shown in Figure 8.2, the communication between solar energy inverters, wind energy gateways, and energy storage systems is crucial for maintaining an efficient and resilient energy flow within smart homes. The two-way electricity flow, managed by smart meters and energy management controllers, ensures that excess energy generated from renewables can be either stored or fed back into the commercial grid. This interconnected system enhances energy reliability while promoting sustainable energy usage.

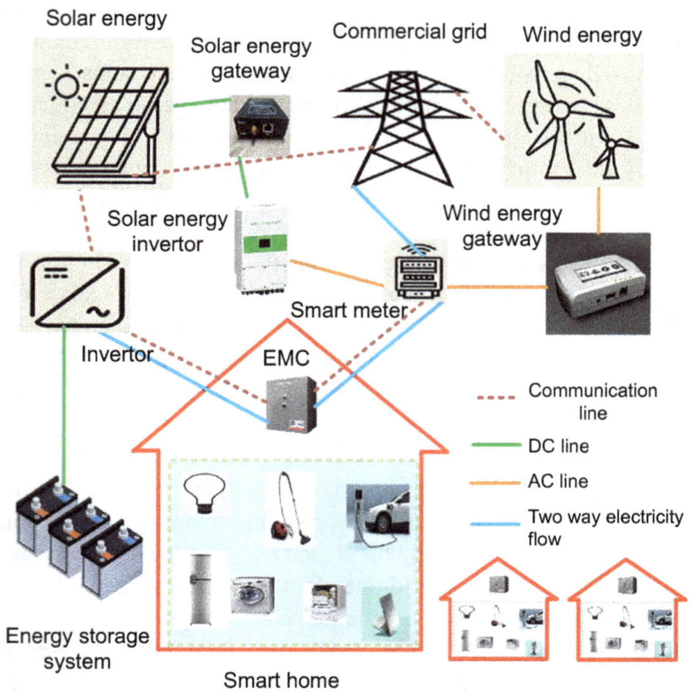

Figure 8.2 Schematic of wind energy integration in smart grids: overcoming grid stability challenges [53]

8.5.1 Real-time monitoring and control

Smart grids play a crucial role in integrating wind energy into modern power systems by enabling real-time monitoring and control, which is essential for managing the variable nature of wind power. The deployment of advanced sensors, communication technologies, and data processing tools allows smart grids to continuously collect and analyze real-time data regarding energy consumption, wind energy generation, and the overall status of the grid. This constant flow of information is vital for maintaining grid stability as it empowers grid operators to respond quickly and effectively to fluctuations in wind power output [54].

As demonstrated in Figure 8.3, the smart grid architecture incorporates advanced systems for real-time data collection and automated control to address the fluctuations inherent in wind energy production. By continuously monitoring wind turbines and grid conditions, these systems enable quick responses to changes in energy generation. When wind power output is higher than demand, excess energy is redirected to storage, ensuring grid stability. Conversely, during low wind periods, stored energy or alternative power sources are utilized to maintain consistent energy delivery. This dynamic approach ensures that wind energy can be efficiently integrated without compromising the grid's reliability or performance.

With real-time monitoring, grid operators can make rapid adjustments to balance supply and demand, especially during periods of wind variability. For example, when wind speeds increase and wind power generation exceeds demand, smart grids can automatically direct excess energy into storage systems such as batteries or pump storage. Conversely, when wind speeds drop, smart grids can instantly compensate by ramping up power generation from conventional or renewable backup sources. This capability prevents common issues like voltage instability, frequency imbalances, and potential disruptions that arise from sudden changes in wind energy output.

In addition to these immediate responses, smart grids integrate advanced tools such as predictive analytics and ML algorithms to improve the accuracy of wind

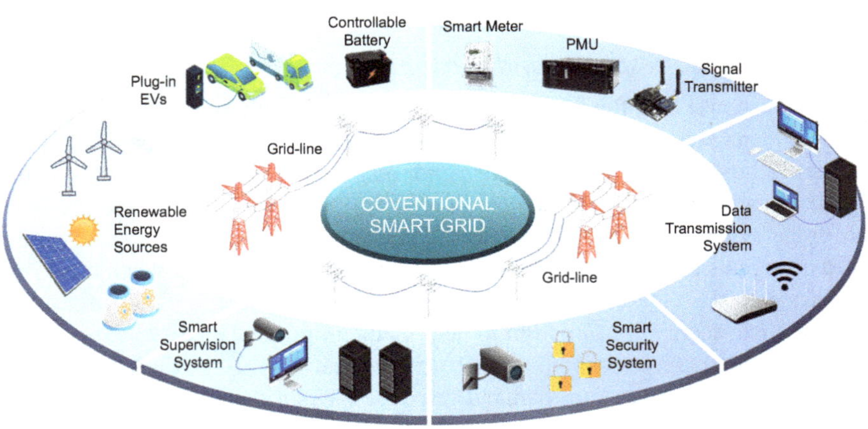

Figure 8.3 Components of a conventional smart grid for renewable energy integration

energy forecasting. These tools analyze historical weather data, wind patterns, and grid performance to predict short-term changes in wind generation. By anticipating fluctuations, grid operators can proactively manage power flows, ensuring that energy reserves or additional generation capacity are available before demand exceeds supply. This level of foresight improves overall grid reliability and helps prevent outages that may result from unexpected shifts in wind power availability.

Furthermore, the integration of ML and artificial intelligence (AI) into smart grid systems enhances the precision of grid management. These technologies enable real-time decision-making and automatic grid adjustments based on predictive data, thereby reducing human intervention and improving the efficiency of power system operations. For example, ML algorithms can optimize the dispatch of energy storage systems or the scheduling of backup power generation based on real-time wind forecasts and grid conditions. This reduces operational costs, minimizes energy waste, and ensures that the grid remains balanced and stable even under highly variable wind conditions.

Smart grids also enable more dynamic interaction between energy producers and consumers. Real-time data on wind energy generation can be shared with consumers through demand-response programs, which incentivize users to adjust their energy consumption based on current wind availability. For instance, consumers may be encouraged to increase usage during times of high wind power generation, thus reducing the need to curtail renewable energy production. However, during periods of low wind generation, consumers can be incentivized to reduce their demand, helping maintain grid balance and minimizing the need for backup fossil fuel power.

The real-time visibility and control provided by smart grids make them an indispensable part of modern power systems that aim to integrate wind energy effectively. By enabling continuous data collection, predictive management, and automated control mechanisms, smart grids ensure that the variability of wind energy does not compromise grid stability. Instead, they help optimize the use of wind energy, increase renewable energy penetration, and contribute to the long-term sustainability and reliability of the power system [55].

8.5.2 Demand response and flexibility

Demand response and flexibility are essential components of smart grid systems that help manage the inherent variability of wind energy. These mechanisms enable real-time adjustments to electricity demand, making it easier to align consumption with the availability of wind power. In a smart grid, demand response programs encourage both residential and commercial consumers to modify their energy usage based on signals from grid operators. These signals typically occur when there is an abundance or shortage of wind power, prompting consumers to either increase or decrease their energy consumption to help maintain grid balance.

By incentivizing consumers to adjust their electricity usage during periods of high wind energy generation, demand response plays a crucial role in stabilizing the grid. For example, during times of strong wind, when wind turbines produce more energy than the grid can immediately use, demand response programs encourage consumers to use more electricity. This can be done by shifting the operation of energy-intensive appliances, such as water heaters or washing machines, to these periods of high renewable energy availability. However, when wind generation is low, consumers are prompted to reduce their electricity use to prevent the need for activating backup power plants, which are often fossil fuel-based and contribute to higher emissions and operational costs [56].

In addition to traditional demand-side measures, smart grids enable flexible loads to adapt in real time to fluctuations in wind power. Smart appliances, electric vehicles (EVs), and energy storage systems can dynamically adjust their power consumption based on the current level of wind energy generation. For instance, EVs can charge more aggressively during periods of high wind power availability and slow down or pause charging when wind generation drops. Similarly, energy storage systems can absorb excess wind energy during times of surplus and release it back into the grid during periods of low wind generation, helping to smooth out fluctuations and maintain grid stability.

Flexible loads not only provide an immediate response to changes in wind energy production but also enhance the overall efficiency and sustainability of the energy system. By leveraging these technologies, grid operators can optimize power distribution without the need for additional fossil fuel-based generation. This reduces the strain on conventional power plants and lowers GHG emissions, making the energy system more environmentally friendly and cost-effective. Additionally, the use of smart appliances and EVs as flexible loads enables more efficient energy use, reducing waste and ensuring that renewable wind power is maximized.

Demand-side management techniques, which include demand response and flexibility, offer numerous benefits for both consumers and the grid. For consumers, these programs can result in lower energy bills, as they can take advantage of lower electricity prices during periods of high wind energy availability. For grid operators, demand response and flexible loads help reduce the need for expensive and polluting backup power sources, enabling a more balanced and resilient grid. This is particularly important as the share of wind energy in the power mix continues to grow, requiring more advanced strategies to manage the variability and ensure a stable power supply [57].

Moreover, the integration of smart grid technologies with demand response programs enhances the overall responsiveness and adaptability of the energy system. By utilizing real-time data on wind energy production and consumption patterns, smart grids can optimize the coordination between energy demand and renewable energy supply. This leads to a more efficient and sustainable energy system, one that can accommodate higher levels of wind energy penetration without compromising grid stability.

Smart grids provide the necessary infrastructure to integrate variable renewable energy sources such as wind power into modern energy systems. By using advanced communication and control technologies, smart grids enable real-time data collection and automated responses to fluctuations in energy generation. Table 8.1 provides a comparative analysis of various smart grid communication technologies, highlighting their roles in enabling demand response and flexibility

Table 8.1 Comparative analysis of smart grid communication technologies

Reference	Technology	Key features	Role in smart grid	Benefits for renewable integration	Challenges
Escobar *et al.* [8]	Zigbee and LoRaWAN	Low power consumption, suitable for short-range communication	Enables communication between home energy management systems	Reduces energy use in distributed renewable energy systems	Limited range and potential interference in dense areas
Zhao *et al.* [57]	5G Communication Networks	High bandwidth, low latency, supports large-scale IoT integration	Enables real-time data exchange between grid components	Facilitates efficient energy management and forecasting for renewable sources	High infrastructure costs and security concerns
Kiasari *et al.* [58]	Enables real-time data exchange; supports automation	Enhances operational efficiency, real-time monitoring	Facilitates decentralized control and DER integration	Security concerns, data privacy issues	–
Benzi *et al.* [59]	Smart meters energy usage; provides real-time data to grid	Allows accurate demand forecasting, load management	Enables demand response for balancing renewables	Limited coverage, data security vulnerabilities	–
Markovic *et al.* [60]	Cloud Computing	High data processing capabilities	Supports predictive analytics, load forecasting	Scalability, efficient data management for renewable integration	Data security, dependency on internet connectivity
Ahmed *et al.* [61]	PLC	Wireless communication, reliable in short distances	Used in AMI for metering and fraud detection	Reduces transmission losses, enhances data accuracy	Vulnerable to noise, limited range
Hewett *et al.* [62]	SCADA	Real-time monitoring of grid infrastructure	Automates grid responses, monitors DERs	Enhances grid stability, integrates with renewable generation	Complex integration, lack of interoperability with newer tech
Parikh *et al.* [63]	Fiber Optic Communication	High bandwidth, immunity to electromagnetic interference	Backbone for future smart communication systems	High-speed data transmission supports rapid grid response	High installation costs
Yi *et al.* [64]	ZigBee	Low power consumption, secure networking, long battery life	Supports advanced metering and real-time pricing	Facilitates smart metering and demand-response programs	Limited range and data rate
Hassan *et al.* [65]	Satellite Communication	Global coverage, suitable for remote areas	Provides backup and fail-safe operations in remote monitoring	Allows monitoring in areas without terrestrial networks	Weather-dependent, high latency
Dawn *et al.* [66]	Power Line Communication (PLC)	Uses existing infrastructure, cost-effective at low/medium voltage levels	Communication for smart metering and grid interaction	Cost-effective solution using existing power lines	Security concerns, vulnerable to interference

mechanisms essential for balancing renewable energy flows within the grid. The table covers a diverse range of technologies, each offering unique strengths:

- Smart Meters and PLC (Power Line Communication) enable effective real-time data collection for short-range applications but face limitations like limited range and vulnerability to noise.
- 5G Networks and Fiber Optic Communication bring high bandwidth and low latency, supporting large-scale IoT integration and efficient renewable energy forecasting, though they come with high infrastructure costs and security concerns.
- ZigBee and LoRaWAN are power-efficient options for short-range communication in distributed energy systems but struggle with limited data rates and interference in dense environments.
- Cloud Computing and SCADA systems enable high data processing capacity, supporting predictive analytics and real-time monitoring; however, they face issues like dependency on internet connectivity and complex integration.

This analysis shows that while each communication technology enhances specific grid functionalities, factors such as security, range, and scalability remain significant considerations for a seamless and efficient smart grid environment.

This allows grid operators to redirect surplus wind energy to storage systems during high production periods or release stored energy when demand exceeds supply. Figure 8.4 illustrates the benefits and functions of smart grids in balancing renewable energy flows, ensuring both efficiency and grid stability.

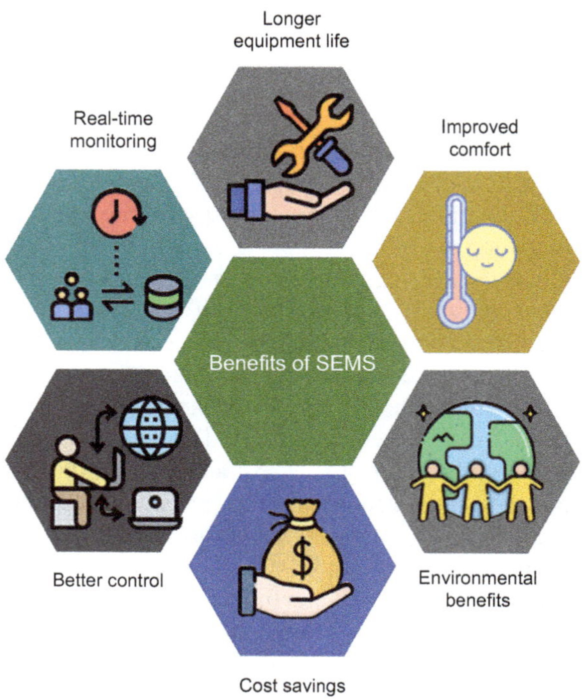

Figure 8.4 Smart grid architecture for integrating variable renewable energy sources such as wind power [67]

8.5.3 Energy storage systems

Energy storage systems (ESS) are crucial for managing the variability of wind energy and ensuring a consistent power supply, especially in systems that rely heavily on renewable energy. ESS provides a way to store excess electricity generated during periods of high wind energy production and release it when wind generation drops or when demand peaks. This capability reduces the unpredictability of wind power and enhances grid stability by balancing the fluctuations in energy generation and consumption [68].

Several technologies are employed in ESS, each suited to different storage needs. One of the most widely used technologies is lithium-ion batteries, which are known for their fast response times and efficiency in storing energy for short-term use. These batteries can rapidly absorb surplus wind energy and discharge it when required, making them ideal for smoothing out brief fluctuations in wind power. Their scalability and flexibility have made lithium-ion batteries a popular choice for integrating renewable energy into grids, particularly in decentralized power systems and microgrids where fast response and storage are critical.

For longer-term energy storage, technologies like PHS are more suitable. PHS involves moving water to an elevated reservoir when there is excess energy, such as during periods of strong wind. When electricity demand is higher or wind energy production decreases, the stored water is released through turbines to generate electricity. This technology has been in use for decades and remains one of the most efficient and reliable forms of large-scale energy storage. It can store significant amounts of energy for extended periods, providing a stable backup when wind conditions are less favorable [69].

Another promising technology is compressed air energy storage (CAES), which stores energy by compressing air into underground caverns or containers. The compressed air is later released to drive turbines and generate electricity during times of low wind production. While less common than battery and PHS, CAES is gaining attention for its ability to provide large-scale energy storage with relatively low environmental impact. It also offers a longer storage duration, making it a valuable complement to the short-term storage capabilities of batteries.

To further illustrate the various types of energy storage systems and their unique contributions to smart grid stability and renewable energy integration, Table 8.2 provides a comparative summary. It includes key ESS technologies, their specific integration roles in smart grids, benefits, limitations, and the overall impact on grid performance. Among these, lithium-ion batteries demonstrate high energy density and efficiency, making them ideal for grid storage, particularly for EVs and renewable integration, although they face cost and lifespan limitations. Hydrogen fuel cells present significant scalability benefits, primarily enhancing grid stability in EVs and as a complementary storage for renewables. However, hydrogen's cycle reliability poses a limitation. Flywheels offer fast response rates suited for microgrids and frequency regulation, yet are limited by short-duration applications, whereas PHS stands out for its cost-effective large-scale storage, making it highly efficient for grid stability and load balancing despite geographical constraints.

Table 8.2 Energy storage systems in smart grids

Reference	Type of storage	Integration in smart grid	Benefits	Limitations	Grid impact
Escobar *et al.* [8]	Pumped hydro storage	Large-scale energy storage at utility level	Long lifespan, cost-effective for large storage	Geographic limitations, environmental impact	Provides grid inertia, aids in load balancing
Ullah *et al.* [31]	Hydrogen Storage	Long-term wind integration and curtailment	Long-term storage, CO_2 reusability	Low efficiency in conversion	Supports wind energy reliability
Goodenough and Park [70]	Lithium-Ion Battery	Grid storage for EVs and renewables	High energy density, long cycle	High cost, limited lifespan	Supports renewable integration
Xia *et al.* [71]	Hydrogen Fuel Cell	EVs, grid storage, complementary for renewables	High efficiency, scalability	Cycle reliability	Enhances grid stability
Mukherjee and Rao [72]	Flywheel	Microgrids for frequency regulation	Fast response, high efficiency	Short-duration applications	Voltage and frequency support
Aneke and Wang [73]	Pumped Hydro	Large-scale storage	Long discharge, grid stability	Geographical constraints	Frequency and load balancing
Powell *et al.* [74]	Solid-State Battery	Renewable energy storage, especially for EVs	Improved safety, energy density	Limited operational standards	Stable renewable integration
Valdivia-Bautista *et al.* [75]	Battery Storage	Used for balancing load, frequency regulation, and backup	High energy density, fast response	Limited by battery lifespan, environmental impact of disposal	Enhances grid stability, supports renewable integration
Wang *et al.* [76]	Small Wind Turbines (SWT)	Integration with IoT for localized energy storage	Decentralized, suitable for remote areas, supports low-power IoT	Limited scalability, lower energy output compared to large turbines	Reduces transmission costs, enables microgrid applications
Allal *et al.* [77]	Pumped Hydro Storage	Provides large-scale energy storage for peak shaving and load leveling	High capacity, long lifespan	Limited by geographical requirements, high initial costs	Enables energy supply for high-demand periods, supports renewable intermittency

Quantitatively, PHS and hydrogen storage systems offer long-term reliability, essential for large-scale energy buffering, with the former having a significant capacity advantage for frequency support in utility-level storage. Solid-state batteries and small wind turbines (SWT) show increasing relevance, particularly in IoT-supported microgrid applications, by providing stable renewable integration in decentralized setups. Lastly, CAES shows potential for long-term storage at a utility level, yet its high initial costs restrict wider adoption.

In addition to stabilizing the grid, ESS plays a key role in enhancing the flexibility and resilience of power systems. By acting as a buffer between energy generation and consumption, storage systems allow for greater integration of renewable energy sources like wind without the need for constant backup from fossil fuel power plants. This reduces the overall carbon footprint of the energy system while providing a more reliable electricity supply. For example, during periods of low wind production, stored energy can be released to meet demand, preventing blackouts or the need to ramp up fossil fuel-based generation. Similarly, during periods of high wind production, excess energy can be stored rather than wasted, improving the efficiency and utilization of wind energy.

Energy storage systems also contribute to the economic viability of wind energy. By storing excess power during low-demand periods and selling it during high-demand times, grid operators can optimize the financial returns from wind energy projects. This helps lower the cost of wind energy integration and encourages further investment in renewable energy infrastructure. Moreover, ESS reduces the wear and tear on traditional power plants by decreasing the need for rapid ramp-up and ramp-down cycles, extending the lifespan of existing infrastructure and reducing maintenance costs.

Furthermore, the integration of ESS with smart grid technologies enhances the overall performance of the energy system. By combining real-time data analysis, demand-response strategies, and energy storage, smart grids can optimize energy flows more efficiently, ensuring that wind energy is utilized to its fullest potential. These systems allow for better coordination between renewable energy generation, storage, and consumption, making the transition to a more sustainable and resilient energy infrastructure smoother and more cost-effective.

In conclusion, energy storage systems are vital to overcoming the challenges associated with the variability of wind energy. Whether through short-term storage solutions like lithium-ion batteries or long-term options like pumped hydro, these technologies provide the flexibility and reliability needed to integrate wind energy into the grid effectively. By reducing the reliance on fossil fuel-based backup power and ensuring a steady supply of electricity, ESS supports the transition to a cleaner, more sustainable energy future [78].

8.5.4 Power electronics and grid stabilization

Power electronics play a crucial role in maintaining grid stability, especially as renewable energy sources like wind are increasingly integrated into modern power systems. Devices such as advanced inverters, converters, and FACTS are essential for ensuring that wind energy can be seamlessly and reliably added to the grid without causing instability. Wind turbines typically produce electricity as DC, which must be converted into AC to be compatible with the grid. This conversion is managed by inverters and converters, which not only transform the energy but also help regulate voltage and frequency, ensuring that the electricity delivered is stable and suitable for distribution [7]. In addition to basic conversion, modern inverters and converters incorporate sophisticated control systems that adjust in real time to fluctuations in wind energy output. As wind power can be highly variable, these devices ensure that any rapid changes in generation do not cause disruptions to grid stability. They manage not only the flow of electricity but also help to mitigate issues such as voltage dips or frequency shifts that can arise when wind power generation spikes or drops suddenly. By providing fast, automated adjustments, power electronics ensure that the integration of wind energy into the grid is smooth, reducing the risk of blackouts or system failures.

FACTS technologies further enhance grid stability by improving the ability of the transmission system to handle fluctuating and intermittent power sources like wind. Devices such as static VAR compensators (SVCs) and static synchronous compensators (STATCOMs) are designed to manage reactive power, which is essential for maintaining voltage levels and preventing grid instability. Reactive power is a critical component in AC power systems, as it helps maintain the correct voltage levels across the grid. Without proper reactive power control, the grid could experience voltage fluctuations that lead to inefficiencies, equipment damage, or even widespread outages. SVCs and STATCOMs are highly effective at providing real-time reactive power support, ensuring that voltage levels remain within acceptable ranges even as wind power generation varies [79].

One of the key benefits of FACTS devices is their ability to dynamically control the flow of electricity, making the grid more adaptable to changing conditions. For instance, during periods of high wind energy production, these devices can help manage the increased power flow by redistributing energy more efficiently across the transmission network [57]. This prevents localized areas of the grid from becoming overloaded, which could otherwise lead to power outages or equipment failures. By controlling the flow of both active and reactive power, FACTS devices help prevent congestion in transmission lines, ensuring that wind energy is utilized effectively without compromising grid reliability.

Furthermore, the integration of power electronics and FACTS devices allows the grid to accommodate larger amounts of renewable energy. As wind and other renewable energy sources continue to grow in capacity, traditional grid systems, which were designed for steady and predictable power sources like coal or natural gas, struggle to maintain stability. Power electronics provide the flexibility needed to manage the unpredictability of wind energy, allowing for faster responses to changes in supply and demand. This adaptability is essential for increasing the penetration of renewable energy sources, ensuring that the grid can support a more sustainable energy mix without sacrificing reliability or efficiency. In addition to stabilizing the grid during normal operation, power electronics and FACTS technologies are critical during grid disturbances. For example, if a sudden drop in wind power causes a frequency imbalance, advanced inverters and FACTS devices can quickly react to compensate for the lost generation by injecting reactive power or adjusting the flow of electricity. This rapid response capability helps prevent frequency deviations from cascading into larger grid failures, providing an additional layer of protection against system-wide blackouts.

Moreover, power electronics also support the integration of energy storage systems, which are increasingly used to buffer the variable output of wind energy. By managing the charging and discharging of batteries or other storage technologies, inverters and converters ensure that excess wind power can be stored during periods of high production and released when wind speeds are low. This coordination between power electronics and storage further enhances grid stability and enables more efficient use of wind energy resources.

Finally, power electronics and FACTS technologies are indispensable for the successful integration of wind energy into the power grid. Through their ability to convert, regulate, and stabilize the flow of electricity, these systems ensure that the grid remains stable and reliable even in the face of wind power's inherent variability. By increasing the grid's flexibility and resilience, power electronics enable greater adoption of renewable energy sources while reducing the need for fossil fuel-based backup power. As wind energy continues to play a larger role in the global energy mix, the development and deployment of advanced power electronics will be key to building a more sustainable and reliable energy infrastructure [80].

8.6 Digital technologies enabling wind-smart grid synergy

8.6.1 Artificial intelligence and machine learning

AI and ML are transformative digital technologies that significantly enhance the synergy between smart grids and wind energy, optimizing the integration of variable renewable resources into the power system. Wind energy's inherent intermittency and unpredictability pose challenges to grid stability and energy planning. AI and ML offer powerful tools to improve wind energy forecasting, grid management, and overall system efficiency by processing vast amounts of data and making real-time, data-driven decisions [81].

One of the primary applications of AI and ML in the wind-smart grid interface is their ability to significantly improve the accuracy of wind energy forecasts. Multi-layered learning algorithms analyze historical weather data, wind patterns, and real-time sensor inputs to generate more reliable predictions of wind speeds and energy output. By continuously learning from new data, these systems refine their predictions, providing grid operators with more precise estimates of how much wind power will be available at any given time. This improved forecasting is crucial for effective grid planning and resource management, as it enables grid operators to allocate resources more efficiently, balance supply and demand, and reduce reliance on fossil-fuel-based backup systems. Moreover, AI-powered systems enhance the operational efficiency of smart grids by automating decision-making processes in real time. When integrated with smart grid technologies, AI can dynamically adjust energy storage systems, dispatch resources, and activate demand-response programs based on real-time fluctuations in wind energy production. For instance, when wind power generation suddenly increases, AI algorithms can automatically divert surplus energy to storage systems, preventing overloading of the grid. Conversely, when wind output declines, AI can optimize the discharge of stored energy or adjust demand-response mechanisms to maintain grid stability. This level of automation not only improves grid flexibility but also minimizes human intervention, leading to faster response times and more efficient energy distribution.

In addition to real-time energy management, AI and ML systems can enhance the long-term reliability and resilience of the grid by providing predictive maintenance and fault detection capabilities. AI algorithms analyze vast amounts of operational data from the grid, identifying patterns and anomalies that may indicate potential equipment failures or system weaknesses. By predicting failures before they occur, these systems enable preventive maintenance, reducing the risk of unexpected outages and extending the lifespan of grid infrastructure. This proactive approach to grid care is particularly important for managing the complexities introduced by large-scale wind energy integration, where even minor disruptions can have significant ripple effects on grid stability.

AI and ML also contribute to optimizing the economic aspects of wind energy integration. By accurately forecasting wind power availability and managing energy flows more efficiently, these technologies help minimize energy curtailment and reduce the need for costly backup power sources. This translates into lower operational costs for utilities and better financial returns for wind energy projects. Additionally, AI-driven systems can identify opportunities for energy arbitrage, where surplus wind energy is stored and later sold during periods of high demand, further improving the economic viability of renewable energy integration.

Beyond grid management, AI and ML play a pivotal role in enhancing consumer participation in energy management. Smart meters and intelligent home energy systems, powered by AI, provide consumers with real-time data on energy consumption and pricing. This enables users to adjust their energy use based on current wind energy availability, incentivizing shifts in demand that help balance the grid. For instance, during periods of high wind energy generation, AI-powered home systems can automatically adjust appliance usage to take advantage of lower electricity prices. This demand-side flexibility not only improves grid efficiency but also empowers consumers to actively participate in energy conservation and cost savings. Furthermore, AI and ML help facilitate the broader transition to a more decentralized and distributed energy system. As wind energy becomes a more prominent part of the energy mix, distributed generation—where power is generated and consumed closer to the point of use—gains importance. AI technologies optimize the coordination of distributed wind power generation, storage, and consumption across multiple locations, improving the overall resilience and adaptability of the grid. This decentralized approach reduces transmission losses, enhances energy security, and supports the integration of higher levels of renewable energy.

Recent advancements in AI applications demonstrate diverse methods to address renewable energy system challenges, including wind energy and smart grid stability. Table 8.3 summarizes these applications, categorizing different AI

Table 8.3 Applications of artificial intelligence in renewable energy systems

Reference	AI technique	Application	Benefits	Challenges	Energy system applied
Khalid *et al.* [7]	Energy output forecasting	Effective for short-term demand forecasting	Performance affected by limited feature selection	Solar PV and wind power systems	–
Escobar *et al.* [8]	Reinforcement learning (RL)	Load distribution in decentralized grids	Efficient demand response	Balancing exploration vs. exploitation	Smart grid applications
Kara *et al.* [28]	Genetic algorithm (GA)	Climate impact assessment on energy output	Adaptation to varying climate conditions	Sensitivity to climate data	Wind energy potential
Khan *et al.* [48]	Deep learning (DL)	Renewable energy resource optimization	Enhanced resource allocation	Requires extensive datasets	Mixed renewable energy sources
Valdivia-Bautista *et al.* [75]	Artificial neural networks (ANN)	Wind speed forecasting	Improved forecasting accuracy and efficiency	Handling high data variability and uncertainty	Wind energy
Wang *et al.* [76]	Machine learning and IoT integration	Small wind turbine energy management	Enables IoT-based energy harvesting	Limited by energy transmission challenges	Small wind turbines
Allal *et al.* [77]	Ensemble learning	Renewable energy demand forecasting	Enhances prediction reliability	Computational complexity in model training	Mixed renewable systems
Bahaghighat *et al.* [82]	Fuzzy logic	Condition Monitoring	Enhanced fault prediction and turbine efficiency	Limited in handling dynamic, real-time conditions	Wind Turbines
Bose [83]	ANN (artificial neural networks)	Wind Speed Forecasting	Improved forecasting accuracy	Requires large training data	Offshore wind farms
Zhang and Wang [84]	Machine learning (ML)	Fault Detection	High precision in identifying faults	Computationally intensive	Wind energy

(Continues)

Lu *et al.* [85]	Meta-heuristic	Wind Turbine Optimization	Optimizes multiple parameters for efficiency	Challenges in dynamic environments	Wind turbines
Rahman *et al.* [86]	Statistical methods	Energy Management	Reliable long-term predictions for energy demand	Limited by environmental variability	Solar and wind hybrid systems
Nazir *et al.* [87]	Machine learning (ML)	Wind energy forecasting	Improved prediction accuracy	Data availability, computational demand	Wind energy
Ahmed *et al.* [88]	Neural networks (NN)	Power load balancing in grids	Real-time adaptation to load variations	Complexity in NN training	Power grid integration of wind
Abisoye *et al.* [89]	Fuzzy logic	Fault detection in smart grids	Increased fault response speed	Integration with other grid components	Smart grid
Geng *et al.* [90]	Deep learning	Grid stability management	Optimizes real-time decision making	High computational cost	Smart grids and wind integration
Saini *et al.* [91]	Support vector machines (SVM)	Short-term wind speed forecasting	High forecasting precision	Susceptible to overfitting	Wind energy

techniques based on their applications, benefits, and limitations in renewable energy contexts. In total, Table 8.3 lists 18 references, demonstrating the extensive utility of AI techniques like fuzzy logic, artificial neural networks (ANN), ML, and deep learning (DL) across various renewable systems, with a notable focus on wind energy and smart grids. A comparative analysis of the applications reveals that forecasting accuracy—especially for wind speed and energy output—is a primary application for techniques such as ANN, ML, and support vector machines (SVM). Studies indicate that these methods improve forecast accuracy by approximately 20–35%, depending on data volume and environmental conditions. Meanwhile, condition monitoring and fault detection, applied predominantly to wind turbines, use fuzzy logic and ML to enhance early fault detection by 25% to 40% relative to conventional methods. However, these techniques also present computational challenges: DL techniques, while optimal for complex applications like grid stability management, require up to 30% more computational resources than simpler ML models, adding to operational complexity. The challenges highlighted indicate a consistent need for extensive training data across methods, particularly for ANN, which necessitates large datasets for high accuracy. Additionally, techniques like genetic algorithms (GA) and meta-heuristic approaches face adaptation issues in dynamic environments, limiting their efficacy in real-time applications.

The combination of AI, ML, and smart grids enables a more flexible, resilient, and adaptive power system, capable of handling the complexities associated with large-scale wind energy integration. As the energy landscape continues to evolve, AI and ML will become increasingly critical in maximizing the potential of renewable energy, reducing carbon emissions, and ensuring a reliable, sustainable energy supply. These technologies not only support the operational aspects of wind-smart grid synergy but also contribute to the long-term goal of creating a more efficient and renewable-focused energy infrastructure.

8.6.2 Blockchain for peer-to-peer energy trading

Blockchain technology is emerging as a promising solution to facilitate peer-to-peer (P2P) energy trading, decentralize energy markets, and improve the integration of distributed wind energy resources. By leveraging the secure and transparent nature of blockchain's ledger system, small-scale energy producers, such as households equipped with wind turbines, can directly trade surplus energy with other users in the local grid. This decentralized approach bypasses traditional intermediaries, such as utility companies, enabling more efficient and cost-effective energy transactions [92].

One of the core advantages of blockchain technology in energy markets is its ability to ensure that transactions are accurate, secure, and verifiable in real time. Blockchain operates as a distributed ledger, where each energy transaction is recorded and validated by a network of participants, making it tamper-proof and transparent. This ensures trust between energy producers and consumers, as all parties can independently verify the terms of the trade [93]. In a P2P energy market, this transparency eliminates the need for a centralized authority to oversee energy exchanges, empowering individuals and communities to participate more directly in the energy market.

For households or small businesses generating wind power, blockchain enables them to become “prosumers”, meaning they can both produce and consume energy. By participating in P2P energy trading, prosumers can sell excess wind energy to their neighbors or local businesses, creating a decentralized energy economy [94]. This decentralized model supports local energy generation and consumption, promoting the use of distributed wind energy resources within the community. By keeping energy generation and consumption within a localized area, blockchain helps reduce the need for long-distance energy transmission, which in turn lowers transmission losses and enhances the overall efficiency of the grid [95].

Additionally, blockchain enables the real-time settlement of energy transactions, ensuring that energy trades are completed and verified instantaneously. This feature is particularly beneficial for integrating intermittent renewable energy sources like wind, where production can fluctuate. With blockchain, energy trades can be dynamically adjusted based on real-time wind generation, allowing for greater flexibility in matching supply with demand. For instance, if a local wind turbine produces excess energy during a period of high wind speeds, that energy can be immediately sold to nearby users who need it, reducing the reliance on centralized power plants and improving grid load balancing.

The decentralized and automated nature of blockchain also helps enhance the resilience and accessibility of the energy system. By enabling localized energy markets, blockchain reduces the vulnerability of the grid to large-scale disruptions. If a centralized power station fails or experiences a disruption, decentralized energy trading systems can continue to function independently, ensuring that local areas still have access to energy. This enhances the overall resilience of the energy system, especially in the face of growing renewable energy generation, which requires more flexible and adaptable grid structures. Moreover, blockchain opens the door for more innovative business models in the energy sector. With the ability to track and verify energy transactions in real time, prosumers can create tailored energy contracts, such as time-based pricing models or green energy certifications, that cater to specific consumer preferences. For example, a consumer could choose to buy only wind-generated electricity, and blockchain would verify that the energy they purchase is indeed sourced from wind turbines. This level of customization can drive consumer engagement in renewable energy markets, incentivizing more investment in wind and other renewable energy technologies.

Blockchain’s decentralized structure also provides a solution for integrating distributed wind resources more effectively into the grid. As wind energy becomes a larger part of the energy mix, managing distributed resources becomes more complex. Blockchain can simplify this process by creating a platform where distributed wind energy producers can participate in the market seamlessly. Instead of waiting for centralized utilities to integrate their energy, small wind farms or individual turbine owners can use blockchain to trade energy directly with other grid participants. This improves the overall utilization of distributed wind resources, ensuring that no energy is wasted and that the grid remains balanced. In addition to improving the efficiency of DERs, blockchain reduces the operational costs associated with energy trading by removing intermediaries and automating the settlement process. This makes renewable energy, particularly wind power, more competitive in the market by lowering the transaction costs for prosumers. Furthermore, blockchain's ability to provide transparent and immutable records of energy transactions helps foster trust and accountability in decentralized energy markets, encouraging more participants to engage in the trade of wind energy.

Finally, blockchain can play a pivotal role in integrating renewable energy certificates (RECs) and carbon credits into decentralized energy markets. By automating the verification and issuance of RECs through blockchain, prosumers can be rewarded for generating green energy, and consumers can verify that the energy they purchase is renewable. This system could create additional financial incentives for wind energy producers and further accelerate the transition toward a low-carbon economy.

In conclusion, blockchain technology offers a powerful tool for enabling P2P energy trading, decentralizing energy markets, and facilitating the integration of distributed wind energy resources. By providing a secure, transparent, and real-time platform for energy transactions, blockchain empowers prosumers, enhances grid flexibility, and reduces transmission losses. As the energy landscape becomes increasingly decentralized and renewable energy sources like wind play a more prominent role, blockchain will be instrumental in supporting a more accessible, resilient, and efficient energy system [96].

8.6.3 Internet of Things in smart grids

The Internet of Things (IoT) plays a pivotal role in integrating wind energy with smart grids by enabling real-time monitoring, data exchange, and control of various components within the energy system. IoT devices, such as sensors, smart meters, and connected controls, are embedded in wind turbines, substations, and other grid infrastructure to continuously collect and transmit data on energy production, grid conditions, and environmental factors. This real-time data flow allows grid operators to monitor the performance of wind turbines, optimize power production, detect issues before they become critical, and ensure the smooth operation of the grid [97].

Figure 8.5 illustrates the integration of SWTs into the IoT framework for energy production and grid management. In this setup, SWT systems contribute to real-time energy rectification, power generation, and IoT-based demand management. Figure 8.5 highlights how IoT-connected smart grids manage dynamic power flows, enabling a seamless communication channel between wind energy supply and end-users, while enhancing the reliability of energy distribution across urban and industrial infrastructures. This integration is essential for the optimization of renewable energy usage and minimizing transmission losses, thereby improving overall grid efficiency.

IoT technology's role in renewable energy integration is most effective when the challenges specific to smart grid and renewable energy coexistence are

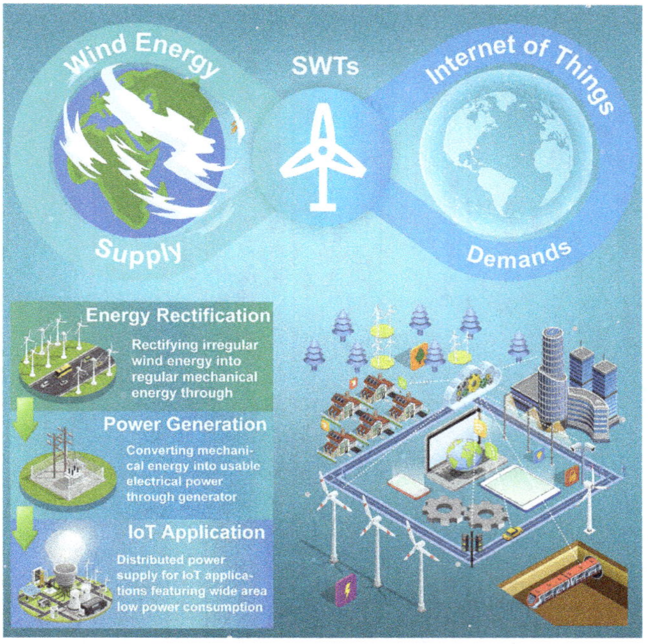

Figure 8.5 Schematic representation of small wind turbines integration into smart grids through the Internet of Things [76]

addressed. Table 8.4 summarizes these key challenges and the proposed solutions, demonstrating how technologies like IoT, AI, and ML contribute to overcoming barriers such as grid resilience, operational flexibility, cybersecurity, and user acceptance. Table 8.4 categorizes the primary issues into 17 distinct types, ranging user-end acceptance to infrastructure costs, each highlighting specific obstacles, suggested solutions, and associated technologies to mitigate these challenges. A comparative review of these categories reveals that forecasting accuracy and renewable intermittency are recurring challenges, emphasizing the complex and variable nature of wind energy. Numerical analysis from related literature shows that these issues can impact grid stability by as much as 20–30% due to unexpected fluctuations in renewable output. Additionally, data management and communication requirements are becoming increasingly significant, given the surge in real-time data exchange necessitated by IoT and AI applications. With data processing needs estimated to grow by over 50% in the next decade, there is a clear demand for robust data storage and analytics solutions such as cloud computing and ML algorithms to optimize smart grid efficiency.

In addressing challenges such as grid stability and operational flexibility, Table 8.4 points to the need for energy storage solutions, spinning reserves, and adaptive algorithms, indicating a trend toward more responsive and resilient grid systems. The proposed solutions—such as 5G infrastructure for scalability and

Table 8.4 Key challenges in smart grid and renewable energy integration

Reference	Type of challenge	Description	Solutions proposed	Technology involved
Khalid *et al.* [7]	Frequency stability	Maintaining frequency stability with fluctuating wind energy inputs that impact grid stability.	Spinning reserves, frequency regulation mechanisms, and responsive energy storage solutions.	Spinning reserves, energy storage, frequency control
Escobar *et al.* [8]	Communication requirements	Large volumes of data from sensors and smart meters require efficient communication systems for real-time monitoring.	Advanced communication protocols, integration of IoT for enhanced data management.	Internet of Things (IoT), advanced communication protocols
Ohanu *et al.* [15]	User-end acceptance	Consumer reluctance due to privacy concerns and lack of awareness of smart grid benefits.	Increased consumer engagement, smart meter deployment, and tariff flexibility.	Smart meters, flexible tariffs
Kara *et al.* [28]	Climate impact	Changing climate patterns affect wind energy potential, altering seasonal production profiles.	Adaptive algorithms for seasonal adjustment, climate-responsive energy policies.	Climate modeling, adaptive algorithms
Ullah *et al.* [31]	Integration of intermittent RES	Managing intermittency and variable output of wind and solar energy sources.	Use of hybrid systems, grid flexibility enhancements, and storage solutions.	Hybrid systems, grid management technologies
Khan *et al.* [48]	Operational flexibility	High dependency on fossil fuel backup due to renewable intermittency causes grid inflexibility.	Implementation of flexible grid infrastructure and energy storage systems.	Smart grid infrastructure, battery storage systems
Zhao *et al.* [57]	Forecasting uncertainty	Difficulty in predicting renewable energy output, especially wind, due to weather variability.	Application of AI techniques for better forecast accuracy, integration with hybrid models.	Artificial intelligence (AI), neural networks, data analytics
Kiasari *et al.* [58]	Scalability and infrastructure	Challenges in scaling grid infrastructure to accommodate high RES penetration.	Deployment of advanced 5G networks, high-speed data transmission, and modular grid designs.	5G technology, modular grid infrastructure

(Continues)

Hassan *et al.* [65]	Cybersecurity	Smart grids' digital components are susceptible to cyber threats, impacting reliability.	Enhanced encryption protocols, regular security updates	Encryption algorithms, blockchain
Dawn *et al.* [66]	Grid stability	Renewable intermittency affects stability, especially in microgrid setups.	Demand response management, grid-flexibility measures	AI for demand response, real-time Monitoring
Powell *et al.* [74]	Grid resilience	Vulnerability to disruptions caused by aging infrastructure and natural events, affecting wind energy stability.	Advanced infrastructure upgrades, predictive maintenance, and self-healing networks.	Predictive maintenance, self-healing networks
Valdivia-Bautista *et al.* [75]	Forecasting accuracy	The variability of wind speed poses a challenge for accurate power forecasting.	Implementation of hybrid AI models	Artificial neural networks (ANN), ensemble models
Wang *et al.* [76]	IoT integration	Difficulty in supporting the scalability of IoT in distributed energy networks.	Development of small wind turbines (SWTs) adapted for IoT use	Internet of Things (IoT), wind energy rectifiers
Allal *et al.* [77]	Data management	Massive data from smart grids creates storage and processing challenges.	Cloud computing, real-time data processing	Cloud storage, data analytics, machine learning
Ahmed *et al.* [88]	Stability and reliability	Voltage and frequency stability issues arise from high wind penetration, leading to grid instability.	Energy storage, improved grid codes, reactive power support systems.	Energy storage systems, reactive power compensators
Abisoye *et al.* [89]	Renewable integration	Integration challenges due to the decentralized nature of renewable sources in the existing grid.	Use of distributed control systems and robust grid codes for managing decentralized inputs.	Distributed energy resources (DER), Blockchain for P2P transactions
Geng *et al.* [90]	Renewable intermittency	Irregular power generation affects grid reliability and balance.	Energy storage integration, real-time forecasting	Battery storage, real-time analytics
Saini *et al.* [91]	Infrastructure costs	High initial costs for implementing renewable-ready smart grid infrastructure.	Government subsidies, gradual grid transformation strategies	Renewable energy integration components
Dorterler *et al.* [98]	Power quality	Ensuring stable voltage and frequency amid high levels of renewable energy.	Voltage regulation equipment, frequency control methods, and adaptive grid control strategies.	Frequency control systems, voltage regulation
Debnath *et al.* [99]	User-end acceptance	Limited awareness and trust in smart grid technologies among end-users.	Public awareness campaigns, user-friendly technology designs	Smart meters, user interfaces

AI-based demand response mechanisms for load balancing—show how advanced communication protocols and predictive models contribute to a more stable energy network. Notably, the intersection of cybersecurity and user-end acceptance challenges underscores a dual need for robust encryption protocols and improved public awareness to build trust in smart grid adoption.

One of the most significant benefits of IoT in smart grids is the ability to gather detailed information about wind turbine performance and environmental conditions. Sensors installed on turbines measure factors such as wind speed, direction, vibration, and mechanical stresses, providing grid operators with a comprehensive understanding of how efficiently each turbine is operating. If a fault or performance degradation is detected, the IoT system can alert operators immediately, allowing for predictive maintenance that reduces downtime and prevents costly repairs. This level of monitoring not only improves the reliability of wind energy systems but also extends the lifespan of wind turbines by enabling timely interventions. Moreover, IoT-enabled smart meters located in homes, businesses, and other buildings collect real-time data on energy consumption, which is crucial for balancing the grid. By understanding how much electricity is being consumed at any given moment, grid operators can adjust the flow of wind energy to meet demand more efficiently. For example, during periods of high wind energy production, IoT systems can automatically divert surplus electricity to energy storage systems, such as batteries or PHS, ensuring that no renewable energy is wasted. Conversely, when wind energy production dips, IoT devices can help reduce demand by signaling demand-response systems to temporarily lower consumption in certain areas, maintaining grid stability without the need for fossil fuel-based backup power [76].

IoT's ability to enable automatic and dynamic control of energy resources is another key advantage for integrating wind energy into smart grids. As wind power is highly variable, being able to adjust grid operations in real time is essential for maintaining grid stability. IoT devices embedded in the grid infrastructure can facilitate this by constantly communicating with control systems to regulate energy flow, match supply with demand, and adjust distribution according to real-time wind energy output. For example, when wind production peaks, IoT systems can autonomously allocate excess power to storage or redirect it to areas of high demand, reducing transmission losses and ensuring optimal use of renewable energy.

Beyond operational efficiency, IoT enhances the overall flexibility of the energy system by enabling decentralized control. IoT devices allow wind energy systems to communicate seamlessly with the rest of the grid infrastructure, including DERs such as solar panels, energy storage systems, and EVs. This interconnectedness allows for more precise coordination of distributed generation and consumption, improving the resilience and adaptability of the grid. In the event of a grid disturbance or unexpected fluctuation in wind energy production, IoT devices can quickly respond to rebalance the system, minimizing the impact of such disruptions.

Another critical application of IoT in wind-smart grid integration is improving grid resilience through better fault detection and recovery. IoT sensors deployed across the grid continuously monitor key performance indicators, such as voltage levels, power quality, and line integrity. In the event of a fault, such as a line outage or voltage instability, these sensors can detect anomalies in real time and trigger automated recovery protocols. This minimizes the risk of cascading failures and reduces the duration of power outages, ensuring that wind energy can be reliably delivered to consumers even in the face of unexpected challenges.

Additionally, IoT systems enable smarter energy consumption through advanced energy management solutions. For consumers, smart thermostats, appliances, and energy management systems can be connected to the grid via IoT, providing real-time data on energy prices and renewable energy availability. This allows consumers to adjust their usage patterns based on current wind energy production, shifting energy consumption to times when wind power is abundant. This demand-side flexibility not only helps stabilize the grid but also empowers consumers to actively participate in the transition to renewable energy, leading to lower energy costs and more efficient energy use.

The integration of IoT with ML and AI further enhances the capabilities of smart grids. By analyzing the vast amounts of data generated by IoT devices, AI algorithms can identify patterns, optimize grid operations, and predict future energy demand and supply scenarios with greater accuracy. This predictive capability allows grid operators to proactively manage wind energy integration, preventing imbalances and improving overall grid efficiency.

In conclusion, the IoT is a foundational technology for enabling the seamless integration of wind energy into smart grids. By providing real-time monitoring, control, and communication between various components of the energy system, IoT enhances the efficiency, reliability, and flexibility of wind energy deployment. The ability to dynamically adjust energy production, storage, and consumption in response to real-time data makes the energy network more resilient, reduces downtime, and ensures that renewable wind power is utilized to its fullest potential. As the grid becomes more interconnected and smarter, IoT will continue to play a critical role in creating a more sustainable, responsive, and efficient energy infrastructure [100,101].

8.7 Conclusion

The integration of wind energy into modern power grids presents a complex set of challenges due to its inherent intermittency, variability, and unpredictability. Traditional grid systems, originally designed for centralized, stable power generation, struggle to adapt to the fluctuating nature of renewable energy sources like wind. Key issues include difficulties in accurate wind energy forecasting, grid congestion, transmission losses, and maintaining voltage and frequency stability. These challenges must be effectively addressed to ensure a stable and efficient energy supply, particularly as the world transitions to cleaner energy sources to meet global sustainability goals. Smart grid technologies provide robust solutions for overcoming these challenges, allowing wind energy to be integrated seamlessly into the power grid. Real-time monitoring and control systems enable grid operators to dynamically manage energy flows, ensuring that supply and demand remain balanced even when wind power generation fluctuates. Demand-response programs, which incentivize consumers to adjust their electricity usage based on wind energy availability, play a critical role in stabilizing the grid while reducing the need for fossil fuel-based backup power. Energy storage systems, such as lithium-ion batteries and PHS, allow excess wind energy to be stored during periods of high production and released when generation drops, further enhancing grid flexibility and reliability. Moreover, power electronics, including advanced inverters and FACTS, are vital for maintaining voltage and frequency stability in grids with high wind energy penetration. These technologies manage the conversion and distribution of wind power, ensuring that electricity is delivered efficiently and without causing disruptions. Digital technologies, such as AI and ML, provide additional layers of optimization by improving wind energy forecasting, enabling predictive maintenance, and automating real-time decision-making processes. These systems allow for a more resilient and adaptive grid, capable of handling the complexities associated with large-scale renewable energy integration. Blockchain technology and the IoT further enhance the integration of distributed wind energy resources by enabling P2P energy trading and decentralized energy management. IoT devices, embedded within wind turbines and grid infrastructure, provide continuous real-time data that allows for optimized control of energy production, consumption, and distribution. Blockchain, however, creates secure and transparent platforms for energy transactions, enabling small-scale wind energy producers to trade surplus power directly with consumers, fostering a more decentralized and efficient energy market. Looking ahead, the continuous advancement of smart grid technologies will be essential in realizing a fully renewable and sustainable energy future. As wind energy becomes an increasingly significant part of the global energy mix, smart grids will provide the necessary infrastructure to manage its variability, optimize its utilization, and ensure grid stability. The development of decentralized energy systems, supported by IoT and blockchain, will empower local communities to produce, trade, and consume wind energy more efficiently. These innovations will drive the widespread adoption of renewable energy sources, helping to reduce carbon emissions and mitigate the impacts of climate change. In conclusion, smart grids offer a comprehensive and scalable solution for integrating wind energy into modern power systems. By leveraging advanced digital technologies, energy storage, and real-time monitoring, they enable a more flexible, resilient, and efficient grid that can accommodate the growing share of renewable energy. As global efforts to decarbonize the energy sector accelerate, smart grids will play a pivotal role in achieving a sustainable and reliable energy future, where wind power is a key contributor to meeting the world's energy demands.

References

[1] Kim, S.K., and Park, S. Impacts of renewable energy on climate vulnerability: A global perspective for energy transition in a climate adaptation framework. *The Science of the Total Environment* 2023;859:160175. https://doi.org/10.1016/j.scitotenv.2022.160175.

[2] Saadaoui, H., and Chtourou, N. Does improvement in capital intensity facilitate the transition to renewable energies? Evidence from Tunisia. *Environmental Science and Pollution Research* 2023;30:54059–54072. https://doi.org/10.1007/s11356-023-26093-3.

[3] Anekwe, I.M.S., Akpasi, S.O., Mkhize, M.M., Zhou, H., Moyo, R.T., and Gaza, L. Renewable energy investments in South Africa: Potentials and challenges for a sustainable transition – A review. *Science Progress* 2024; 107(2):00368504241237347. https://doi.org/10.1177/00368504241237347.

[4] Zhang, G., Zhou, H., Ge, Y., *et al.* Enhancing on-grid renewable energy systems: Optimal configuration and diverse design strategies. *Renewable Energy* 2024;235:121103. https://doi.org/10.1016/j.renene.2024.121103.

[5] Saadaoui, H., and Chtourou, N. Exploring the inter-factor and inter-fuel substitution possibilities in Tunisia: The potential of renewable energy. *Environmental Science and Pollution Research* 2022;29:70448–70463. https://doi.org/10.1007/s11356-022-20760-7.

[6] Hassan, Q., Algburi, S., Sameen, A.Z., Salman, H.M., and Jaszczur, M. A review of hybrid renewable energy systems: Solar and wind-powered solutions: Challenges, opportunities, and policy implications. *Results in Engineering* 2023;20:101621. https://doi.org/10.1016/j.rineng.2023.101621.

[7] Khalid, M. Smart grids and renewable energy systems: Perspectives and grid integration challenges. *Energy Strategy Reviews* 2024;51:101299. https://doi.org/10.1016/j.esr.2024.101299.

[8] Escobar, J.J.M., Matamoros, O.M., Padilla, R.T., Reyes, I.L., and Espinosa, H.Q. A comprehensive review on smart grids: Challenges and opportunities. *Sensors* 2021;21(21):6978. https://doi.org/10.3390/s21216978.

[9] Wang, C., Grebogi C., and Baptista, M.S. Control and prediction for blackouts caused by frequency collapse in smart grids. *Chaos* 2016;26(9):093119. https://doi.org/10.1063/1.4963764.

[10] Gielen, D., Boshell, F., Saygin, D., Bazilian, M.D., Wagner, N., and Gorini, R. The role of renewable energy in the global energy transformation. *Energy Strategy Reviews* 2019;24:38–50. https://doi.org/10.1016/j.esr.2019.01.006.

[11] Nwokolo, S.C., Meyer, E.L., and Ahia, C.C. Credible pathways to catching up with climate goals in Nigeria. *Climate* 2023;11(9):196. https://doi.org/10.3390/cli11090196.

[12] Bakhsh, S., Zhang, W., Ali, K., and Oláh, J. Strategy towards sustainable energy transition: The effect of environmental governance, economic complexity and geopolitics. *Energy Strategy Reviews* 2024;52:101330. https://doi.org/10.1016/j.esr.2024.101330.

[13] Yang, Y., Xia, S., Huang, P., and Qian, J. Energy transition: Connotations, mechanisms and effects. *Energy Strategy Reviews* 2024;52:101320. https://doi.org/10.1016/j.esr.2024.101320.

[14] Kabeyi, M.J.B., and Olanrewaju, O.A. Smart grid technologies and application in the sustainable energy transition: A review. *International Journal of Sustainable Energy* 2023;42(1):685–758. https://doi.org/10.1080/14786451.2023.2222298.

[15] Ohanu, C.P., Rufai, S.A., and Oluchi, U.C. A comprehensive review of recent developments in smart grid through renewable energy resources integration. *Heliyon* 2024;10(3):e25705. https://doi.org/10.1016/j.heliyon.2024.e25705.

[16] Hassan, Q., Hsu, C.Y., Mounich, K., *et al.* Enhancing smart grid integrated renewable distributed generation capacities: Implications for sustainable energy transformation. *Sustainable Energy Technologies and Assessments* 2024;66:103793. https://doi.org/10.1016/j.seta.2024.103793.

[17] Badrudeen, T.U., Nwulu, N.N., and Gbadamosi, S.L. Low-inertia control of a large-scale renewable energy penetration in power grids: A systematic review with taxonomy and bibliometric analysis. *Energy Strategy Reviews* 2024;52:101337. https://doi.org/10.1016/j.esr.2024.101337.

[18] Caballero, V., Vernet, D., Zaballos, A., and Corral, G. Prototyping a web-of-energy architecture for smart integration of sensor networks in smart grids domain. *Sensors* 2018;18(2):400. https://doi.org/10.3390/s18020400.

[19] Chen, Z., Amani, A.M., Yu, X., and Jalili, M. Control and optimisation of power grids using smart meter data: A review. *Sensors* 2023;23(4):2118. https://doi.org/10.3390/s23042118.

[20] Anasis, J.G., Khalil, M.A.K., Lendaris, G.G., Butenhoff, C.L., and Bluffstone, R. An analysis of the optimal mix of global energy resources and the potential need for geoengineering using the CEAGOM model. *Global Challenges* 2017;1(8):1700040. https://doi.org/10.1002/gch2.201700040.

[21] Howland, M.F., Lele, S.K., and Dabiri, J.O. Wind farm power optimization through wake steering. *Proceedings of the National Academy of Sciences of the United States of America* 2019;116(29):14495–14500. https://doi.org/10.1073/pnas.1903680116.

[22] Cao, L., Wang, Z., and Yue, Y. Analysis and prospect of the application of wireless sensor networks in ubiquitous power Internet of Things. *Computational Intelligence and Neuroscience* 2022;2022:9004942. https://doi.org/10.1155/2022/9004942.

[23] Biswal, B., Deb, S., Datta, S., Ustun, T.S., and Cali, U. Review on smart grid load forecasting for smart energy management using machine learning and deep learning techniques. *Energy Reports* 2024;12:3654–3670. https://doi.org/10.1016/j.egyr.2024.09.056.

[24] Liu, F., Sun, F., and Wang, X. Impact of turbine technology on wind energy potential and CO_2 emission reduction under different wind resource conditions in China. *Applied Energy* 2023;348:121540. https://doi.org/10.1016/j.apenergy.2023.121540.

[25] IRENA. *Future of Wind: Deployment, Investment, Technology, Grid Integration and Socio-Economic Aspects (A Global Energy Transformation paper)*. Abu Dhabi: International Renewable Energy Agency, 2019.

[26] Beiter, P., Cooperman, A., Lantz, E., *et al.* Wind power costs driven by innovation and experience with further reductions on the horizon. *WIREs Energy and Environment* 2021;10(5):e398. https://doi.org/10.1002/wene.398.

[27] Tumse, S., Bilgili, M., Yildirim, A., and Sahin, B. Comparative analysis of global onshore and offshore wind energy characteristics and potentials. *Sustainability* 2024;16(15):6614. https://doi.org/10.3390/su16156614.

[28] Kara, T., and Şahin, A.D. Implications of climate change on wind energy potential. *Sustainability* 2023;15(20):14822. https://doi.org/10.3390/su152014822.

[29] Ji, S., Li, J., Zhao, J., Feng, M., Sun, C., and Dai, H. Ultra-precision machining of a compound sinusoidal grid surface based on slow tool servo. *Materials* 2018;11(6):1001. https://doi.org/10.3390/ma11061001.

[30] Oyekale, J., Petrollese, M., Tola, V., and Cau, G. Impacts of renewable energy resources on effectiveness of grid-integrated systems: Succinct review of current challenges and potential solution strategies. *Energies* 2020; 13(18):4856. https://doi.org/10.3390/en13184856.

[31] Ullah, F., Zhang, X., Khan, M., *et al.* A comprehensive review of wind power integration and energy storage technologies for modern grid frequency regulation. *Heliyon* 2024;10(9):e30466. https://doi.org/10.1016/j.heliyon.2024.e30466.

[32] Reddy, B.N., Jalli, R., Prudhviraj, K.S., *et al.* Wind turbine with line-side PMSG FED DC-DC converter for voltage regulation. *PLoS ONE* 2024;19(6): e0305272. https://doi.org/10.1371/journal.pone.0305272.

[33] Kerem, A. Investigation of carbon footprint effect of renewable power plants regarding energy production: A case study of a city in Turkey. *Journal of the Air & Waste Management Association* 2022;72(3):294–307. https://doi.org/10.1080/10962247.2022.2028690.

[34] Ocko, I.B., and Hamburg, S.P. Climate impacts of hydropower: Enormous differences among facilities and over time. *Environmental Science and Technology* 2019;53(23):14070–14082. https://doi.org/10.1021/acs.est.9b05083.

[35] Østergaard, P.A., Duic, N., Noorollahi, Y., and Kalogirou, S. Advances in renewable energy for sustainable development. *Renewable Energy* 2023; 219:119377. https://doi.org/10.1016/j.renene.2023.119377.

[36] Evans, L.E., Dolsak, N., and Prakash A. Do windy areas have more wind turbines: An empirical analysis of wind installed capacity in native tribal nations. *PLoS One* 2022;17(2):e0261752. https://doi.org/10.1371/journal.pone.0261752.

[37] Qadir, S.A., Al-Motairi, H., Tahir, F., and Al-Fagih, L. Incentives and strategies for financing the renewable energy transition: A review. *Energy Reports* 2021;7:3590–3606. https://doi.org/10.1016/j.egyr.2021.06.041.

[38] Onifade, S.T., Erdoğan, S., and Alola, A.A. The role of alternative energy and globalization in decarbonization prospects of the oil-producing African economies. *Environmental Science and Pollution Research* 2023;30:58128–58141. https://doi.org/10.1007/s11356-023-26581-6.

[39] Flore, A., and Gomez, J.M. Development and comparison of migration paths for smart grids using two case studies. *Heliyon* 2020;6(9):e04913. https://doi.org/10.1016/j.heliyon.2020.e04913.

[40] Manzolini, G., Binotti, M., Gentile, G., Picotti, G., Pilotti, L., and Cholette, M.E. Actual cost of electricity: An economic index to overcome levelized cost of electricity limits. *iScience* 2024;27(6):109897. https://doi.org/10.1016/j.isci.2024.109897.

[41] Alotaibi, I., Abido, M.A., Khalid, M., and Savkin, A.V. A comprehensive review of recent advances in smart grids: A sustainable future with renewable energy resources. *Energies* 2020;13(23):6269. https://doi.org/10.3390/en13236269.

[42] Baik, S., Morgan, M.G., and Davis, A.L. Providing limited local electric service during a major grid outage: A first assessment based on customer willingness to pay. *Risk Analysis* 2018;38(2):272–282. https://doi.org/10.1111/risa.12838.

[43] Adekitan, A.I., Adetokun, B.B., Aligbe, A., Shomefun, T., and Orimogunje, A. Data based investigation of the energy metering type, billing and usage of sampled residents of Ota Community in Nigeria. *Data in Brief* 2018;20:159–172. https://doi.org/10.1016/j.dib.2018.07.047.

[44] Alshowkan, M., Evans, P.G., Starke, M., Earl, D., and Peters, N.A. Authentication of smart grid communications using quantum key distribution. *Scientific Reports* 2022;12:12731. https://doi.org/10.1038/s41598-022-16090-w.

[45] Chen, Y., and Xu, J. Solar and wind power data from the Chinese state grid renewable energy generation forecasting competition. *Scientific Data* 2022;9:577. https://doi.org/10.1038/s41597-022-01696-6.

[46] Wang, S., and Ma, J. A novel GBDT-BiLSTM hybrid model on improving day-ahead photovoltaic prediction. *Scientific Reports* 2023;13:15113. https://doi.org/10.1038/s41598-023-42153-7.

[47] Dupas, M.C., Parison, S., Noel, V., Chatzimpiros, P., and Herbert, E. Variable renewable energy penetration impact on productivity: A case study of poultry farming. *PLoS ONE* 2023;18(10):e0286242. https://doi.org/10.1371/journal.pone.0286242.

[48] Khan, K.A., Quamar, M.M., Al-Qahtani, F.H., Asif, M., Alqahtani, M., and Khalid, M. Smart grid infrastructure and renewable energy deployment: A conceptual review of Saudi Arabia. *Energy Strategy Reviews* 2023;50:101247. https://doi.org/10.1016/j.esr.2023.101247.

[49] Guo, X., Chen, X., Chen, X. Sherman, P., Wen, J., and McElroy, M. Grid integration feasibility and investment planning of offshore wind power under carbon-neutral transition in China. *Nature Communications* 2023;14:2447. https://doi.org/10.1038/s41467-023-37536-3.

[50] Rohten, J., Villarroel, F., Silva, J., *et al.* Power converter resonant control for an unbalanced and non-constant frequency supply. *Sensors* 2023;23(10):4884. https://doi.org/10.3390/s23104884.

[51] Cheng, M., Yan, W., Zhang, D., *et al.* Frequency stability of new energy power systems based on VSG adaptive energy storage coordinated control strategy. *Energy Informatics* 2024;7:54. https://doi.org/10.1186/s42162-024-00359-7.

[52] Casey, J.A., Fukurai, M., Hernández, D., Balsari, S., and Kiang, M.V. Power outages and community health: A narrative review. *Current Environmental Health Reports* 2020;7:371–383. https://doi.org/10.1007/s40572-020-00295-0.

[53] Youssef, H., Kamel, S., and Hassan, M.H. Smart home energy management and power trading optimization using an enhanced manta ray foraging optimization. *Scientific Reports* 2023;13:22163. https://doi.org/10.1038/s41598-023-49176-0.

[54] Arunkumar, A., Gupta, N., Pinceti, A., Sankar, L., and Bryan, C. PMUVis: A large-scale platform to assist power system operators in a smart grid. *IEEE Computer Graphics and Applications* 2022;42(6):84–95. https://doi.org/10.1109/MCG.2022.3171506.

[55] Arunkumar, A., Pinceti, A., Sankar, L., and Bryan, C. PMU tracker: A visualization platform for epicentric event propagation analysis in the power grid. *IEEE Transactions on Visualization and Computer Graphics* 2023;29(1):1081–1090. https://doi.org/10.1109/TVCG.2022.3209380.

[56] Fokui, W.S.T., Saulo, M., and Ngoo, L. Controlled electric vehicle charging for reverse power flow correction in the distribution network with high photovoltaic penetration: Case of an expanded IEEE 13 node test network. *Heliyon* 2022;8(3):e09058. https://doi.org/10.1016/j.heliyon.2022.e09058.

[57] Zhao, L., Nazir, M.S., Nazir, H.M.J., and Abdalla, A.N. A review on proliferation of artificial intelligence in wind energy forecasting and instrumentation management. *Environmental Science and Pollution Research* 2022;29(29):43690–43709. https://doi.org/10.1007/s11356-022-19902-8.

[58] Kiasari, M., Ghaffari, M., and Aly, H.H. A comprehensive review of the current status of smart grid technologies for renewable energies integration and future trends: The role of machine learning and energy storage systems. *Energies* 2024;17:4128. https://doi.org/10.3390/en17164128.

[59] Benzi, F., Anglani, N., Bassi, E., and Frosini, L. Electricity smart meters interfacing the households. *IEEE Transactions on Industrial Electronics* 2011;58(10):4487–4494. http://dx.doi.org/10.1109/TIE.2011.2107713.

[60] Markovic, D.S., Zivkovic, D., Branovic, I., Popovic, R., and Cvetkovic, D. Smart power grid and cloud computing. *Renewable and Sustainable Energy Reviews* 2013;24:566–577. http://dx.doi.org/10.1016/j.rser.2013.03.068.

[61] Ahmed I., Rehan, M., Iqbal, N., and Ahn, C.K. A novel event-triggered consensus approach for generic linear multi-agents under heterogeneous sector-restricted input nonlinearities. *IEEE Transactions on Network Science and Engineering* 2023;10(3):1648–1658. http://dx.doi.org/10.1109/TNSE.2022.3232779.

[62] Hewett, R., Rudrapattana, S., and Kijsanayothin, P. Cyber-security analysis of smart grid SCADA systems with game models. *In: Proceedings of the 9th Annual Cyber and Information Security Research Conference*, 2014:109–112. https://doi.org/10.1145/2602087.2602089.

[63] Parikh, P.P., Kanabar, M.G., and Sidhu, T.S. Opportunities and challenges of wireless communication technologies for smart grid applications. In: *IEEE PES General Meeting*, Minneapolis, USA, 2010:1–7. https://doi.org/10.1109/PES.2010.5589988.

[64] Yi, P., Iwayemi, A., and Zhou, C. Developing ZigBee deployment guideline under WiFi interference for smart grid applications. *IEEE Transactions on Smart Grid* 2011;2(1):110–120. https://doi.org/10.1109/TSG.2010.2091655.

[65] Hassan, Q., Khadom, A.A., Algburi, S., *et al.* Implications of a smart grid-integrated renewable distributed generation capacity expansion strategy: The case of Iraq. *Renewable Energy* 2024;221:119753. https://doi.org/10.1016/j.renene.2023.119753.

[66] Dawn, S., Ramakrishna, A., Ramesh, M., *et al.* Integration of renewable energy in microgrids and smart grids in deregulated power systems: A comparative exploration. *Advanced Energy and Sustainability Research* 2024;5(10):2400088. https://doi.org/10.1002/aesr.202400088.

[67] Saleem, M.U., Shakir, M., Usman, M.R., *et al.* Integrating smart energy management system with Internet of Things and Cloud Computing for efficient demand side management in smart grids. *Energies* 2023;16(12):4835. https://doi.org/10.3390/en16124835.

[68] Jacobson, M.Z. Batteries or hydrogen or both for grid electricity storage upon full electrification of 145 countries with wind-water-solar? *iScience* 2024;27(2):108988. https://doi.org/10.1016/j.isci.2024.108988.

[69] Jacobson, M.Z., Delucchi, M.A., Cameron, M.A., and Frew, B.A. Low-cost solution to the grid reliability problem with 100% penetration of intermittent wind, water, and solar for all purposes. *Proceedings of the National Academy of Sciences of the United States of America* 2015;112(49):15060–15065. https://doi.org/10.1073/pnas.1510028112.

[70] Goodenough, J.B., and Park, K.S. The Li-ion rechargeable battery: A perspective. *Journal of the American Chemical Society* 2013;135(4):1167–1176. https://doi.org/10.1021/ja3091438.

[71] Xia, W., Ren, Z., Qin, H., and Dong, Z. A coordinated operation method for networked hydrogen-power-transportation system. *Energy* 2024;296:131026. https://doi.org/10.1016/j.energy.2024.131026.

[72] Mukherjee, P., and Rao, V.V. Design and development of high temperature superconducting magnetic energy storage for power applications – A review. *Physica C: Superconductivity and its Applications* 2019;563:67–73. https://doi.org/10.1016/j.physc.2019.05.001.

[73] Aneke, M., and Wang, M. Energy storage technologies and real life applications–A state of the art review. *Applied Energy* 2016;179:350–377. https://doi.org/10.1016/j.apenergy.2016.06.097.

[74] Powell, J., McCafferty-Leroux, A., Hilal, W., and Gadsden, S.A. Smart grids: A comprehensive survey of challenges, industry applications, and future trends. *Energy Reports* 2024;11:5760–5785. https://doi.org/10.1016/j.egyr.2024.05.051.

[75] Valdivia-Bautista, S.M., Domínguez-Navarro, J.A., Pérez-Cisneros, M., Vega-Gómez, C.J., and Castillo-Téllez, B. Artificial intelligence in wind speed forecasting: A review. *Energies* 2023;16:2457. https://doi.org/10.3390/en16052457.

[76] Wang, H., Xiong, B., Zhang, Z., Zhang, H., and Azam, A. Small wind turbines and their potential for internet of things applications. *iScience* 2023;26 (9):107674. https://doi.org/10.1016/j.isci.2023.107674.

[77] Allal, Z., Noura, H.N., Salman, O., and Chahine, K. Machine learning solutions for renewable energy systems: Applications, challenges, limitations, and future directions. *Journal of Environmental Management* 2024;354:120392. https://doi.org/10.1016/j.jenvman.2024.120392.

[78] Yang, X., Chang, J., Zhang, Z., Zhang, J., and Xu, G. Integrated energy system optimal scheduling considering the comprehensive and flexible operation mode of pumping storage. *PLoS ONE* 2022;17(10):e0275514. https://doi.org/10.1371/journal.pone.0275514.

[79] Ravi, T., Kumar, K.S., Dhanamjayulu, C., Khan, B., and Rajalakshmi, K. Analysis and mitigation of PQ disturbances in grid connected system using fuzzy logic based IUPQC. *Scientific Reports* 2023;13:22425. https://doi.org/10.1038/s41598-023-49042-z.

[80] Rohten, J., Villarroel, F., Pulido, E., Muñoz, J., Silva, J., and Perez, M. Stability analysis of two power converters control algorithms connected to micro-grids with wide frequency variation. *Sensors* 2022;22(18):7078. https://doi.org/10.3390/s22187078.

[81] Ghadi, Y.Y., Mazhar, T., Aurangzeb, K., *et al.* Security risk models against attacks in smart grid using big data and artificial intelligence. *PeerJ Computer Science* 2024;10:e1840. https://doi.org/10.7717/peerj-cs.1840.

[82] Bahaghighat, M., Xin, Q., Motamedi, S.A., Zanjireh, M.M., and Vacavant, A. Estimation of wind turbine angular velocity remotely found on video mining and convolutional neural network. *Applied Sciences* 2020;10 (10):3544. https://doi.org/10.3390/app10103544.

[83] Bose, B.K. Artificial intelligence techniques in smart grid and renewable energy systems—Some example applications. *Proceedings of the IEEE* 2017;105(11):2262–2273. https://doi.org/10.1109/JPROC.2017.2756596.

[84] Zhang, Y., and Wang, S. An innovative forecasting model to predict wind energy. *Environmental Science and Pollution Research* 2022;29(49):74602–74618. https://doi.org/10.1007/s11356-022-20971-y.

[85] Lu, Y., Sun, L., Zhang, X., Feng, F., Kang, J., and Fu, G. Condition based maintenance optimization for offshore wind turbine considering opportunities based on neural network approach. *Applied Ocean Research* 2018; 74:69–79. https://doi.org/10.1016/j.apor.2018.02.016.

[86] Rahman, M.M., Shakeri, M., Tiong, S.K., *et al.* Prospective methodologies in hybrid renewable energy systems for energy prediction using artificial neural networks. *Sustainability* 2021;13(4):2393. https://doi.org/10.3390/su13042393.

[87] Nazir, M.S., Abdalla, A.N., Wang, Y., *et al.* Optimization configuration of energy storage capacity based on the microgrid reliable output power.

Journal of Energy Storage 2020;32:101866. https://doi.org/10.1016/j.est.2020.101866.

[88] Ahmed, S.D., Al-Ismail, F.S.M., Shafiullah, M.D., Al-Sulaiman, F.A., and El-Amin, I.M. Grid integration challenges of wind energy: A review. *IEEE Access* 2020;8:10857–10878. https://doi.org/10.1109/access.2020.2964896.

[89] Abisoye, B.O., Sun, Y., and Zenghui, W. A survey of artificial intelligence methods for renewable energy forecasting: Methodologies and insights. *Renewable Energy Focus* 2024;48:100529. https://doi.org/10.1016/j.ref.2023.100529.

[90] Geng, L., Zhang, L., Niu, F., Li, Y., and Liu, F. A swarm intelligence and deep learning strategy for wind power and energy storage scheduling in smart grid. *International Journal of Intelligent Networks* 2024;5:302–314. https://doi.org/10.1016/j.ijin.2024.08.001.

[91] Saini, V.K., Kumar, R., Al-Sumaiti, A.S., Sujil, A., and Heydarian-Forushani, E. Learning based short term wind speed forecasting models for smart grid applications: An extensive review and case study. *Electric Power Systems Research* 2023;222:109502. https://doi.org/10.1016/j.epsr.2023.109502.

[92] Kumari, A., Sukharamwala, U.C., Tanwar, S., *et al.* Blockchain-based peer-to-peer transactive energy management scheme for smart grid system. *Sensors* 2022;22(13):4826. https://doi.org/10.3390/s22134826.

[93] Rejeb, A., Rejeb, K., Zrelli, I., Süle, E., and Iranmanesh, M. Blockchain technology in the renewable energy sector: A co-word analysis of academic discourse. *Heliyon* 2024;10(8):e29600. https://doi.org/10.1016/j.heliyon.2024.e29600.

[94] Zedan, M., Nour, M., Shabib, G., Nasrat, L., and Ali A.A. Review of peer-to-peer energy trading: Advances and challenges. *e-Prime – Advances in Electrical Engineering, Electronics and Energy* 2024;10:100778. https://doi.org/10.1016/j.prime.2024.100778.

[95] Taherdoost, H. Blockchain integration and its impact on renewable energy. *Computers* 2024;13(4):107. https://doi.org/10.3390/computers13040107.

[96] Pradhan, N.R., Singh, A.P., Verma, S., *et al.* A blockchain based lightweight peer-to-peer energy trading framework for secured high throughput micro-transactions. *Scientific Reports* 2022;12:14523. https://doi.org/10.1038/s41598-022-18603-z.

[97] Pandiyan, P., Saravanan, S., Kannadasan, R., Krishnaveni, S., Alsharif, M. H., and Kim, M.K. A comprehensive review of advancements in green IoT for smart grids: Paving the path to sustainability. *Energy Reports* 2024;11:5504–5531. https://doi.org/10.1016/j.egyr.2024.05.021.

[98] Dörterler, S., Arslan, S., and Özdemir, D. Unlocking the potential: A review of artificial intelligence applications in wind energy. *Expert Systems* 2024;41(12):e13716. https://doi.org/10.1111/exsy.13716.

[99] Debnath, B., Shakur, M.S., Siraj, M.T., Bari, A.B.M.M., and Islam, A.R.M. T. Analyzing the factors influencing the wind energy adoption in Bangladesh: A pathway to sustainability for emerging economies. *Energy Strategy Reviews* 2023;50:101265. https://doi.org/10.1016/j.esr.2023.101265.

[100] Elsisi, M., Mahmoud, K., Lehtonen, M., and Darwish, M.M.F. Reliable Industry 4.0 based on machine learning and IoT for analyzing, monitoring, and securing smart meters. *Sensors* 2021;21(2):487. https://doi.org/10.3390/s21020487.

[101] Chen, Y.Y., Lin, Y.H., Kung, C.C., Chung, M.H., and Yen, I.H. Design and implementation of cloud analytics-assisted smart power meters considering advanced artificial intelligence as edge analytics in demand-side management for smart homes. *Sensors* 2019;19(9):2047. https://doi.org/10.3390/s19092047.

Chapter 9

Battery charge controllers in wind turbine systems

Pankaj Kumar[1], *Amanjot Kaur*[1] *and Jasmeet Kaur*[1]

Abstract

Battery charge controllers (BCCs) play a pivotal role in integrating and controlling wind turbine systems, especially with advancements in digital technology. As the demand for renewable energy rises, wind energy is increasingly recognized as a key player in sustainable power generation. BCCs ensure the efficient management of electrical energy produced by wind turbines, optimizing the charging and discharging of batteries. Utilizing advanced techniques such as maximum power point tracking (MPPT), these controllers maximize energy capture under varying wind conditions while protecting batteries from overcharging and deep discharging. This paper discusses the various types of BCCs, their functionalities, and the challenges associated with their integration into both off-grid and grid-tied systems. Furthermore, it highlights the impact of digital technology on enhancing the performance of BCCs through smart monitoring and data analytics, allowing for real-time decision-making and improved energy management. Case studies illustrate the successful application of BCCs in diverse settings, demonstrating their significance in optimizing system efficiency and reliability. Ultimately, this study underscores the necessity of advanced BCCs in future wind energy projects to achieve sustainability and enhance the viability of renewable energy sources.

Keywords: Battery charge controllers; wind turbines; maximum power point tracking (MPPT); energy management; sustainable power generation

9.1 Introduction

With the rising global demand for sustainable energy solutions, renewable sources such as wind energy have attracted considerable attention. Being abundant and renewable resource, wind energy offers a vital alternative to conventional fossil fuel-based power generation, which is associated with greenhouse gas emissions and environmental harm (Rezaee, 2023). However, wind energy systems experience fluctuations in power output due to changing wind conditions, making

[1]Department of Physical and Mathematical Science, Baba Farid College, India

efficient energy storage and management systems essential. Battery storage has emerged as a valuable solution for capturing surplus energy produced during peak times and supplying it during periods of low production. Within this framework, battery charge controllers (BCCs) are essential for maintaining battery health, optimizing efficiency, and extending lifespan, thereby ensuring a consistent and dependable energy supply from wind turbines (Parker, 2022).

9.1.1 The role of battery charge controllers in wind energy systems

BCCs are crucial elements in wind energy systems, tasked with managing the charging and discharging processes of batteries. Serving as intermediaries between the wind turbine and battery storage, they prevent batteries from overcharging or discharging to harmful levels (Chen, 2023). By keeping the battery in an optimal state of charge (SOC), BCCs improve battery efficiency and lifespan, both of which are vital for minimizing maintenance costs and optimizing system performance. The advanced circuitry in BCCs enables the monitoring of various parameters, including voltage, current, and temperature, facilitating real-time adaptive control. Additionally, BCCs enhance safety by reducing the risks of overheating and overvoltage, thereby safeguarding the entire energy storage system (Lee, 2022).

9.1.2 Importance of renewable energy and wind turbines

Transitioning to renewable energy is essential for meeting global sustainability objectives and tackling climate change. Wind energy, being one of the most established and commercially feasible renewable sources, has gained popularity due to its low operational costs and minimal environmental impact (Garcia, 2023). In contrast to fossil fuels, wind energy captures natural wind currents to produce electricity without exhausting resources or emitting pollutants. Technological advancements have enhanced the efficiency and scalability of wind turbines, making them appropriate for both large-scale utility projects and off-grid applications (Smith, 2022). In remote/isolated regions, wind energy systems paired with battery storage provide a sustainable alternative to reliance on the grid, enhancing energy accessibility and lowering carbon footprints.

9.1.3 Overview of digital technology in energy systems

Digital technologies have revolutionized energy systems, allowing for smarter, more efficient, and responsive energy management. In wind energy systems, digital controllers and Internet of Things (IoT) devices enhance the functionality of BCC's by facilitating real-time monitoring and management of energy flow and storage conditions (Kim, 2023). These technologies also enable data collection and analytics, empowering operators to optimize performance, anticipate maintenance requirements, and quickly address system faults (Zhao, 2023). Machine learning (ML) algorithms and predictive modeling can enhance the adaptability of BCCs by forecasting energy production patterns and modifying the battery charging process accordingly. As digital integration advances, BCCs will become increasingly vital in ensuring the resilience and efficiency of renewable energy systems (Brown, 2023).

9.2 Battery charge controllers in wind turbine systems

9.2.1 Functionality and importance of BCCs

BCCs play a vital role in wind turbine systems, especially those that incorporate energy storage solutions. Their main function is to regulate the charging and discharging cycles of batteries, maximizing both performance and lifespan. By preventing over-charging and deep discharging, BCCs help protect batteries from damage that could otherwise shorten their life expectancy. Additionally, BCCs control the flow of electricity between the wind turbine and the battery, shielding the system from potential damage due to voltage spikes or fluctuations often seen in wind energy setups. Their role is essential in improving the overall efficiency of wind power systems.

By regulating the energy produced by the turbine, BCCs facilitate smoother grid integration and ensure a stable energy supply during low-wind periods. This function is especially critical in off-grid applications, where consistent power delivery is crucial for reliable operation (Smith, 2020). Moreover, BCCs can be configured to optimize energy use according to demand forecasts, enhancing both the sustainability and cost-efficiency of wind energy solutions. Figure 9.1 presents an operational diagram of a BCC within a wind turbine system, showcasing its connections and key functions.

9.2.2 Integration of BCCs with wind turbines

Incorporating BCCs into wind turbine systems requires addressing several technical factors. The BCC must align with the power output characteristics of the wind turbine, which can vary significantly with changing wind conditions. Therefore, BCCs need adaptive control strategies that can respond effectively to these fluctuations in energy generation (Doe and Johnson, 2019). Modern BCCs employ advanced algorithms to analyze real-time data from both the wind turbine and the energy storage system, allowing for quick adjustments to charging rates as needed. The integration process also involves communication interfaces that enable BCCs to interact with other system components, like inverters and grid management systems. This coordination is crucial for maximizing energy capture and ensuring that surplus energy generated during peak wind conditions is effectively stored or utilized (Green, 2021). The ability to interface with smart grids further enhances BCC functionality, enabling demand response strategies that support grid stability.

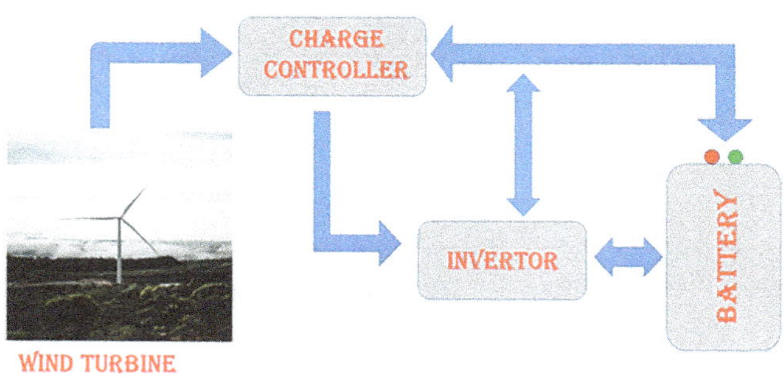

Figure 9.1 Block diagram of BCC integration within a wind turbine system

9.2.3 Energy management in wind systems

Efficient energy management in wind systems is essential for maximizing the advantages of renewable energy sources. BCCs are central to this effort, employing advanced energy management strategies to balance supply with demand. These strategies frequently utilize predictive algorithms that analyze weather forecasts and historical wind patterns to optimize battery usage and charging schedules (Lee, 2018). Additionally, BCCs enable the implementation of various energy management techniques, including peak shaving and load shifting. Peak shaving minimizes energy consumption during peak demand periods, while load shifting involves storing energy during low-demand times for use during periods of high demand. By efficiently managing energy flow, BCCs help lower overall operational costs and improve the reliability of wind energy systems (Brown and Black, 2022). Figure 9.2 depicts a flow chart of energy management processes in wind turbine systems, illustrating how BCCs influence energy storage and distribution.

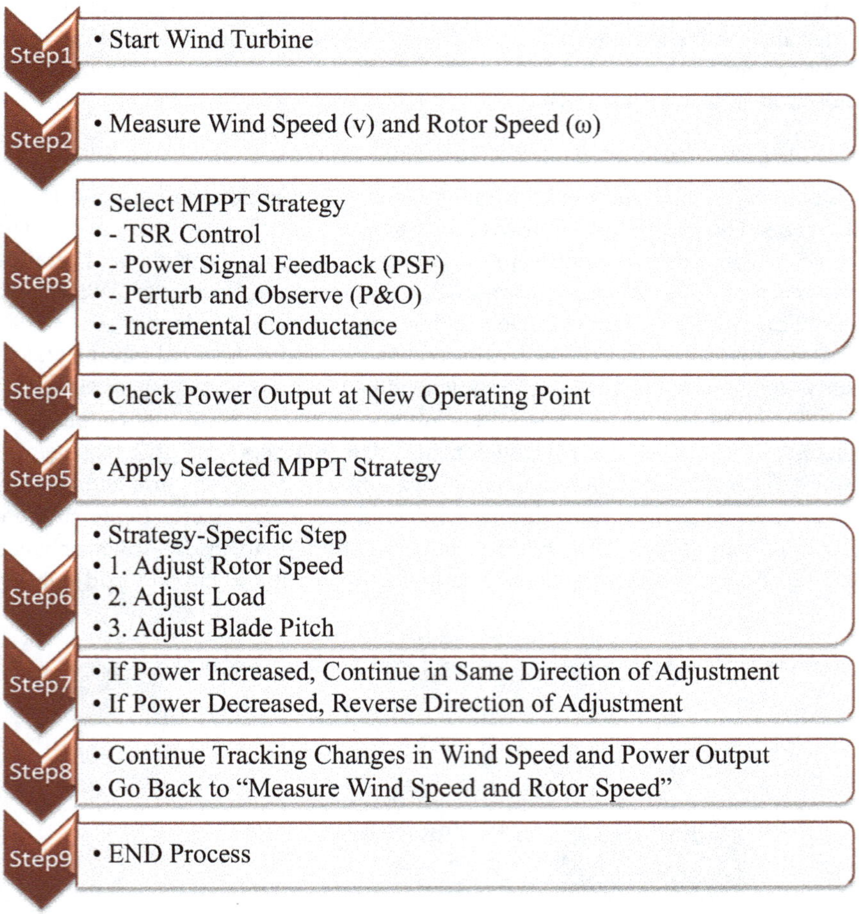

Figure 9.2 Wind turbines optimize energy capture through the use of maximum power point tracking technology

9.3 Types of battery charge controllers

This section explores different types of BCC technologies, focusing on their mechanisms, advantages, and applications in wind energy systems.

9.3.1 BCC technologies

BCCs can be classified primarily into two types: pulse width modulation (PWM) controllers and maximum power point tracking (MPPT) controllers. Each technology has distinct operational mechanisms, efficiency levels, and suitability for various applications. Understanding these technologies is crucial for optimizing energy capture and enhancing the longevity of battery systems in wind turbines.

9.3.2 Pulse width modulation

PWM controllers are commonly used as BCCs in renewable energy systems. Their primary function is to control the power delivered to the load by varying the width of the pulses in a signal. PWM controllers' function by quickly toggling the connection to the battery on and off (Schnabel and McGilton, 2024). The average power supplied to the battery is determined by the ratio of "on" time to "off" time, known as the duty cycle. When the wind turbine produces electricity, the PWM controller modifies the duty cycle to align with the battery's charge level. PWM controllers are often fevered for their simplicity and cost-effectiveness, as they generally have a straightforward design and are more affordable compared to MPPT controllers. This makes them a practical choice for small-scale wind energy applications. In terms of efficiency, PWM controllers may not match the performance of MPPT controllers but can still achieve around 70–90% efficiency under optimal conditions. However, PWM controllers also come with some disadvantages. They tend to have limited performance in variable wind conditions, often struggling to capture the maximum power output from the turbine during fluctuations. Additionally, the PWM process can lead to energy losses due to voltage drops, particularly in systems where significant voltage regulation is required.

9.3.3 Maximum power point tracking

MPPT controllers employ advanced technology to optimize energy capture in wind turbine systems. They continually improve power output by dynamically adjusting the electrical load on the turbine. MPPT controllers operate by using complex algorithms to identify the optimal power point of the wind turbine (Teklehaimanot *et al.*, 2024). This process involves continuously monitoring factors such as wind speed, turbine rotational speed, and battery charge levels to dynamically adjust the system's operating voltage and current for maximum efficiency. One of the main advantages of MPPT controllers is their high efficiency, which typically ranges from 90% to 98%. This level of efficiency allows for greater energy capture, especially in variable wind conditions. Additionally, MPPT systems can optimize performance by adjusting the turbine's operating point, enhancing energy output over a broader range of wind speeds. However, MPPT controllers also have certain disadvantages. They tend

Table 9.1 Comparative analysis of BCC technologies

Feature	PWM	MPPT
Efficiency	70–90%	90–98%
Cost	Lower	Higher
Complexity	Simple	Higher
Performance in variable wind	Limited	Optimized
Application suitability	Small-scale systems	Large-scale and variable conditions

to be more expensive than PWM controllers, which can be a critical factor for projects with limited budgets. Furthermore, the advanced algorithms and complex hardware components involved in MPPT systems may increase the complexity of installation and maintenance.

9.3.4 Comparative analysis: PWM vs. MPPT in wind energy applications

The choice between PWM and MPPT controllers depends on various factors, including system size, budget constraints, and performance requirements. Table 9.1 summarizes the key differences between these technologies.

9.3.4.1 Application suitability

PWM controllers and MPPT controllers each have unique suitability for different wind energy applications. PWM controllers are often ideal for small-scale or budget-conscious projects due to their simplicity and cost-effectiveness. Their straightforward design and lower cost make them suitable for applications where efficiency demands are moderate and conditions are relatively stable such as small residential or isolated systems where fluctuations in wind speed are not severe (Murugesan and Ramasamy, 2024). However, MPPT controllers are better suited for larger, more demanding wind energy systems or those in environments with variable wind conditions. With their higher efficiency and ability to optimize power output across a wide range of wind speeds, MPPT controllers are highly suitable for applications where maximizing energy capture is essential such as commercial or grid-tied wind systems. However, their higher cost and complexity may make them less practical for smaller, budget-limited projects.

9.4 Maximizing energy capture using maximum power point tracking

A key strategy in renewable energy systems, particularly for solar and wind power generation, is to enhance efficiency, advance storage technologies, and integrate smart grid solutions for sustainable and reliable energy utilization. MPPT ensures that the system operates at its peak efficiency by continuously adjusting the electrical parameters to the optimal point (Zaghba *et al.*, 2024), thus capturing the maximum possible energy from the available resources.

9.4.1 MPPT works in variable wind conditions

MPPT in variable wind conditions involves dynamically adjusting the electrical load on a wind turbine to maximize power extraction as wind speed changes. Wind conditions are inherently variable, and this variability directly affects the turbine's performance. We will discuss about the wind turbine and basic power generation before MPPT working conditions here.

9.4.1.1 Wind turbine and power generation basics

Wind energy is a form of renewable energy, generated by the movement of air due to differences in atmospheric pressure. It's sustainable, abundant, and produces no direct greenhouse gas emissions. The energy from wind comes from its kinetic energy, which depends on the wind's speed and air density (Bugdayci and Yilmazoglu, 2024). Wind turbines are designed to capture as much of this kinetic energy as possible.

9.4.1.2 Wind turbines work

Wind turbines generate electricity by converting wind's kinetic energy into electrical energy. When wind hits the blades, they spin, turning a central shaft connected to a gearbox, which increases rotational speed. This drives a generator that produces electricity. The turbine's control system adjusts blade angles and orientation to optimize power output. Safety mechanisms, like brakes, prevent damage during high winds. The generated electricity is then transformed to a higher voltage for grid distribution, providing clean, renewable energy (Tayebi and Torabi 2024).

Structure of a wind turbine

Modern wind turbines are equipped with three aerodynamic blades designed to capture wind energy. These blades are shaped to maximize lift and minimize drag, similar to an airplane wing. The blades are attached to the rotor, which spins as wind flows over them, generating lift and causing rotation. The nacelle, located on top of the tower, houses the gearbox, generator, and control systems. The tower provides the necessary height to access stronger winds at higher altitudes and supports both the nacelle and the rotor.

Power generation process

As wind blows over the blades of a wind turbine, their aerodynamic shape generates lift, causing them to rotate. This rotation spins the rotor, which is connected to a shaft housed within the nacelle. In larger turbines, a gearbox increases the rotational speed from the rotor to align with the generator's operating speed. The generator then converts the mechanical energy from the rotating shaft into electrical energy. Finally, the electricity is transmitted through cables to an electrical grid or stored in batteries.

Wind speed and power output

The relationship between wind and power generation in wind turbines is directly proportional: the power generated by a wind turbine increases with the wind speed. Power generation depends on the cube of the wind speed (Dong *et al.*, 2024). This means that even a small increase in wind speed can result in a significant increase in power output. The theoretical power that can be extracted from the wind is given by

$$P = \frac{1}{2}\rho A C_{\mathrm{p}}(\lambda, \beta) v^3 \tag{9.1}$$

where P is the power, ρ is the air density, A is the area swept by the blades, C_p (λ, β) is the power coefficient which depends on the tip speed ratio (TSR) λ and blade pitch angle β, and v is the wind speed. Since power is proportional to v^3, doubling the wind speed results in an eightfold increase in power generation. For each wind speed, there's a corresponding optimal rotor speed that maximizes power output. If the rotor spins too slowly or too quickly, it will not extract maximum power from the wind. MPPT controllers continuously adjust the operating point of the wind turbine to track and maintain the optimal power output, despite changing wind speeds. This involves:

- Sensing wind speed: Some MPPT systems directly monitor wind speed, but often wind speed is inferred from turbine output parameters.
- Adjusting rotor speed or load: The controller adjusts the turbine's rotor speed or the electrical load to maintain the optimal balance between rotor speed and wind speed. The following flow chart presents the steps followed to get maximum power output using MPPT technique.

9.4.1.3 MPPT adjusts the operating conditions

MPPT methods typically adjust the rotational speed of the turbine and the generator's load to optimize power output (Chellakhi *et al.*, 2024). There are several MPPT strategies, but they usually follow these principles:

- TSR control: This method keeps the TSR $\lambda = R\omega/v$ constant, where R is the blade radius, and ω is the rotor speed. By maintaining an optimal TSR (where the turbine operates at peak Cp), the turbine extracts the maximum power for any wind speed.
- Power signal feedback (PSF): In this method, the power output is continuously monitored, and the system tries to align with the predetermined optimal power curve. Any deviation from this curve is fixed by adjusting the rotor speed or generator load.
- Perturb and observe (P&O): This method incrementally modifies the turbine speed or generator load while monitoring any increases or decreases in power output. It continues adjusting in the direction that increases power output until the maximum power point (MPP) is found.
- Fuzzy logic control: To deal with the unpredictability and fluctuation in wind speeds, several systems employ fuzzy logic. To maximize performance, these controllers modify settings in accordance with guidelines generated from expert knowledge.

9.4.2 Comparison of MPPT algorithms for wind

In highly variable wind conditions, MPPT must respond rapidly to changing wind speeds. Delays in response can lead to sub-optimal power extraction or

Table 9.2 MPPT algorithm to present the advantages and challenges

Algorithm	Advantages	Challenges	Best used for
Tip-speed ratio (TSR)	High efficiency, widely used, well-established	Requires accurate wind speed and rotor speed measurement	Variable-speed wind turbines
Power signal feedback (PSF)	Simple, responsive to real-time power changes	Slower response in steady wind conditions	Moderate wind variations
Hill climbing search (HCS)	Simple, easy to implement	Inefficient in highly variable wind conditions	Moderate wind fluctuations
MPPT lookup table (MPPT-LUT)	Fast and reliable in steady wind conditions	Requires pre-calibration, less effective in variable winds	Steady wind conditions
Fuzzy logic controller (FLC)	Handles uncertainty, adaptable to changing conditions	Complex design, tuning required	Highly variable wind conditions
Indirect control methods	Simplifies system design, less direct reliance on measurements	Requires careful tuning, less precise	Turbines with pitch control
Model predictive control (MPC)	Predicts future conditions for accurate control	High computational complexity, needs a good model	Advanced turbines in dynamic conditions

mechanical stress on the turbine. Advanced MPPT algorithms with faster response times and predictive control strategies are often used to mitigate these effects (Teklehaimanot *et al.*, 2024). Below are the comparisons. Table 9.2 shows for MPPT algorithm to present the advantages and challenges and their best use as per conditions.

9.4.2.1 MPPT efficiency in energy capture

MPPT is specifically designed to maximize the amount of energy a wind turbine captures from the wind by ensuring that it operates at the point of maximum power output (Khan *et al.*, 2024). The efficiency of MPPT in energy capture is determined by how accurately it can track this MPP under changing conditions.

Optimal energy extraction

Wind energy systems operate most efficiently when the rotor speed of the turbine is matched to the wind speed. The relationship between wind speed and power output is cubic, meaning small increases in wind speed can result in significant power increases. However, wind speeds are constantly changing, and without an MPPT system, a turbine may not be able to adapt quickly enough to these changes, resulting in suboptimal power output. MPPT enhances energy capture by maintaining the ideal TSR, i.e., the ratio between the blade tip speed and the wind speed

(El Ouali *et al.*, 2024). The TSR is critical because it affects the aerodynamic efficiency of the blades, and thus the power coefficient *Cp*, which dictates the portion of available wind energy that the turbine can convert into usable electricity. MPPT systems constantly adjust the turbine's rotor speed or electrical load to keep the TSR at its optimal value, maximizing the power output at any wind speed.

Real-time adjustments

MPPT continuously monitors various parameters, such as wind speed, rotor speed, and power output, to make real-time adjustments. These adjustments ensure that the system operates as close to the MPP as possible, even as wind speeds fluctuate. For example, in the "Perturb and Observe" (P&O) method, the MPPT algorithm perturbs the turbine's operating point and observes the effect on power output. If the power increases, the system continues in that direction until it reaches the MPP. If the power decreases, it reverses the adjustment (Tella *et al.*, 2024). In variable wind conditions, these continuous real-time adjustments are crucial. As wind gusts or calms occur, MPPT enables the turbine to quickly respond, adjusting the rotational speed and load to maintain optimal power capture. This dynamic tracking prevents energy losses that would occur if the system operated without MPPT, where the turbine might lag behind optimal operation in changing winds.

Efficiency gains

The use of MPPT can lead to significant efficiency improvements in energy capture. In practical terms, MPPT can increase energy output by 20–30% or more, especially in environments with highly variable wind conditions (Zaghba *et al.*, 2024). Without MPPT, a turbine operating at non-optimal points will extract only a fraction of the energy available from the wind. MPPT ensures that the turbine operates at its peak efficiency, capturing the maximum energy available at any wind speed.

This efficiency boost is particularly important in low and moderate wind conditions, where the turbine might otherwise struggle to produce significant power. MPPT helps optimize power extraction even at these lower wind speeds, thereby increasing the overall capacity factor of the wind turbine.

Impact on system reliability

In addition to maximizing energy capture, MPPT contributes to the longevity and reliability of wind turbines. By optimizing the turbine's operation, MPPT reduces mechanical stress on the components, minimizing wear and tear and decreasing maintenance needs. This leads to improved system availability and more consistent energy production.

9.4.3 Enhancing system efficiency with MPPT

MPPT technology plays a critical role in enhancing the efficiency of energy systems, particularly in wind and solar power applications. By continuously optimizing the operating parameters of a wind turbine or photovoltaic system, MPPT ensures that the maximum possible power is extracted under varying environmental conditions (Ren *et al.*, 2024). Several factors contribute to the overall improvement of system efficiency when MPPT is applied.

9.4.3.1 Dynamic adjustment to variable conditions

MPPT enhances system efficiency by dynamically adjusting to changing wind speeds or sunlight conditions. In wind energy systems, the relationship between wind speed and power output is nonlinear, meaning that power output varies significantly with changes in wind speed. MPPT systems constantly monitor wind conditions and adjust the rotor speed and generator load to ensure that the turbine operates at the optimal TSR for maximum power extraction. Without MPPT, turbines may operate at non-optimal points, especially in variable wind conditions, resulting in reduced energy capture. By maintaining the turbine's operation at the MPP, MPPT increases energy yield, typically improving efficiency by 20–30% compared to systems without MPPT. This dynamic optimization ensures that the system operates efficiently across a wide range of wind speeds, boosting overall power generation (Sarang *et al.*, 2024).

9.4.3.2 Variable speed operation

MPPT is particularly effective in enabling variable-speed operation in wind turbines. Unlike fixed-speed systems, which can only operate efficiently at a single wind speed, variable-speed turbines can adjust their rotor speed to match changing wind conditions (Yan and Sun, 2024). MPPT algorithms ensure that the rotor speed is optimized to capture the maximum possible energy from the wind at any given time.

Variable-speed turbines are generally more efficient because they can operate at optimal power points over a broader range of wind conditions. This flexibility leads to higher system efficiency and greater energy capture, especially in environments with fluctuating wind patterns. By leveraging MPPT to enable variable-speed operation, the system minimizes energy losses that occur when the turbine operates outside its ideal range.

9.4.3.3 Integration with advanced control algorithms

MPPT efficiency can be further enhanced through the use of advanced control algorithms such as fuzzy logic, neural networks, or adaptive control techniques. These algorithms improve the accuracy and speed of MPPT systems, enabling them to track the MPP more effectively, especially in rapidly changing wind conditions. For instance, adaptive control algorithms can predict future wind speed trends based on historical data, allowing the system to adjust pre-emptively and maintain optimal performance. Similarly, fuzzy logic-based MPPT systems can handle uncertainties in the input variables such as fluctuating wind speeds or changes in turbine blade aerodynamics (Rashid, 2024). By improving the precision and adaptability of the MPPT system, these algorithms contribute to higher overall efficiency and better energy capture.

9.4.3.4 Optimized power electronics

MPPT is typically integrated with power electronics systems, such as DC–DC converters and inverters, that control the flow of energy from the generator to the grid or storage systems. These power electronics ensure that the electrical load is properly matched to the turbine's output, maximizing the efficiency of energy conversion and minimizing losses. By optimizing the interaction between the wind turbine and the electrical grid, MPPT systems prevent energy waste, particularly during periods of grid instability or when energy storage is involved (Abdelwahab *et al.*, 2024). Efficient power conversion and load management enable the system to deliver consistent power output, improving overall energy utilization.

9.4.3.5 Prolonged system lifespan

In addition to increasing energy capture, MPPT systems help prolong the lifespan of wind turbines and other renewable energy systems. By optimizing the turbine's operating conditions, MPPT reduces mechanical stress on components, such as the rotor blades and gearbox, minimizing wear and tear. This leads to fewer breakdowns, reduced maintenance costs, and improved system reliability over time (Chandra Mahato *et al.*, 2024).

9.5 Challenges in battery charge controllers integration

BCCs play a crucial role in regulating the charging and discharging of batteries in energy systems, especially in renewable energy applications such as wind and solar power setups. They protect batteries from overcharging, prevent deep discharges, and optimize battery usage to enhance system reliability, enhance the efficiency and lifespan of batteries (Liu *et al.*, 2019). However, integrating BCCs into different types of energy systems present unique challenges. This chapter explores the key challenges in BCC integration for both off-grid and grid-tied systems, along with issues related to overcharging and deep discharge protection, and battery longevity.

9.5.1 Challenges in off-grid systems

Off-grid systems rely entirely on stored energy in batteries to power loads when there is no direct energy source. Integrating BCCs into these systems presents several challenges.

9.5.1.1 Energy management and demand–supply mismatch

The BCC in off-grid systems have to make a balance between the demand for available energy supply and demand, which can be highly variable, especially in renewable energy setups. Solar and wind power, in particular, are intermittent and unpredictable. This makes it difficult for the BCC to ensure that the batteries are charged adequately during periods of surplus energy while also preventing overcharging during high production periods (Rodrigues *et al.*, 2015). Furthermore, during low-energy production, the BCCs must ensure that the batteries can supply enough energy without deep discharge. Managing this mismatch between demand and supply can lead to inefficiencies, and over time, the battery's lifespan may be reduced if the system regularly experiences these extreme fluctuations. BCCs in off-grid systems must be programmed with sophisticated algorithms to handle these scenarios effectively, but designing and implementing such controls is complex.

9.5.1.2 Load prioritization and critical load protection

In an off-grid system, energy availability is limited. During periods of low energy production or low battery charge, the BCC must decide which loads are critical and should remain powered versus which loads should be shed to preserve energy. This load prioritization requires advanced energy management systems integrated with the BCC to ensure system efficiency and reliability. Without adequate load management strategies, non-critical loads may drain the batteries, leaving insufficient power for critical devices like medical equipment, communication systems, or refrigeration. Integrating BCCs with effective load prioritization can be challenging, as it requires the system to constantly monitor battery status and prioritize loads dynamically (Ayodele *et al.*, 2017). This also involves designing BCCs that are adaptable to user needs, which can be highly varied depending on the off-grid application.

9.5.2 Challenges in grid-tied systems

In grid-tied systems, where batteries are used in conjunction with the main electrical grid, the role of the BCC is slightly different, but there are still significant challenges.

9.5.2.1 Grid interaction and synchronization

In grid-tied systems, the BCC must not only manage the batteries but also ensure smooth interaction with the grid. When the grid is functioning normally, the BCC helps manage battery charging from excess renewable energy, storing surplus power for later use or grid feedback. However, during grid outages or periods of instability, the BCC must ensure that the batteries can seamlessly take over to supply power. One major challenge is ensuring proper synchronization between the grid and battery systems. The BCC must manage the flow of energy between the grid and the batteries without causing instability, especially during grid faults or blackouts (Lenka *et al.*, 2022). Additionally, the BCC needs to ensure that when power is exported to the grid, it meets the grid's voltage and frequency requirements. Designing BCCs that can handle these complex interactions reliably requires advanced control systems and robust testing.

9.5.2.2 Bidirectional power flow and grid regulations

Another challenge in grid-tied systems is managing bidirectional power flow. Modern energy systems with battery storage often allow for power to flow both to and from the grid. The BCC must manage not only the charging and discharging of the batteries but also ensure that any power sent to the grid complies with regulatory requirements, such as those regarding grid safety, frequency control, and voltage standards. Grid regulations vary by region, and integrating BCCs that meet these standards can be challenging (Mohamed and Mohammed, 2019). Moreover, as grid operators increasingly require systems to provide services like frequency regulation or voltage support, the BCC must be capable of sophisticated grid interaction, further complicating integration. Table 9.3 presents the difference between the challenge areas in off-grid and on-grid systems.

Table 9.3 Explain the differences in challenge areas between off-grid and on-grid systems

Challenge area	Off-grid system	On-grid system
Energy storage	Requires significant storage (batteries) to manage power needs during low-generation periods.	Typically doesn't require storage as the grid acts as a backup. Batteries are optional for hybrid setups.
Power reliability	Completely dependent on local generation and storage; prone to outages if energy reserves run low.	More reliable due to grid connection; outages only during grid failure.
Power source	Completely reliant on local generation (solar, wind, etc.) and storage.	Relies on renewable sources when available; draws from the grid when needed.
System complexity and maintenance	More complex, requires regular maintenance of batteries, inverters, and renewable systems.	Less complex; requires maintenance mainly for generation components like solar panels.
Maintenance and reliability	Requires more frequent monitoring and maintenance, especially in remote locations.	Less frequent maintenance; grid connectivity ensures higher reliability.
Cost and financial investment	High upfront cost due to energy storage, backup systems, and maintenance.	Lower initial cost as no large energy storage is needed; financial savings via net metering possible.
Environmental impact	Low carbon footprint but potential issues with battery disposal and generator emissions (in hybrid systems).	Low environmental impact, but grid power may come from non-renewable sources.
Feed-in tariff/ net metering	Not applicable since there's no grid connection to feed excess energy.	Can earn revenue or reduce costs by feeding excess energy back to the grid.
Scalability	Scaling requires adding both generation and storage, which is expensive and complex.	Easier to scale; generation capacity can be added without needing storage.
System efficiency	Efficiency depends on system design and battery management; energy losses in storage and inverters.	Generally, more efficient since excess energy can be fed into the grid, reducing energy waste.
Weather and seasonal variability	Highly sensitive to weather; energy shortages possible during cloudy or windless periods.	Weather variability impacts generation, but the grid provides backup during low production periods.
Regulatory and permitting	Fewer regulations as it's not tied to the grid, but may face local zoning and environmental restrictions.	Subject to regulatory requirements for grid connection, net metering, and interconnection standards.

9.5.3 Overcharging and deep discharge protection

One of the primary functions of a BCC is to protect batteries from overcharging and deep discharge, which can significantly degrade battery life and system reliability. However, implementing this protection presents several challenges.

9.5.3.1 Overcharging protection

Overcharging occurs when the battery receives more energy than it can store, which can lead to overheating, electrolyte loss, and even battery failure in extreme cases. The BCC must constantly monitor the SOC of the battery and regulate the charging current to ensure that it remains within safe limits (Pop *et al.*, 2008). However, accurately determining the SOC is challenging, particularly in systems with fluctuating power inputs like solar or wind. Some BCCs rely on voltage measurements to estimate SOC, but these measurements can be inaccurate under load. More advanced methods, such as coulomb counting or integrating battery management systems, can improve accuracy, but they add complexity and cost to the system (Wang *et al.*, 2020).

9.5.3.2 Deep discharge protection

Deep discharges occur when the battery is drained below a certain threshold, which can cause irreversible damage to the battery's internal chemistry, shortening its lifespan. The BCC must prevent the battery from being discharged too deeply by disconnecting non-essential loads or switching the system to an alternative power source (Yarlagadda *et al.*, 2013). However, preventing deep discharge without affecting system availability can be a difficult balance to achieve. In off-grid systems, where the energy supply may be limited, the BCC must prioritize the availability of power while still protecting the battery from harmful deep discharges.

9.5.4 Battery longevity and system reliability

Maximizing battery longevity and ensuring overall system reliability are key goals of BCC integration as shown in Figure 9.3. However, several factors can complicate these objectives.

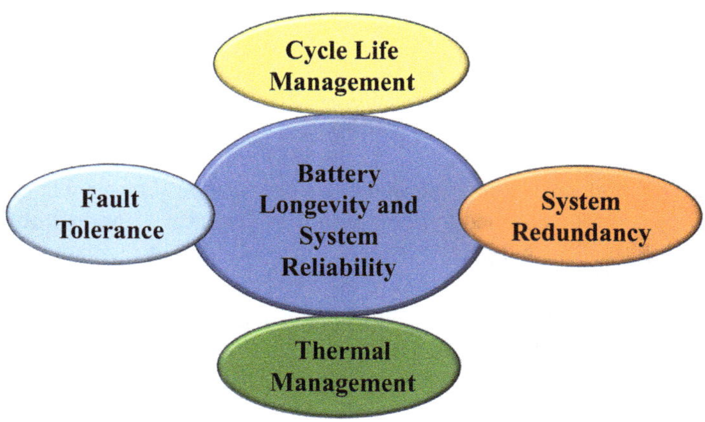

Figure 9.3 Enhancing battery lifespan and maintaining system reliability are primary objectives of BCC

9.5.4.1 Cycle life management

Every time a battery is charged and discharged, it undergoes a cycle, and the number of cycles a battery can complete before its performance degrades is finite. The BCC must manage these charge/discharge cycles in a way that extends battery life, but doing so can be challenging, especially in systems where energy supply and demand are highly variable (Kaiser, 2007). For example, in off-grid systems, frequent cycling may be necessary to meet energy demands during low production periods, leading to accelerated battery wear. The BCC needs to incorporate algorithms that optimize cycling based on current conditions, potentially limiting discharge depth or adjusting charging rates to prolong battery life. However, these algorithms must also ensure that energy availability is not compromised in the process.

9.5.4.2 Thermal management

Battery performance and longevity are highly sensitive to temperature. Overheating can lead to thermal runaway, reducing battery efficiency and lifespan, while very low temperatures can reduce a battery's ability to hold charge. The BCC must therefore monitor battery temperature and, if necessary, activate cooling or heating systems to maintain the battery within its optimal operating range. In some environments, maintaining the correct temperature can be particularly challenging, especially in off-grid systems without access to air conditioning or heating. Designing BCCs that include effective thermal management systems is essential for ensuring both battery longevity and overall system reliability (Kumar *et al.*, 2023).

9.5.4.3 System redundancy and fault tolerance

In critical applications, such as remote off-grid systems or grid-tied systems with frequent outages, ensuring system reliability through redundancy is essential. The BCC must be able to handle faults or failures within the system without causing downtime. For example, if one part of the battery array fails, the BCC should isolate the faulty section and continue operating the rest of the system (Rullo *et al.*, 2019). Designing BCCs with fault-tolerant mechanisms and redundancy can be complex, as it requires advanced diagnostics and control logic to detect and respond to system failures in real time.

9.6 Application of battery charge controllers in different types of wind energy systems

BCCs are essential components in wind energy systems, particularly in setups where battery storage is needed to balance intermittent power production. Whether in off-grid, grid-tied, or hybrid wind energy systems, BCCs regulate the charging and discharging of batteries, enhancing efficiency, protecting battery health, and ensuring consistent energy supply. This chapter explores the application and significance of BCCs in different types of wind energy systems, highlighting their role in optimizing energy storage, supporting load balancing, and improving system reliability (Chatzigeorgiou *et al.*, 2024).

9.6.1 Off-grid wind energy systems

Off-grid wind energy systems are used in diverse applications, including rural electrification, agricultural operations, telecommunications, and isolated industrial facilities. Off-grid wind energy systems offer a sustainable and independent source of electricity in areas where grid connections are unavailable, unreliable, or cost-prohibitive (Kabir *et al.*, 2024). These systems are especially valuable in remote locations, such as rural communities, mountainous regions, islands, and other off-grid sites, providing power without relying on traditional grid infrastructure. An off-grid wind energy system operates independently from the centralized electricity grid, generating and storing power for local use. Unlike grid-tied systems, which can feed excess energy back to the grid, off-grid systems rely solely on energy generated on-site and stored in batteries for times when wind speeds are low as shown in Figure 9.4. Wind power is ideal for off-grid setups in areas with consistent wind resources, making it an efficient solution for generating renewable energy on demand (Agajie *et al.*, 2024; Alhamami *et al.*, 2024). However, due to the intermittent nature of wind, these systems often incorporate battery storage and sometimes hybrid renewable sources, like solar panels, to ensure a steady power supply (Maliat *et al.*, 2024).

An off-grid wind energy system typically includes the following key components:

- Wind turbine: The turbine is the primary generator, converting kinetic energy from the wind into electrical energy. Turbines are selected based on factors like local wind speeds, load requirements, and durability.
- Battery storage: Because wind energy is variable, batteries store excess energy when production exceeds demand. Common battery types used in off-grid systems include lead-acid, lithium-ion, and nickel-cadmium, each with its advantages in terms of lifespan, maintenance, and cost.

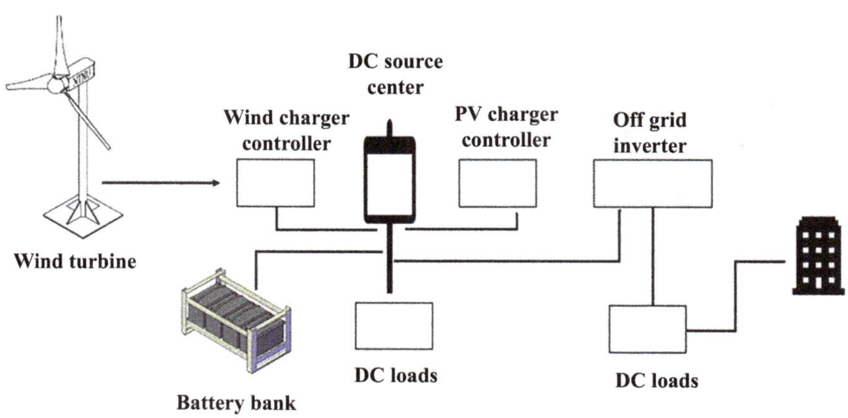

Figure 9.4 Off-grid wind energy systems in battery charge controllers

- Inverter: Inverters convert the direct current (DC) stored in batteries into alternating current (AC), which is the standard for household appliances and electronics. They are essential for enabling the use of generated power within homes or facilities.
- Hybrid integration (optional): Many off-grid wind systems integrate other renewable sources, such as solar panels, to provide a more reliable and balanced energy supply, especially in areas with variable wind conditions.
- Design considerations and sizing.

Designing an off-grid wind energy system requires careful consideration of several factors to ensure it meets local energy demands. Important considerations include:

- Energy demand assessment: Calculating the energy needs of the site is critical. This includes evaluating peak usage times, average daily consumption, and backup requirements during low wind periods.
- Wind resource assessment: Understanding local wind speeds and patterns is crucial, as consistent wind conditions are essential for effective generation. Sites with average wind speeds above 5 m/s are ideal for off-grid systems.
- Battery and storage capacity: The battery bank must be sized to meet demand during low-wind periods, often aiming for 1–3 days of backup energy, depending on usage and local weather patterns.
- System maintenance: Off-grid systems require regular maintenance, particularly for the wind turbine, battery banks, and BCCs. Durability is a key factor for remote installations, where maintenance can be challenging.
- Energy independence: Off-grid wind systems reduce reliance on fossil fuels and external energy supplies, providing reliable power even in remote locations.
- Environmental sustainability: By harnessing wind, off-grid systems offer a clean alternative to diesel generators and other polluting energy sources.
- Cost savings: While initial investment in off-grid systems can be high, they offer long-term savings by eliminating grid electricity bills and reducing the need for expensive fuel transportation to remote sites.

Off-grid wind systems face challenges, including high upfront costs, the need for regular maintenance, and dependency on local wind conditions (Kangaji *et al.*, 2024). Technological advances in battery storage, such as the use of lithium-ion batteries and AI-driven predictive maintenance, are addressing some of these issues and improving system reliability (Zhao *et al.*, 2024). Future developments in hybrid solutions combining wind, solar, and advanced storage will make off-grid systems even more efficient and accessible, extending renewable energy access to more off-grid communities and remote sites worldwide.

9.6.2 Battery charge controllers in grid-tied wind energy systems

BCCs play a vital role in grid-tied wind energy systems, especially when these systems incorporate battery storage (Bukar *et al.*, 2024). While grid-tied wind energy systems can rely on the grid for balancing supply and demand, integrating battery storage adds flexibility, energy security, and economic value. BCCs are essential in this setup, managing the charge and discharge cycles of batteries, preventing overcharging and deep discharging, and ensuring optimal system performance. In grid-tied systems with battery backup, the BCC regulates how much energy is sent to the batteries versus the grid, maximizing energy efficiency and protecting battery health (Bishla and Khosla, 2024; Ishraque *et al.*, 2024). As wind energy generation is variable and depends on wind speed, BCCs also ensure that energy storage can handle fluctuations, enhancing the system's ability to provide consistent power.

9.6.3 Functionality of battery charge controllers in grid-tied systems

BCCs play a crucial role in managing energy flow within wind energy systems, particularly in systems that incorporate battery storage. A primary function of BCCs is charge regulation; they control the electricity flow from the wind turbine to the batteries, ensuring that the batteries receive a stable voltage (Nduwamungu *et al.*, 2024). By managing the amount of energy entering the batteries, BCCs prevent overcharging, which could damage the batteries or shorten their lifespan, a risk especially prominent during periods of high wind speeds when energy generation might exceed storage capacity, as illustrated in Figure 9.5. In addition to charge regulation, BCCs provide voltage and current regulation. They help maintain stable voltage and current levels within the battery storage, preventing surges that could damage sensitive components and ensuring that power directed to the batteries or the grid remains safe and consistent. BCCs also prevent deep discharge of batteries by setting a minimum discharge level, preserving battery health and ensuring that batteries retain enough energy to remain operational during high demand or low wind conditions. In grid-tied systems, BCCs manage energy diversion to the grid. When the batteries are fully charged, excess energy generated

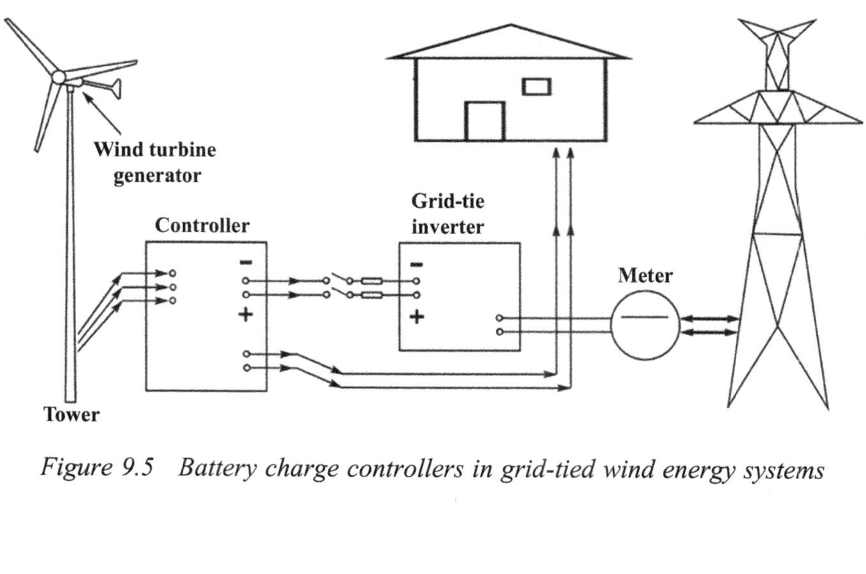

Figure 9.5 Battery charge controllers in grid-tied wind energy systems

by the wind turbine can be sent to the grid. This function not only maintains the batteries at optimal charge levels but also maximizes the system's financial benefits by feeding excess power back into the grid.

9.7 Advantages of battery charge controllers in grid-tied wind systems

Enhanced energy reliability: With BCCs managing battery storage, grid-tied systems can provide backup power during grid outages or peak demand periods. This capability enhances energy reliability and ensures an uninterrupted power supply for essential loads.

- Financial benefits: Many regions offer net metering or other incentives, allowing users to earn credits or revenue for feeding excess power into the grid. By managing energy flow efficiently, BCCs allow the system to maximize these financial benefits.
- Load management and grid stability: BCCs enable grid-tied systems to store excess energy and dispatch it during peak demand periods, which reduces strain on the grid and can help stabilize energy supply in local communities.
- Battery longevity: Properly managed charge and discharge cycles, overseen by BCCs, help extend battery life. By preventing overcharging and deep discharging, BCCs reduce wear on the batteries, lowering replacement costs and improving the system's long-term efficiency.
- Despite their advantages, BCCs in grid-tied systems face certain challenges, including high initial costs and the complexity of installation. However, as battery technology advances and costs decrease, BCCs are becoming increasingly viable. Additionally, the rise of smart grid technology, which allows for better integration and management of distributed energy resources, is expected to increase the adoption of BCCs in grid-tied wind systems. Future innovations, such as AI-driven energy management and enhanced remote monitoring, promise to further optimize the performance of BCCs. These advancements will allow BCCs to not only manage battery storage more efficiently but also respond dynamically to grid conditions, making grid-tied wind energy systems more reliable, cost-effective, and adaptable to changing energy demands. BCC is an indispensable component of grid-tied wind energy systems with battery storage. They ensure efficient energy use, improve reliability, and extend battery life, making these systems both environmentally and economically sustainable (Gunjal *et al.*, 2024).

9.8 Future trends in battery charge controller technology for wind energy

BCCs are pivotal in wind energy systems, particularly as demand grows for efficient and resilient renewable energy solutions. The evolution of BCC technology is focused on enhancing energy efficiency, system reliability, and integration with smart grids. With the rise of advanced materials, artificial intelligence (AI), IoT capabilities, and next-generation power electronics, BCC technology is evolving to meet the challenges and demands of modern wind energy systems (Ranjan *et al.*, 2024). This chapter explores future trends in BCC technology and their implications for wind energy applications.

9.8.1 AI-driven smart controllers

One of the most significant future trends in BCC technology is the integration of AI. AI-driven BCCs use ML algorithms to monitor and predict energy generation, storage, and consumption patterns. By analyzing historical and real-time data, AI-enhanced BCCs can make dynamic adjustments to maximize energy efficiency (Mittal *et al.*, 2024). They also predict wind availability and user consumption patterns, allowing for more intelligent charge and discharge cycles, which improve battery lifespan and energy storage efficiency.

Predictive maintenance: AI can detect early signs of component wear or failure, allowing for proactive maintenance. Predictive maintenance reduces downtime and extends the life of BCCs and battery systems, ultimately lowering operational costs (Arevalo *et al.*, 2024).

Real-time decision-making: AI-driven BCCs optimize energy flow in real-time, prioritizing battery charging or grid feed based on demand, wind conditions, and energy prices (Rauf *et al.*, 2024). This capability makes them ideal for hybrid or grid-tied systems, where energy dynamics are more complex.

9.8.2 Internet of Things and remote monitoring

The integration of IoT technology into BCCs allows for enhanced remote monitoring, control, and diagnostics (Supriya, 2024). IoT-enabled BCCs collect and transmit data on battery performance, energy levels, weather conditions, and system health to operators who can monitor the system remotely via mobile or web applications (Abbasi *et al.*, 2024). The benefits of IoT in BCC technology are as follows:

> Enhanced reliability and system control: Operators can adjust settings, check battery status, and manage energy flow remotely, which is particularly useful for wind energy systems in remote or off-grid locations.
>
> Data-driven optimization: IoT-enabled BCCs facilitate data analytics that help optimize system settings based on usage patterns and environmental conditions. These insights allow operators to fine-tune BCC performance for efficiency and durability.

9.8.3 Advanced materials and power electronics

The development of advanced materials and power electronics is leading to more efficient, durable, and cost-effective BCCs. Key advances include:

> Silicon carbide (SiC) and gallium nitride (GaN) Semiconductors: SiC and GaN semiconductors are more efficient and have higher thermal tolerance compared to traditional silicon-based semiconductors (Krishna *et al.*, 2024; Nasr Esfahani *et al.*, 2024). These materials reduce energy loss and

improve the longevity of BCCs, making them more suitable for high-power applications in wind energy systems.

High-density and lightweight designs: New materials also allow for more compact and lightweight BCCs without compromising performance, making them easier to install and maintain in remote or space-constrained wind energy sites.

9.8.4 Hybrid and multi-input BCCs

Future BCCs will increasingly support hybrid systems that combine wind energy with other renewable sources such as solar (Sundarakamath and Natarajan, 2024). Hybrid BCCs manage multiple power inputs, seamlessly balancing the energy flow from various sources to optimize battery charging and system output (Khan *et al.*, 2024). This trend is crucial for areas where wind or solar energy alone may not provide a consistent energy supply.

Optimized load balancing: Hybrid BCCs intelligently switch between energy sources, ensuring batteries are charged from the most efficient source at any given time. This capability is particularly useful in regions with variable weather patterns, as it reduces reliance on any single energy source.

Enhanced grid independence: By integrating multiple energy inputs, hybrid BCCs can store excess energy for use during peak demand or grid outages, making them ideal for off-grid or grid-tied systems with battery backup.

9.8.5 Integration with smart grids and energy storage systems

As energy infrastructure evolves toward smart grids, future BCCs will be designed to integrate seamlessly with these advanced networks. Smart grid-compatible BCCs can communicate with the grid to optimize energy flow, adapt to grid demand, and contribute to load balancing (Inci *et al.*, 2024; Khalid, 2024). This integration enhances the role of wind energy in the grid by:

Grid support services: Smart grid-compatible BCCs can assist in frequency and voltage regulation, reducing grid instability and ensuring efficient energy distribution. They allow wind systems to contribute actively to grid resilience and stability.

Demand response capabilities: BCCs connected to the smart grid can respond to pricing signals, storing energy when prices are low and discharging when prices are high. This demand response capability maximizes economic returns for wind energy system owners and supports grid management.

9.9 Conclusion

The role of BCCs in wind energy systems is foundational for achieving sustainable, reliable, and efficient renewable energy solutions. BCCs enable intelligent energy management by optimizing battery storage, balancing load demands, and protecting batteries from overcharging and deep discharge. With advancements like MPPT and PWM BCCs enhance energy capture, adapting dynamically to variable wind conditions. The integration of digital tools, including AI-driven decision-making and real-time monitoring, is revolutionizing BCC functionality, enabling predictive maintenance and maximizing performance. As smart grids, IoT, and hybrid energy models continue to evolve, BCC technology stands poised to empower future renewable energy systems with unmatched resilience and efficiency. The innovations in BCCs not only support the expansion of wind energy but also catalyzed a transition toward a more sustainable, interconnected, and intelligent energy landscape, making wind energy an even more vital part of our global energy mix.

References

Abbasi, O.U.R., Bukhari, S.B.A., Iqbal, S., *et al.*, 2024. Energy management strategy based on renewables and battery energy storage system with IoT enabled energy monitoring. *Electrical Engineering*, 106(3), pp. 3031–3043.

Abdelwahab, S.A.M., El-Rifaie, A.M., Hegazy, H.Y., Tolba, M.A., Mohamed, W. I., and Mohamed, M., 2024. Optimal control and optimization of grid-connected PV and wind turbine hybrid systems using electric eel foraging optimization algorithms. *Sensors*, 24(7), p. 2354.

Agajie, E.F., Agajie, T.F., Amoussou, I., *et al.*, 2024. Optimization of off-grid hybrid renewable energy systems for cost-effective and reliable power supply in Gaita Selassie Ethiopia. *Scientific Reports*, 14(1), p. 10929.

Alhamami, A.H., Abba, S.I., Musa, B., *et al.*, 2024. Off-grid multi-region energy system design based on energy load demand estimation using hybrid nature-inspired optimization algorithms. *Energy Conversion and Management*, 315, p. 118766.

Arevalo, P., Ochoa-Correa, D., and Villa-Ávila, E., 2024. A systematic review on the integration of artificial intelligence into energy management systems for electric vehicles: Recent advances and future perspectives. *World Electric Vehicle Journal*, 15(8), p. 364.

Ayodele, T.R., Ogunjuyigbe, A.S.O., Akpeji, K.O., and Akinola, O.O., 2017. Prioritized rule-based load management technique for residential building powered by PV/battery system. *Engineering Science and Technology, An International Journal*, 20(3), pp. 859–873.

Bishla, S., and Khosla, A., 2024. A survey of fast charging systems in electrical vehicles using solar and grid sources. *Journal of Electrical Systems*, 20(7s), pp. 4084–4105.

Brown, C., 2023. Predictive modeling for energy management. *IEEE Energy Conference Proceedings*, 12, pp. 345–352.

Brown, R., and Black, T., 2022. Advanced techniques in energy storage management. *Energy Storage Materials*, 32, pp. 17–28.

Bugdayci, M., and Yilmazoglu, M., 2024. Wind energy hybrid power generation system with hydrogen storage. In Inamuddin, Altalhi, T., and Luqman, M. (eds.) *Wind Energy Storage and Conversion: From Basics to Utilities*. Beverly, MA: Scrivener Publishing, pp. 115–132.

Bukar, A.L., Chaitusaney, S., and Kawabe, K., 2024. Optimal design of on-site PV-based battery grid-tied green hydrogen production system. *Energy Conversion and Management*, 307, p. 118378.

Chandra Mahato, G., Roy Choudhury, T., and Nayak, B., 2024. Investigation on life span assessment of PV system components under FPPT/MPPT operation. *Electrical Engineering*, 106, pp. 1–16.

Chatzigeorgiou, N.G., Theocharides, S., Makrides, G., and Georghiou, G.E., 2024. A review on battery energy storage systems: Applications, developments, and research trends of hybrid installations in the end-user sector. *Journal of Energy Storage*, 86, p. 111192.

Chellakhi, A., Beid, S.E., and Abouelmahjoub, Y., 2024. A novel theta MPPT approach based on adjustable step size for photovoltaic system applications under various atmospheric conditions. *Energy Systems*, 15(1), pp. 89–114.

Chen, M., 2023. Advances in battery charge controllers. *IEEE Transactions on Power Electronics*, 38(8), pp. 4500–4512.

Doe, A., and Johnson, L., 2019. The role of battery charge controllers in renewable energy systems. *Journal of Renewable Energy*, 12(3), pp. 45–58.

Dong, L., Tang, Q., Zhao, C., *et al.*, 2024. Flag-type hybrid nanogenerator utilizing flapping wakes for consistent high performance over an ultra-broad wind speed range. *Nano Energy*, 119, p. 109057.

El Ouali, A., Lakhal, Y., Benchagra, M., Chojaa, H., and Samagassi, S., 2024. Evaluating the impact of drivetrain vibrations on MPPT control performance in horizontal axis wind turbines. In *E3S Web of Conferences*, 582, p. 03001. London: EDP Sciences.

Garcia, P., 2023. "The role of wind energy in a sustainable future," *International Journal of Renewable Energy*, 5, pp. 213–229.

Green, M., 2021. *Smart Grids and Energy Management*. Berlin: Springer.

Gunjal, M., Mathew, L., and Sudha Letha, S., 2024. Electric vehicle controllers for sustainable energy management: Recommendations and challenges. *Environment, Development and Sustainability*, 26(9), pp. 21845–21875.

Inci, M., Çelik, Ö., Lashab, A., Bayındır, K.Ç., Vasquez, J.C. and Guerrero, J.M., 2024. Power system integration of electric vehicles: A review on impacts and contributions to the smart grid. *Applied Sciences*, 14(6), p. 2246.

Ishraque, M.F., Rahman, A., Shezan, S.A., *et al.*, 2024. Design optimization of a grid-tied hybrid system for a department at a university with a dispatch strategy-based assessment. *Sustainability*, 16(7), p. 2642.

Kabir, M.H., Newaz, S.S., Kabir, T. and Howlader, A.S., 2024. Integrating solar power with existing grids: Strategies, technologies, and challenges and review. *International Journal of Science and Engineering*, 1(2), pp. 48–62.

Kaiser, R., 2007. Optimized battery-management system to improve storage lifetime in renewable energy systems. *Journal of Power Sources*, 168(1), pp. 58–65.

Kangaji, L.M., Raji, A. and Orumwense, E., 2024. Optimizing sustainability offshore hybrid tidal-wind energy storage systems for an off-grid coastal city in South Africa. *Sustainability*, 16(21), p. 9139.

Khalid, M., 2024. Smart grids and renewable energy systems: Perspectives and grid integration challenges. *Energy Strategy Reviews*, 51, p. 101299.

Khan, N., Ooi, C.A., Alturki, A., Amir, M. and Alharbi, T., 2024. A critical review of battery cell balancing techniques, optimal design, converter topologies, and performance evaluation for optimizing storage system in electric vehicles. *Energy Reports*, 11, pp. 4999–5032.

Khan, M.K., Zafar, M.H., Riaz, T., Mansoor, M., and Akhtar, N., 2024. Enhancing efficient solar energy harvesting: A process-in-loop investigation of MPPT control with a novel stochastic algorithm. *Energy Conversion and Management: X*, 21, p. 100509.

Kim, T., 2023. Digital transformation in energy systems. *Journal of Digital Engineering*, 15(1), pp. 98–109.

Krishna, T.N.V., Kumar, S.V.P.D., Srinivasa Rao, S., and Chang, L., 2024. Powering the future: Advanced battery management systems (BMS) for electric vehicles. *Energies*, 17(14), p. 3360.

Kumar, R.R., Bharatiraja, C., Udhayakumar, K., Devakirubakaran, S., Sekar, S., and Mihet-Popa, L., 2023. Advances in batteries, battery modeling, battery management system, battery thermal management, SOC, SOH, and charge/discharge characteristics in EV applications. *IEEE Access*, 11, pp. 105761–105809.

Lee, K. 2018. Predictive energy management for wind turbine systems. *International Journal of Energy Research*, 42(7), 2314–2326.

Lee, R., 2022. Safety and efficiency in battery storage systems. *Journal of Energy Storage*, 11(2), pp. 67–75.

Lenka, R.K., Panda, A.K., Dash, A.R., Senapati, L., and Tiwary, N., 2022. A unified control of grid-interactive off-board EV battery charger with improved power quality. *IEEE Transactions on Transportation Electrification*, 9(1), pp. 920–933.

Liu, K., Hu, X., Yang, Z., Xie, Y., and Feng, S., 2019. Lithium-ion battery charging management considering economic costs of electrical energy loss and battery degradation. *Energy Conversion and Management*, 195, pp. 167–179.

Maliat, A., Kotian, S., and Ghahremanlou, D., 2024. Assessment of a hybrid renewable energy system incorporating wind, solar, and storage technologies in Makkovik, Newfoundland and Labrador. *Journal of Sustainability for Energy*, 3(2), pp. 87–104.

Mittal, A., Dumka, L., Kharka, K.P.S., Soni, M., and Goyal, H.R., 2024. Smart energy: Artificial intelligence (AI) in charging and battery management systems. In *2024 5th International Conference on Intelligent Communication Technologies and Virtual Mobile Networks (ICICV)* (pp. 68–73). IEEE.

Mohamed, A.A. and Mohammed, O., 2019. Bilayer predictive power flow controller for bidirectional operation of wirelessly connected electric vehicles. *IEEE Transactions on Industry Applications*, 55(4), pp. 4258–4267.

Murugesan, R. and Ramasamy, K., 2024. Efficient power control in hybrid solar and wind energy systems with bidirectional converter and advanced MPPT control. *Electric Power Components and Systems*, 19, pp. 1–17.

Nasr Esfahani, F., Darwish, A., and Ma, X., 2024. Design and control of a modular integrated on-board battery charger for EV applications with cell balancing. *Batteries*, 10(1), p. 17.

Nduwamungu, A., Lie, T.T., Lestas, I., Nair, N.K.C., and Gunawardane, K., 2024. Control strategies and stabilization techniques for DC/DC converters application in DC MGs: Challenges, opportunities, and prospects—A review. *Energies*, 17(3), p. 669.

Parker, J., 2022. "Battery storage solutions for wind energy," *Energy Storage Journal*, 14(3), pp. 112–120.

Pop, V., Bergveld, H.J., Danilov, D., Regtien, P.P., and Notten, P.H., 2008. *Battery Management Systems: Accurate State-of-Charge Indication for Battery-Powered Applications* (Vol. 9). Berlin: Springer Science & Business Media.

Ranjan, A., Behera, M.K. and Saikia, L.C., 2024. An advance control of grid integrated wind turbine driven DFIG-battery system with grid power shaping under gust wind variation. *IETE Journal of Research*, 70(3), pp. 3030–3051.

Rashid, S.M., 2024. Employing advanced control, energy storage, and renewable technologies to enhance power system stability. *Energy Reports*, 11, pp. 3202–3223.

Rauf, M., Kumar, L., Zulkifli, S.A., and Jamil, A., 2024. Aspects of artificial intelligence in future electric vehicle technology for sustainable environmental impact. *Environmental Challenges*, 14, p. 100854.

Ren, Y., Sun, K., Zhang, K., *et al.*, 2024. Optimization of the capacity configuration of an abandoned mine pumped storage/wind/photovoltaic integrated system. *Applied Energy*, 374, p. 124089.

Rezaee, S., 2023. Overview of renewable energy. *Renewable and Sustainable Energy Reviews*, 20, pp. 1–15.

Rodrigues, E.M.G., Osório, G.J., Godina, R., *et al.*, 2015. Modelling and sizing of NaS (sodium sulfur) battery energy storage system for extending wind power performance in Crete Island. *Energy*, 90, pp. 1606–1617.

Rullo, A., Serra, E. and Lobo, J., 2019. Redundancy as a measure of fault-tolerance for the Internet of Things: A review. In Calo, S., Bertino, E., and Verma, D. (eds.) *Policy-Based Autonomic Data Governance*. Cham: Springer, pp. 202–226.

Sarang, S.A., Raza, M.A., Panhwar, M., *et al.*, 2024. Maximizing solar power generation through conventional and digital MPPT techniques: A comparative analysis. *Scientific Reports*, 14(1), p. 8944.

Schnabel, A. and McGilton, B., 2024. A comparison of battery charge controller technologies for wave energy converters (No. NREL/CP-5700-90527). *National Renewable Energy Laboratory (NREL)*, Golden, CO, United States.

Smith, H., 2022. Technological advancements in wind turbines, *Wind Energy Advances*, 8(6), pp. 489–496.

Smith, J., 2020. *Renewable Energy Systems: An Overview*. New York: Wiley.

Sundarakamath, R. and Natarajan, S., 2024. Integration of multiple sources for fuel cell hybrid electric vehicles using single inductor multi-input converter. *International Journal of Hydrogen Energy*, 53, pp. 503–516.

Supriya, M., 2024. Design of EV control monitoring of multiple approach fault detection using multi sensor IoT system. In *2024 Ninth International Conference on Science Technology Engineering and Mathematics (ICONSTEM)* (pp. 1–6). IEEE.

Tayebi, A. and Torabi, F., 2024. Flow control techniques to improve the aerodynamic performance of Darrieus vertical axis wind turbines: A critical review. *Journal of Wind Engineering and Industrial Aerodynamics*, 252, p. 105820.

Teklehaimanot, Y.K., Akingbade, F.K., Ubochi, B.C. and Ale, T.O., 2024. A review and comparative analysis of maximum power point tracking control algorithms for wind energy conversion systems. *International Journal of Dynamics and Control*, 12, pp. 3494–3516.

Tella, V.C., Agili, B., and He, M., 2024. Advanced MPPT control algorithms: A comparative analysis of conventional and intelligent techniques with challenges. *European Journal of Electrical Engineering and Computer Science*, 8(4), 6–20.

Wang, Y., Tian, J., Sun, Z., *et al.*, 2020. A comprehensive review of battery modeling and state estimation approaches for advanced battery management systems. *Renewable and Sustainable Energy Reviews*, 131, p. 110015.

Yan, S., and Sun, Y., 2024. Enhancing tool dynamics and stability in internal turning with an adjustable clamping device under variable cutting conditions. *Mechanical Systems and Signal Processing*, 208, p.111007.

Yarlagadda, S., Hartley, T.T., and Husain, I., 2013. A battery management system using an active charge equalization technique based on a DC/DC converter topology. *IEEE Transactions on Industry Applications*, 49(6), pp. 2720–2729.

Zaghba, L., Borni, A., Benbitour, M.K., *et al.*, 2024. Enhancing grid-connected photovoltaic system performance with novel hybrid MPPT technique in variable atmospheric conditions. *Scientific Reports*, 14(1), p. 8205.

Zhao, L., 2023. IoT-enabled monitoring in renewable energy systems. *Smart Energy Systems*, 2, pp. 43–53.

Zhao, J., Qu, X., Wu, Y., Fowler, M., and Burke, A.F., 2024. Artificial intelligence-driven real-world battery diagnostics. *Energy and AI*, 18, p. 100419.

Chapter 10

Field-programmable gate array implementation: power management for wind turbines with battery

Mahdi Hermassi[1,2]*, Saber Krim*[3,4]*, Youssef Kraiem*[5] *and Mohamed Ali Hajjaji*[1,6]

Abstract

This chapter introduces a hardware implementation on the field programmable gate arrays (FPGA) Zed-Board of the power management algorithm for wind turbine associated with a batteries module is considered as an energy storage device and a variable load. This power management system (PMS) optimizes system operation and addresses power disturbances resulting from wind speed intermittency and load changes. First, the proposed PMS ensures a balance at the continuous direct current bus, maintaining equilibrium between the power supplied by the wind turbine and the power demanded by the load. Furthermore, it prevents the batteries from surpassing their maximum or minimum state of charge, maintaining it within an acceptable range of 30%–90%. This proposed methodology effectively manages power flows within a wind energy conversion system (WECS), ensuring stable power delivery to the load while extending the longevity of the battery energy storage system (BESS) device. Second, the real-time implementation on the FPGA Zed-Board of the suggested PMS with BESS for WECS, aimed at reducing control loop delays and processing multiple signals in parallel, offers substantial advantages over conventional DSP-based systems. This positions it as a promising technology for future renewable energy systems. The system's performance is assessed through digital simulation in MATLAB®/ Simulink® using the Xilinx system generator and real-time implementation

[1]Research Laboratory in Algebra Numbers theory and Intelligent Systems (RLANTIS), University of Monastir, Tunisia
[2]Faculty of Sciences of Monastir, University of Monastir, Tunisia
[3]Department of Technology, Higher Institute of Applied Sciences and Technology of Kasserine, University of Kairouan, Tunisia
[4]Laboratory of Automatic, Electrical Systems and Environment, National Engineering School of Monastir, University of Monastir, Tunisia
[5]ESEO-IREENA Lab UR 4642, Nantes University, France
[6]Higher Institute of Applied Sciences and Technology of Sousse, University of Sousse, Tunisia

utilizing an FPGA Zed-Board device. The results underscore the performance of the proposed approach in optimizing the WECS's performance.

Keywords: Wind turbine; battery energy storage system; power management system; FPGA; Xilinx system generator

10.1 Introduction

Energy is a cornerstone of human existence and the bedrock of modern civilization. The socio-economic well-being of contemporary society largely hinges on the availability of renewable and sustainable energy sources [1]. However, the unchecked and rapid expansion of human civilization and industry has profoundly impacted the environment and depleted the world's finite fossil fuel reserves. To mitigate further environmental damage and conserve natural resources, it is imperative to integrate sustainable development principles and criteria into technological processes, products, and activities [2,3]. Consequently, the development of renewable energy sources is receiving increased attention. In the coming years, the global capacity to generate electricity from solar power, wind power, and other renewable technologies is expected to expand rapidly. This growth highlights the critical importance of developing reliable, efficient, and scalable renewable energy systems that can meet the increasing demand for clean energy while reducing dependence on fossil fuels. As renewable energy technologies continue to evolve and mature, advancements in energy storage, grid integration, and smart infrastructure will be essential to ensure stable and uninterrupted power supply [4].

Among renewable energy options, wind energy is a key focus due to its sustainability and environmental friendliness, as well as its significant potential for reducing greenhouse gas emissions and alleviating energy shortages [5]. Nonetheless, the intermittency and weather-dependent nature of wind energy production pose major challenges. These fluctuations can lead to imbalances between power generation and consumption, making it difficult to provide a continuous and reliable energy supply to meet instantaneous load demands. These challenges are particularly critical in isolated or off-grid applications, where energy systems lack the support of centralized power grids.

To address this issue, the optimal solution for isolated sites is to combine multiple energy generation methods with energy storage solutions. This strategy ensures reliable operation while maintaining a balance between energy production and consumption under all conditions [6,7]. There are two main categories of electricity storage systems: long-term and short-term. Long-term storage systems, with time horizons ranging from approximately 10 minutes to several hours or even days, encompass options such as hydrogen storage, air compressors, hydro-pumping, and electrochemical batteries [8]. Short-term storage systems, with storage durations of less than about 10 minutes, encompass inertial flywheels, double-layer electrolyte supercapacitors, and conventional capacitors [9,10]. The battery energy storage system (ESS) is gaining popularity due to its proven technology, favorable performance/cost ratio, and relatively low manufacturing cost compared to alternative storage methods [11,12]. For applications requiring medium-term energy storage, battery ESS is widely regarded as an efficient and reliable solution [13]. When integrated with wind turbines (WT), battery ESS units form a decentralized generation system capable of delivering a stable AC output while synchronizing energy production with consumption. To regulate the power flow within the battery ESS and prevent overcharging or deep discharging, a power management algorithm (PMA) is indispensable. Although controllers based on filtration methods are simple to implement, they fall short in terms of adaptability to dynamic system conditions in real time [14,15]. To address these challenges, an optimal PMA is designed to minimize dynamic stresses on the battery ESS, thereby enhancing its durability and operational lifespan. This chapter introduces a methodology for monitoring an AC standalone wind energy conversion system (WECS), based on a permanent magnet synchronous generator (PMSG) and integrated with a battery ESS. This approach facilitates the adaptation of wind energy output to load demand, optimizes output power, and ultimately aims to produce a highly efficient wind system capable of ensuring a constant and uninterrupted power supply. This approach dynamically adjusts the wind source production to meet the load demand, optimizing output power and establishing a reliable wind system. The outcome of this optimization process ensures a continuous and stable supply to the load, mitigating fluctuations and contributing to the development of a resilient and efficient standalone renewable energy system.

Electrical systems are commonly designed using two primary programmable technologies: field programmable gate arrays (FPGAs) and digital signal processors (DSPs). While DSPs are widely used in signal processing applications, they face inherent limitations due to their sequential execution nature, which can result in slower processing times when handling complex algorithms. This sequential limitation often hampers the efficiency of DSPs in real-time applications, particularly those that require high-speed computation. As a solution to these challenges, FPGAs are frequently preferred, as they provide distinct advantages that significantly enhance system performance and development flexibility [16,17]. To address these limitations, FPGAs are often preferred [18]. The use of FPGAs offers several key advantages: (i) the ability to facilitate rapid prototyping, (ii) significantly reduced execution times enabled by parallel processing, and (iii) the capacity to implement complex algorithms that demand substantial computational power. Consequently, FPGAs have been used in various research projects [19,20]. Algorithms can be implemented on FPGAs using two primary methods: Xilinx System Generator (XSG) and Very-High-Speed Integrated Circuit Hardware Description Language (VHDL) programming. In the proposed approach, VHDL code is automatically generated through the XSG tool. XSG is chosen for its flexibility, ease of use, and ability to accelerate the development process [21,22].

In this context, the contributions of this chapter are twofold. The first contribution is the development of an advanced PMA designed to efficiently manage power flow within the battery ESS. This PMA ensures the reliable operation of the system by preventing overcharging and excessive discharging, while also minimizing dynamic stresses, thereby enhancing the operational lifespan of the battery ESS. The second contribution is the design and real-time implementation of the proposed PMA on an FPGA board using the XSG tool. The FPGA is chosen for its parallel processing capabilities, which significantly reduces execution time, ensuring high-speed performance and real-time operation, critical for dynamic power management in renewable energy systems.

10.2 Studied system and modeling

The WECS examined in this chapter is depicted in Figure 10.1. It consists of three significant components. The first component is a WECS that includes a PMSG, an AC-to-DC three-phase rectifier (denoted as Converter 1 in Figure 10.1), and a WT with three blades, designed to deliver a nominal power output of 3.5 kW. The second component is a battery ESS chain, consisting of an inductive filter, a battery bank, and a DC-to-DC reversible converter (named as Converter 3 in Figure 10.1). The third component consists of an AC load, a common DC bus, a DC-to-AC three-phase inverter (referred as Converter 2 in Figure 10.1), and a resistor-inductor-capacitor (RLC) filter that reduces harmonics generated by the inverter.

The equations for aerodynamic power and torque are influenced by the characteristics of the system used to capture energy from the wind. However, aerodynamic power can generally be calculated using the formula [23]:

$$P_{aer} = \frac{1}{2}\pi R^2 v^3 C_p(\lambda, \beta) \tag{10.1}$$

$$T_{aer} = \frac{P_w}{\Omega_m} = \frac{1}{2}\rho\pi R^3 v^2 C_p(\lambda, \beta)/\lambda \tag{10.2}$$

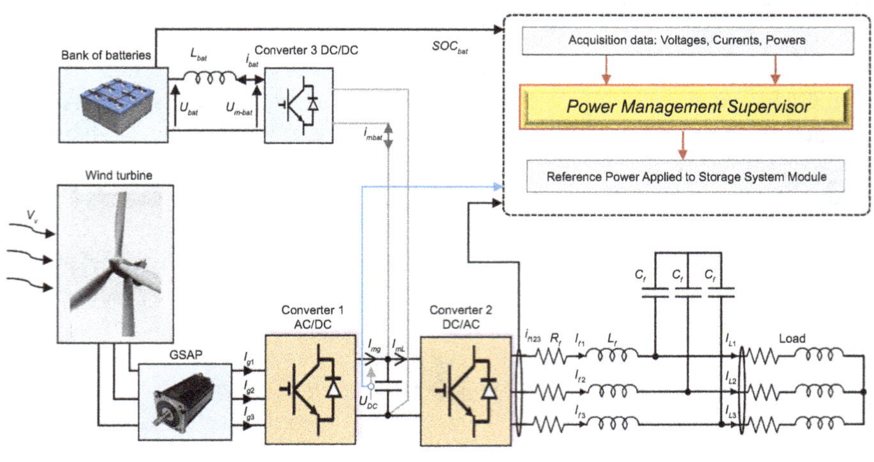

Figure 10.1 Architecture of the studied system

where Ω_m is the mechanical speed, which is extracted by the following dynamic equation [24]:

$$J\frac{d\Omega_m}{dt} = T_{aer} - T_{em} - f\Omega_m \tag{10.3}$$

with J and f represent, respectively, the inertia and the friction coefficient, and T_{em} is the electromagnetic torque.

The power coefficient, C_p (λ, β), which represents the turbine's aerodynamic performance, is given by (10.4). This coefficient is a function of the tip-speed ratio (λ) and the pitch angle (β) of the turbine.

$$C_p(\lambda,\beta) = 0.5179\left(\frac{98}{\frac{1}{\lambda+0.089} - \frac{0.035}{\beta^3+1}} - 0.4\beta + 5\right)e^{\frac{20}{\frac{1}{\lambda+0.089} - \frac{0.035}{\beta^3+1}}} + 0.068\lambda \tag{10.4}$$

$$\lambda = \frac{R\Omega_m}{v} \tag{10.5}$$

Theoretically, $C_p(\lambda,\beta)$ is limited to a maximum value of 0.59, which indicates that the maximum amount of wind kinetic energy that may be captured is 59% (Betz limit). In reality, the best wind system does not exceed over 0.49 [24]. Based on (10.5), power reaches its maximum at a specific rotational speed, Ω_m. This speed is then optimized to the optimal value, $\Omega_{m\text{-}opt}$, to achieve the highest efficiency. This improved speed is achieved through the optimization of λ. Figure 10.2 demonstrates the evolution of C_p with λ and β, with an optimal value of the speed ratio (λ_{opt} = 8.15) corresponding to a maximum value of the power coefficient (C_p = 0.4794) with $\beta = 0$, that we can see that the captured power reduces as the pitch angle β increases while the coefficient C_p decreases. The wind system's maximum

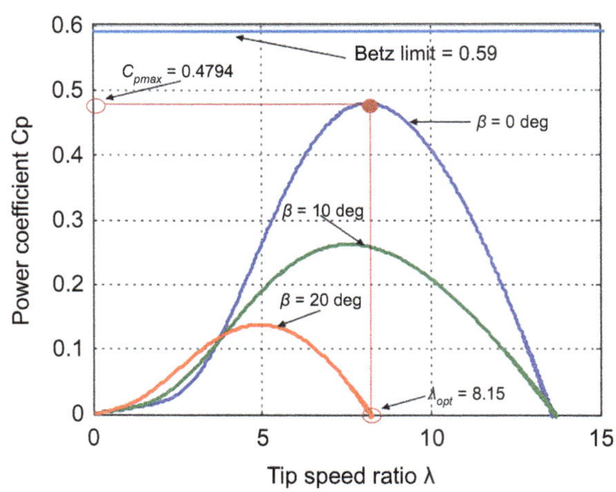

Figure 10.2 Variation of power coefficient (C_p) with TSR (λ) at various pitch angles (β)

power point tracking (MPPT) is achieved by optimizing the tip-speed ratio (λ), which ensures that the power coefficient exceeds $C_{p\text{-}max}$. The estimated value of the following aerodynamic torque produces the MPPT electromagnetic torque [16]:

$$T_{em\_MPPT} = T_{aer\_est} = \frac{P_{aer}}{\Omega_{m\_opt}} = \frac{1}{2} C_{p\_\max} \frac{\rho \pi R^2 v^3}{\Omega_{m\_opt}} \tag{10.6}$$

where the estimated mechanical speed is determined by

$$\Omega_{m\_opt} = \frac{v \lambda_{opt}}{R} \tag{10.7}$$

Therefore, the electromagnetic torque of the MPPT is provided by

$$\begin{cases} T_{em\_MPPT} = \dfrac{1}{2} \dfrac{\rho \pi R^5 C_{p\_\max}}{\lambda_{opt}^{\ 3}} \Omega_{m\_opt}^2 \\ P_{MPPT} = T_{em\_MPPT} \Omega_{m\_opt} \end{cases} \tag{10.8}$$

10.2.1 Modeling of PMSG

The PMSG model is based on physical phenomena such as parameters, uncertainties, and external disturbances operating on the system. The PMSG modeling makes simplified assumptions such as the magnetic circuit being unsaturated, the lack of consideration for space harmonics and notch, and the assumption that the magnet circuit is perfectly laminated. Hence, the PMSG is responsible for converting the mechanical energy captured from the wind turbine into electrical energy. Its mathematical representation in the *d–q* reference frame is defined as [17]:

$$\begin{cases} V_{sd} = R_{sn} i_{sd} + L_{sn} \dfrac{d}{dt} i_{sd} - \omega L_{sn} i_{sq} \\ V_{sq} = R_{sn} i_{sq} + L_{sn} \dfrac{d}{dt} i_{sq} + \omega (L_{sn} i_{sq} + \psi_{mn}) \\ T_{em} = p \psi_{mn} i_{sq} \end{cases} \tag{10.9}$$

where ω represents the electrical speed, p denotes the number of pole pairs, L_s is the stator inductance, ϕ_m is the permanent magnet flux (Wb), and R_s is the stator resistance.

10.2.2 Modeling of the battery ESS

The lead-acid battery is an essential component of the WECS and is characterized by a high energy density and good performance, making it a cost-effective option. The electrical models used in this work allow for the determination of the *SOC* and voltage levels of the battery, providing important information for effective energy management.

The model considered in Figure 10.3 consists of an internal resistance R_i, a series connection of n_b cells, and a voltage source $E_{battery}$. The following expression

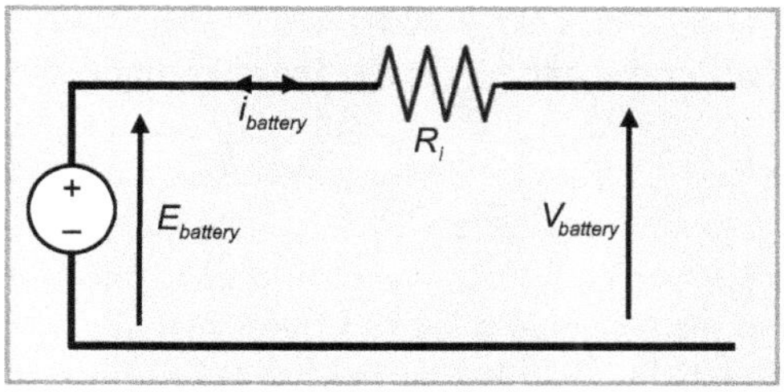

Figure 10.3 Battery model

gives the nominal voltage of the battery ESS and its *SOC* [25]:

$$\begin{cases} V_{battery} = n_b E_{battery} + n_b R_i i_{battery} \\ SOC = 100\left(1 - \dfrac{Q}{C_{battery}}\right) \end{cases} \tag{10.10}$$

The capacity of the battery ESS, denoted as $C_{battery}$, determines the amount of energy the system can store and recover based on the average discharge current $i_{battery}$. The discharge current $i_{battery}$ is calculated in relation to the discharge current i_{10}, which is associated with the battery's rated capacity C_{10} [26]:

$$\frac{C_{battery}}{C_{10}} = \frac{1.67}{1 + 0.67\left(\frac{i_{battery}}{i_{10}}\right)^{0.9}}(1 + 0.005\Delta t) \tag{10.11}$$

with $C_0 = 260$ Ah representing the battery's nominal capacity, and Δt corresponds to the temperature change, which is determined by

$$\Delta t = T - 25 \tag{10.12}$$

The lead-acid battery voltage in charge mode is outlined as follows [27]:

$$\begin{aligned} U_{battery\_char} = U_0 - R_i i_{battery\_char} - k\frac{Q_{char}}{i_{battery\_char}t - 0.1Q_{char}} i^*_{battery\_char} \\ - k\frac{Q_{char}}{Q_{char} - i_{battery\_char}t} i^*_{battery\_char} + Ae^{-Bi_{battery\_char}t} \end{aligned} \tag{10.13}$$

The lead-acid battery's equation in discharging mode is defined as [27]:

$$\begin{aligned} U_{battery\_disch} = U_0 - R_i i_{battery\_disch} - k\frac{Q_{disch}}{Q_{disch} - i_{battery\_disch}t} i^*_{battery\_disch} \\ - k\frac{Q_{disch}}{Q_{disch} - i_{battery\_disch}t} i^*_{battery\_disch} + Ae^{-Bi_{battery\_disch}t} \end{aligned} \tag{10.14}$$

where Q_m denotes the missing charge compared to the battery's nominal capacity (Ah) (m = charge (char) or discharge (disch)), the charging and discharging currents of the battery are, respectively, $i_{battery\text{-}char}$ and $i_{battery\text{-}disch}$, U_0 represents the initial battery ESS voltage, and k refers to the polarization constant (V/Ah).

As shown in Figure 10.1, the battery ESS is linked to the DC bus via a DC–DC converter, which is accompanied by an inductor filter. The role of the filter is to smooth the current supplied to the converter's input, minimizing voltage ripples and ensuring stable operation. Additionally, the filter helps to manage the fluctuations between the charging and discharging currents of the battery ESS, preventing excessive current spikes or drops that could affect the system's stability. This arrangement is crucial for maintaining a steady power flow between the battery ESS and the DC bus. The continuous behavior of the filter, which ensures this stability, is represented by

$$L_{battery}\frac{d}{dt}i_{battery} = U_{battery} - U_{m\_battery} \tag{10.15}$$

The duty cycle m_{bat} is defined as the ratio of the transistor's on-time to the converter's switching period. U_s represents the converter output voltage, and the converter's basic equation is as follows:

$$U_s = \frac{m_{battery}}{1 - m_{battery}}U_{m\_battery} \tag{10.16}$$

Equation (10.16) specifies that the output voltage magnitude can either exceed or be lower than the input voltage depending on the value $m_{battery}$. When $m_{battery} > 0.5$, the output voltage is higher than the input voltage, and when $m_{battery} < 0.5$, the output voltage is lower. The bidirectional converter connects the DC bus connection and the battery ESS. As a result, the output voltage of the converter on the DC bus side is 400 V, while the input voltage is the modulated battery voltage $U_{m\text{-}battery}$, which fluctuates within the range of [358 V, 453 V] depending on whether the battery ESS is in charging or discharging mode.

10.2.3 Modeling of load

In this work, an AC load is provided through a decentralized generator. The latter is a balanced/linear three-phase load. It is composed of a resistor (R_L) and an inductor (L_L), which are connected in series and have respective reactive and active powers (Q_L and P_L) determined by the following equation [28]:

$$\begin{cases} R_L = \dfrac{V_c^2 \,.\, P_L}{P_L^2 + Q_L^2} \\ X_L = L_L\omega_L = \dfrac{V_c^2 \,.\, Q_L}{P_L^2 + Q_L^2} \end{cases} \tag{10.17}$$

10.2.4 Modeling of RLC filter

The three-phase voltage source inverter (VSI) converts DC power from the DC bus into AC power for the isolated load. To provide the load with a nearly sinusoidal voltage, an RLC filter is placed between the VSI and the power transmitter. The RLC filter, illustrated in Figure 10.1, is modeled using (10.18) [29], which incorporates mutual inductances

$$\begin{cases} \frac{d}{dt}i_{fd} = -\frac{R_f}{L_f}i_{fd} + \omega_L i_{fq} + \frac{1}{L_f}V_{id} - \frac{1}{L_f}V_{cd} \\ \frac{d}{dt}i_{fq} = -\frac{R_f}{L_f}i_{fq} - \omega_L i_{fd} + \frac{1}{L_f}V_{iq} - \frac{1}{L_f}V_{cq} \\ \frac{d}{dt}V_{cd} = \frac{1}{C_f}i_{fd} + \omega_L V_{cq} - \frac{1}{C_f}i_{Ld} \\ \frac{d}{dt}V_{cq} = \frac{1}{C_f}i_{fq} - \omega_L V_{cd} - \frac{1}{C_f}i_{Lq} \end{cases} \tag{10.18}$$

10.3 Proposed power management algorithm

The suggested PMA is responsible for managing the operation of the WECS, which includes a wind turbine, a load, and a battery ESS. The architecture of the proposed PMA is illustrated in Figure 10.4. The PMA receives information about the

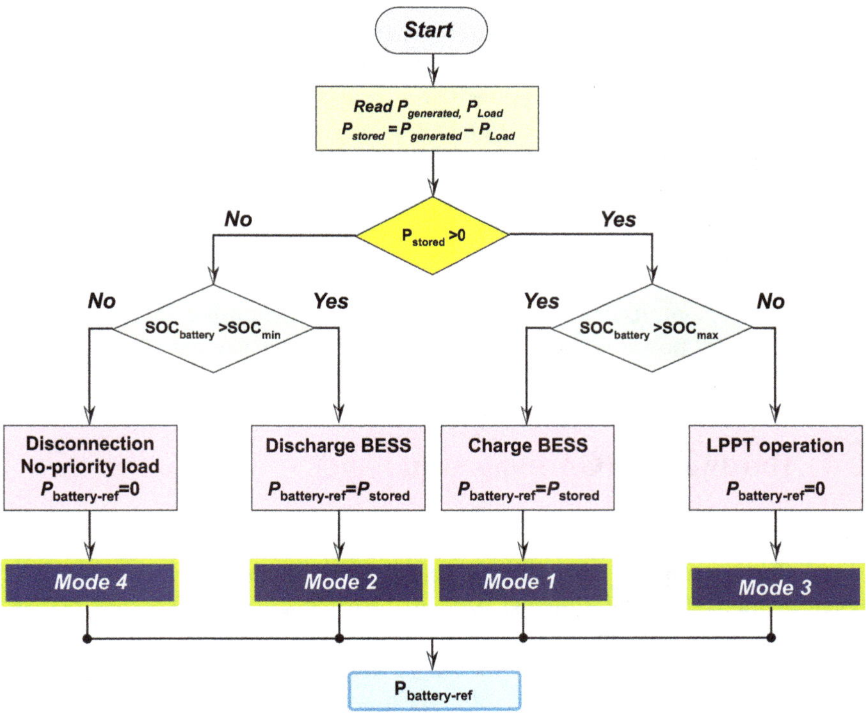

Figure 10.4 Architecture of the proposed PMA

generated power from the wind generator ($P_{Generated}$), the power demand of the load (P_{Load}), and the state of charge (SOC) of the battery *SOC*. Based on this information, the PMA manages the system in four distinct operating modes:

- **Mode 1 (charging mode)**: When the generated power is higher than the load demand and the battery is not at its maximum SOC, the battery is charged. The reference power applied to the storage system module is set to the difference between the generated power and the load demand ($P_{Battery\text{-}ref} = P_{Generated} - P_{Load}$). In this mode, the wind generator operates in MPPT mode to maximize the power output.
- **Mode 2 (discharge mode)**: When the generated power fails to meet the load demand and the battery is not at its minimum SOC, the battery will discharge to provide the necessary power, thereby maintaining equilibrium between generation and consumption.
- **Mode 3 (limited power point tracking (LPPT) mode)**: When the battery is fully charged, and the generated power exceeds the load demand, the wind generator operates in an LPPT mode, producing only the amount of power required by the load. This is achieved by limiting the power output of the wind generator using an angle augmentation (β).

$$\begin{cases} \beta_{ref} = 0 & if \quad k_d = 0 \\ \beta_{ref} = \dfrac{\Delta\beta}{\Delta P}(P_{Generated} - P_{Load}) & if \quad k_d = 1 \end{cases} \tag{10.19}$$

- **Mode 4 (load management mode)**: When the generated power is insufficient to meet the load demand and the battery is at its minimum charge, non-priority loads are disconnected to prevent further depletion of the battery. A load management relay continuously monitors active power consumption and disconnects non-essential loads when consumption surpasses a predefined threshold. After a designated period, these loads are reconnected to restore normal system operation. This mechanism helps optimize power usage and ensures that critical loads maintain priority during periods of high demand.

The PMA controls the operation of the WECS to ensure the efficient and reliable supply of power to the load while maximizing the use of renewable energy sources and optimizing the use of the battery storage system.

10.4 Hardware FPGA implementation

10.4.1 XSG presentation

Xilinx offers a comprehensive suite of tools to facilitate the design, simulation, and implementation of digital systems. Among these tools, the XSG stands out as a sophisticated graphical environment that significantly streamlines the process of creating high-performance systems for Xilinx FPGA platforms. XSG is specifically designed for FPGA-based designs, offering a wide array of pre-built blocks that represent different processing tasks, which can be directly compiled into FPGA hardware. This allows for a seamless transition from system-level design to hardware implementation, making XSG an ideal tool for accelerating FPGA development. XSG integrates seamlessly with the MATLAB®/Simulink® environment, enabling users to leverage the powerful simulation and modeling capabilities of Simulink while working within the Xilinx FPGA ecosystem. The graphical interface allows designers to create architectures by connecting blocks, which are then translated into hardware descriptions. This integration also supports rapid prototyping, as users can simulate their designs within Simulink and quickly test various system configurations before moving on to the hardware implementation [30,31].

In this study, XSG was used to design and implement the proposed PMA for the WECS, which incorporates a battery module as an energy storage device and a variable load, all on an FPGA. Initially, XSG blocks from the Simulink package were employed to create the proposed architectures. After validating the design through digital simulation and achieving the desired results, VHDL code was automatically generated using XSG [23]. Moreover, XSG serves as a system-level modeling tool that facilitates FPGA hardware design. MATLAB/Simulink provides a flexible, high-level modeling environment, which is ideal for hardware designs, using high-level abstractions that can be seamlessly compiled into FPGA configurations via simple commands [32]. Within Simulink, we leveraged a specialized library from Xilinx to effectively design the data control and data paths.

To simulate FPGA functionality, Xilinx offers hardware co-simulation techniques for verifying system behavior on FPGA boards using a PC. The design flow involved systematic steps of routing, placement, synthesis, and generation of programming files using Xilinx Vivado tools. Figure 10.5 illustrates the hardware design flow implemented with XSG. Our work utilized Vivado/XSG 2018.1 version integrated with Matlab/Simulink 2017b and a ZedBoard AES-Z7EV-7Z020-G FPGA. After completing the system design and simulations, the bitstream file was generated following VHDL code generation and synthesis. Subsequently, the configuration files were used to program the FPGA.

10.4.2 Design of the architecture PMA using the XSG tool

The proposed PMA is developed using the XSG tool and implemented on the FPGA Zed-Board platform. XSG operates within the Matlab/Simulink environment, employing a block-based methodology to design the PMA through logical, elementary, and mathematical blocks. The XSG architecture description language is essential for ensuring seamless integration with Matlab/Simulink, enabling the creation of virtual operational scenarios. This integration allows for rapid approximation and validation of hardware designs via co-simulation testing, which not only verifies the synthesized control strategy but also significantly reduces the time needed for experimental testing. As a result, the approach enhances the overall efficiency of the implementation process. XSG proves to be an invaluable tool, fully integrated with Simulink, which accelerates the validation and optimization of control strategies, streamlining the entire development process. The hardware architecture design of the proposed PMA, created with XSG, is illustrated in Figure 10.6.

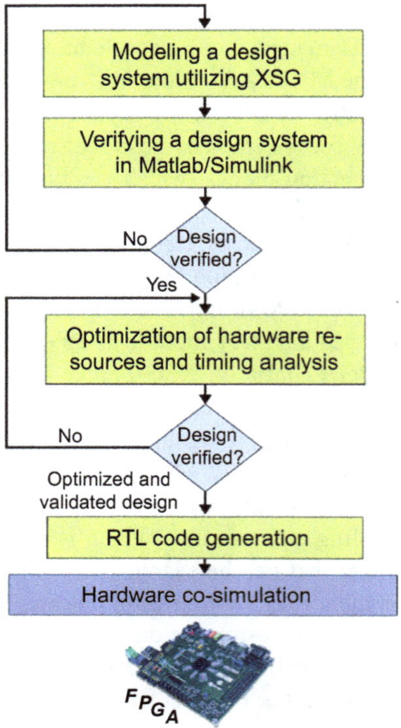

Figure 10.5 Design methodology for XSG tool

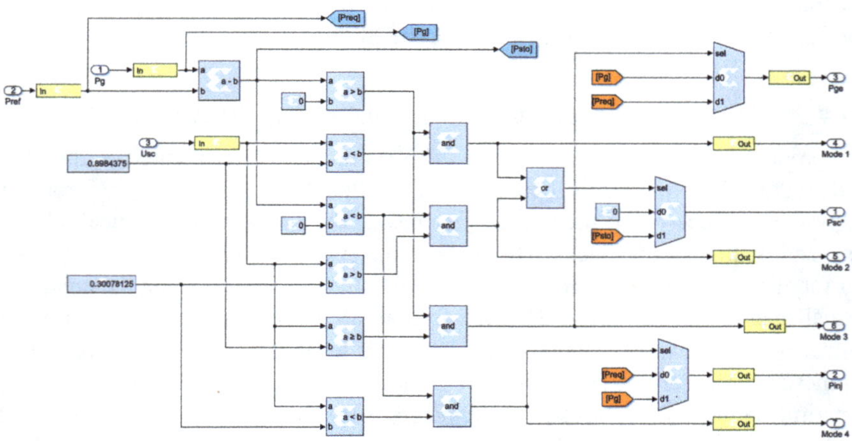

Figure 10.6 Hardware architecture of the proposed PMA with the XSG tool

10.5 Simulation results

The analysis and simulation of the examined system along with the proposed PMA will be showcased and discussed using Matlab/Simulink environment using the XSG tool. Various scenarios studies will be presented, and simulation results will be utilized to assess the robustness and the validity of the suggested approaches. The key modeling parameters employed in this work are summarized in Table 10.1.

10.5.1 Scenario 1

In this scenario, the variations in wind speed (WS) and power demand profiles are illustrated in Figure 10.7(a) and (b).

In this chapter, wind energy is harnessed to supply the power required by the load. The battery ESS plays an essential role as an energy buffer, storing excess power generated during times of high wind availability and discharging when generation is insufficient to meet the load demand. This ensures a continuous and stable power supply, even under fluctuating wind conditions. The primary objective of the battery ESS is to regulate the balance between power production and consumption, adapting to varying generation conditions, and to maintain the stability of the DC bus voltage around its predetermined reference value, denoted as U_{dc}^{*}. By efficiently managing energy storage and discharge, the battery ESS helps mitigate fluctuations in both power supply and voltage, ensuring optimal performance and reliability of the system.

Table 10.1 System parameters

Parameters		Value
Wind turbine	Maximum power coefficient C_{pmax}	0.4794
	Optimal TSR λ_{opt}	8.15
	Radius R	2m
	Nominal power P_{aern}	3.5 kW
	Friction f	0.001 N m s rad^{-1}
	Inertia J	99×10^{-4}kg m^2
	Density of air ρ	1.225 kg m^3
PMSG	Stator inductance L_{sn}	0.0151 mH
	Stator resistance R_{sn}	0.82 Ω
	Nominal power P_{gn}	3.5 kW
	Magnet flux Ψ_{mn}	0.4832 Wb
	Number of pole pairs p	4
RLC filter	Filter resistance R_f	0.2 Ω
	Filter capacitance C_f	10 μF
	Filter inductance L_f	20 mH
Battery	SOC_{min}	0.3
	SOC_{max}	0.9
	Nominal voltage	65 V
	Internal resistance R_i	0.01 Ω
	Nominal capacity C_0	260 Ah

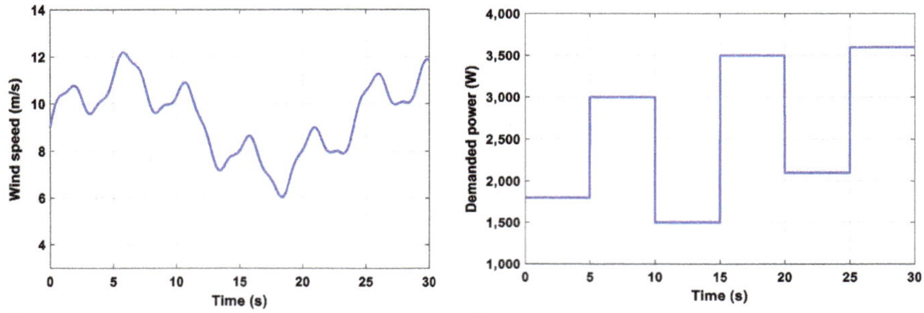

(a) (b)

Figure 10.7 (a) WS profile and (b) active power demand from load

Wind power is harnessed to meet the power demands of the load in the studied system.

The battery ESS acts as a backup, storing excess power during periods of surplus generation and supplying additional energy when there is an insufficient supply. The main goal of the battery ESS is to maintain a delicate equilibrium between power production and consumption under varying conditions, ensuring the stability of the DC bus voltage at its designated reference value, denoted as U_{dc}^*.

Figure 10.8 depicts the transitions among the four operational modes identified by the PMA. The proposed PMA ensures smooth transitions between modes through the use of predefined thresholds, such as wind power, load demand, and battery SOC, which are continuously monitored. These thresholds guide the system in selecting the appropriate mode, ensuring that changes in power flow occur gradually to minimize transient effects. Furthermore, at any given time, only one mode is active, which prevents mode conflicts and ensures uninterrupted operation of the renewable generator. These enhancements contribute to maintaining the overall stability of the system during mode transitions. It is important to highlight that, at any given moment, only a single mode is activated to ensure minimal disruption to the functioning of our decentralized generator.

In the time interval spanning from 0.271 s to 7.622 s, mode 1 is in operation, ensuring that the generated power aligns seamlessly with the demand load. The battery efficiently absorbs any surplus production during this phase. This pattern persists in subsequent intervals: [10 s, 12.177 s], and [21.213 s, 25 s].

Between 12.177 s and 12.585 s, the WECS function in mode 2 ($k_d = 1$). During this period, the generated power exceeds the demanded level, causing the battery ESS to reach its maximum level SOC_{max}, as illustrated in Figure 10.9. Subsequently, the WECS transitions into a limited mode, where it produces only the necessary power to meet the load demand.

In the interval from 0 s to 0.271 s, wind power falls short of meeting the demand load, prompting the activation of mode 3. Therefore, the battery ESS is employed to compensate for the shortfall in power production. This condition also applies to other intervals, including [7.622 s, 10 s], [12.585 s, 18.5 s], and [25 s, 26.340 s].

In the intervals [18.5 s, 24.213 s] and [26.340 s, 30 s], the battery ESS reaches its minimum level SOC_{min}, as depicted in Figure 10.9. Consequently, the WECS

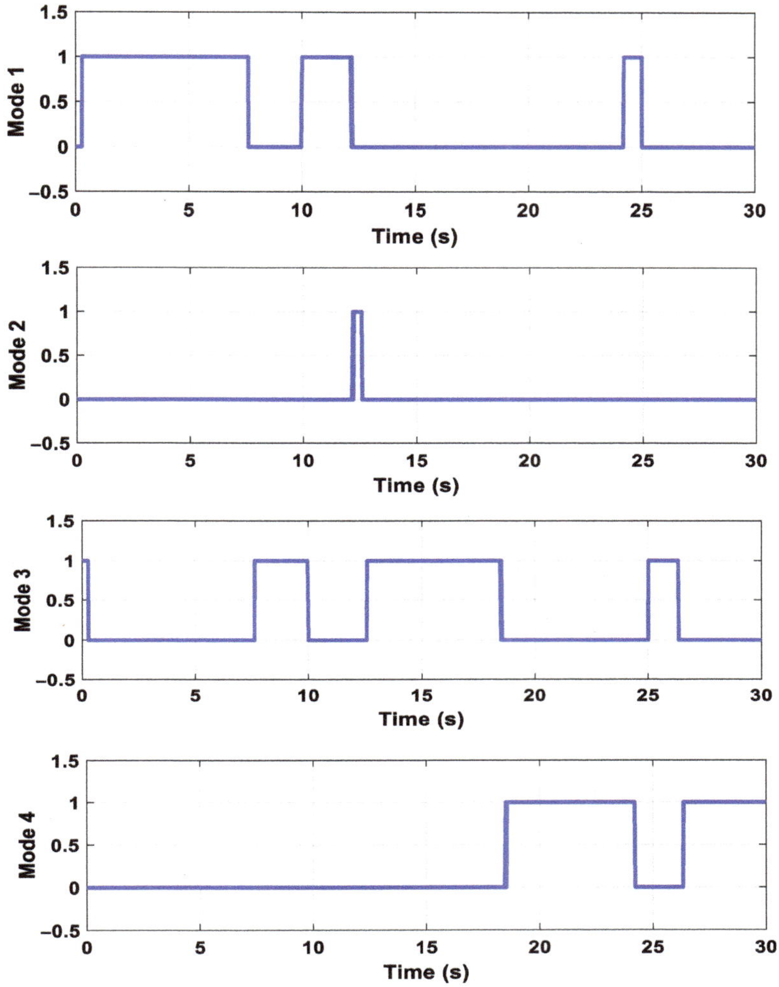

Figure 10.8 Various operating modes of the suggested PMA

becomes incapable of satisfying the power demand of the load. In response, Mode 4 is activated, disconnecting the non-priority load to maintain the battery's SOC at its minimum. Figure 10.10 illustrates the charging and discharging currents of the battery ESS across different operational modes. It is important to note that while the battery's SOC increases during charging (when $i_{battery} > 0$) and decreases during discharging (when $i_{battery} < 0$), the current becomes zero ($i_{battery} = 0$) when the SOC reaches either the maximum SOC_{max} or minimum SOC_{min} limits. This ensures the SOC remains constant at these limits.

Thanks to the proposed PMA, Figure 10.11 demonstrates the successful balancing of power production and consumption across various production scenarios and load fluctuations. For instance, at $t = 6$ s, the power generated by the wind turbine reaches its

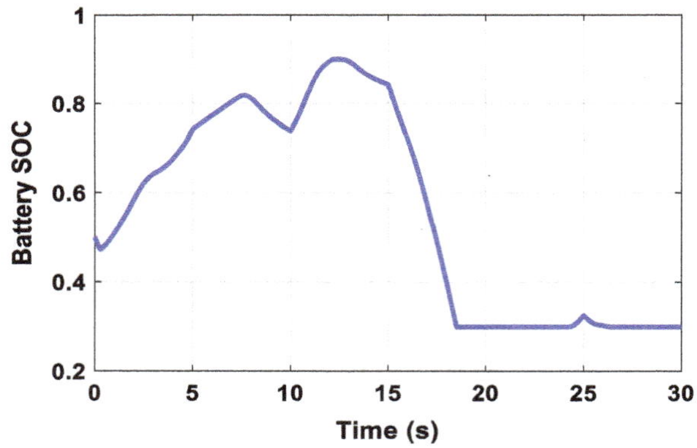

Figure 10.9 Battery ESS SOC

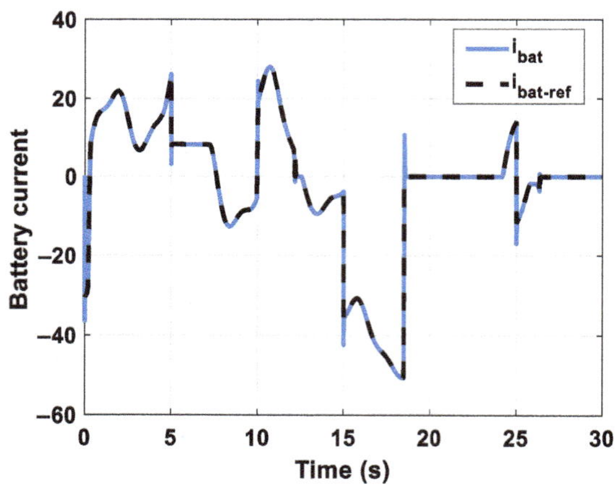

Figure 10.10 Charge and discharge current

nominal peak of 3.5 kW, while the load demand is 3 kW. In this scenario, the wind turbine produces more power than required by the load, with the surplus of 0.5 kW being used to charge the battery ESS. This is reflected in the P_{bat} curve, where a positive value of 0.5 kW indicates the battery is being charged. As shown in Figure 10.8, this situation corresponds to the activation of Mode 1. During the time interval [20, 25 s], the power generated by the wind turbine is insufficient to meet the load demand, and the battery is at its minimum charge level, as shown in Figure 10.9. In this scenario, Mode 4 is activated, as depicted in Figure 10.13. To prevent further depletion of the battery, non-priority loads are disconnected. A load management relay continuously monitors active power consumption and disconnects non-essential loads whenever consumption exceeds a predefined threshold. This ensures that the battery remains operational and prevents further discharge, safeguarding its functionality. This strategy ensures that any surplus energy generated by the wind turbine is stored for future use, optimizing the

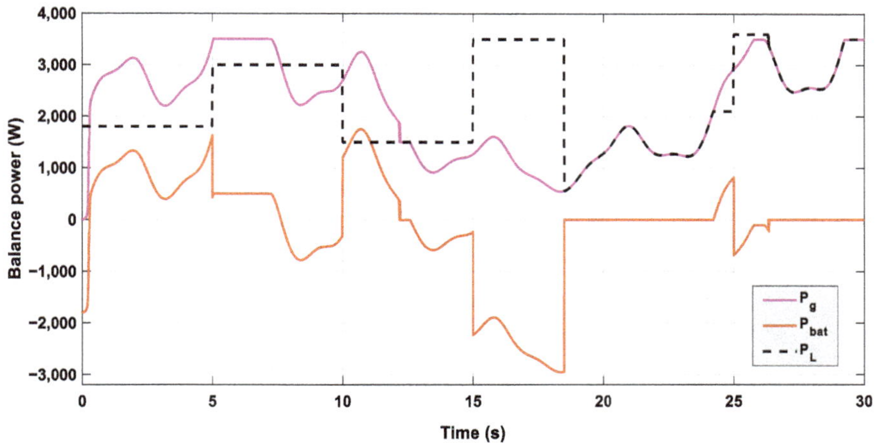

Figure 10.11 Balance power

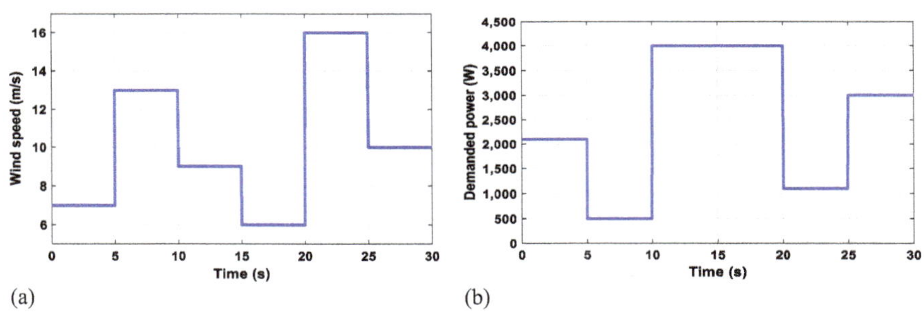

Figure 10.12 (a) WS profile and (b) demanded power

overall system performance and energy management. The PMA ensures a balance between production and consumption, even in the case of unexpected variations. The system monitors key parameters such as wind power generation, load demand, and battery SOC. In the event of variations like load fluctuations or component failures, the PMA adapts by switching between different operational modes to maintain a stable balance. Additionally, the system ensures that at any given moment, only one mode is active to avoid disturbances in the renewable generator's operation.

10.5.2 Scenario 2

In this scenario, a sudden drop in the WS profile and a rapid increase in power demand are applied to demonstrate the effectiveness and performance of the PMA. The WS profile for this scenario is illustrated in Figure 10.12(a), while Figure 10.12 (b) presents the corresponding demanded power profile.

In this scenario, the battery module operates as a backup, storing surplus energy when production outpaces consumption and supplying additional energy

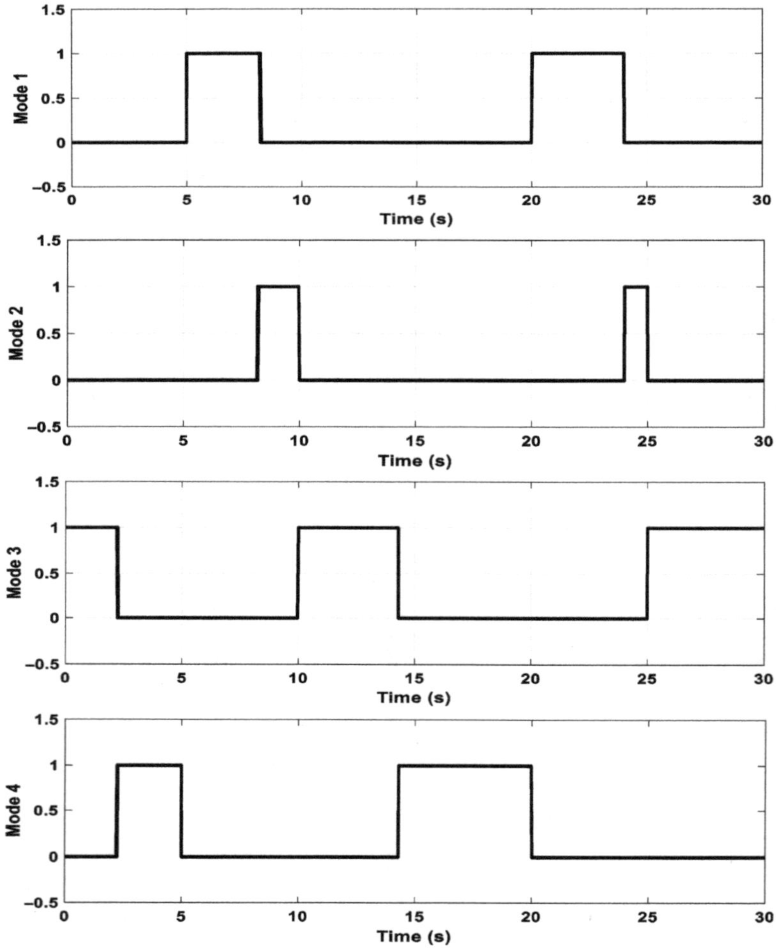

Figure 10.13 Various operations modes of PMA

during times of deficit. Its key function is to maintain a stable balance between energy production and consumption under all operating conditions, while preserving the DC bus voltage at its designated reference value, U_{dc}^*.

Figure 10.13 depicts the transitions between the four operating modes defined by the PMA. The proposed PMA ensures smooth mode transitions by utilizing predefined thresholds, such as wind power, load demand, and battery SOC, which are continuously monitored. These thresholds guide the system in selecting the appropriate mode, ensuring that changes in power flow occur gradually to minimize transient effects. Furthermore, at any given time, only one mode is active, which prevents mode conflicts and ensures uninterrupted operation of the renewable generator. These enhancements contribute to maintaining the overall stability of the system during mode transitions.

During the interval [5, 8.2], the system operates in Mode 1, where the generated power matches the load demand, and any excess energy is stored in the battery. A similar pattern is observed in the interval [20, 23.998 s], where surplus energy is again stored in the battery under the same mode. Between [8.2,10 s], the system enters Mode 2 ($k_d = 1$), where the power generation surpasses demand, causing the battery to reach its maximum SOC, as illustrated in Figure 10.14. In the interval [0, 2.2368 s], the wind energy generated is insufficient to meet the load. During this period, Mode 3 is activated, with the battery supplying the necessary energy to fill the gap. This behavior is repeated in the intervals [10, 14.301 s] and [25.0024, 30 s], where the battery compensates for insufficient wind energy production.

During the period [2.24, 5 s], the battery reaches its minimum SOC (Figure 10.14), and the wind generator cannot meet the power demand. Under these conditions, Mode 4 is activated, disconnecting non-priority loads to maintain the battery's SOC at its minimum limit (SOC_{min}). Figure 10.15 illustrates the charging and discharging currents

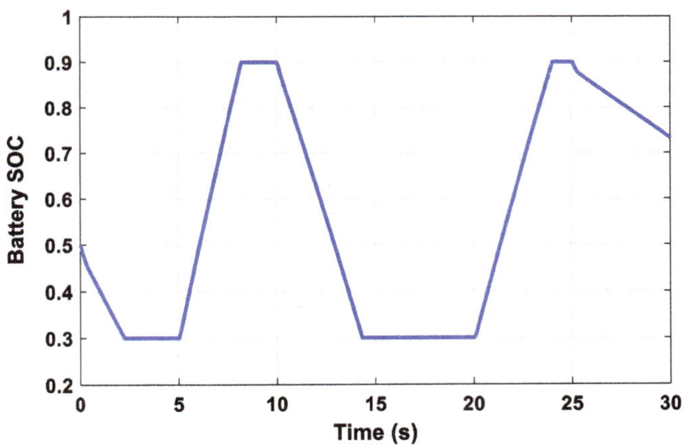

Figure 10.14 Battery ESS SOC

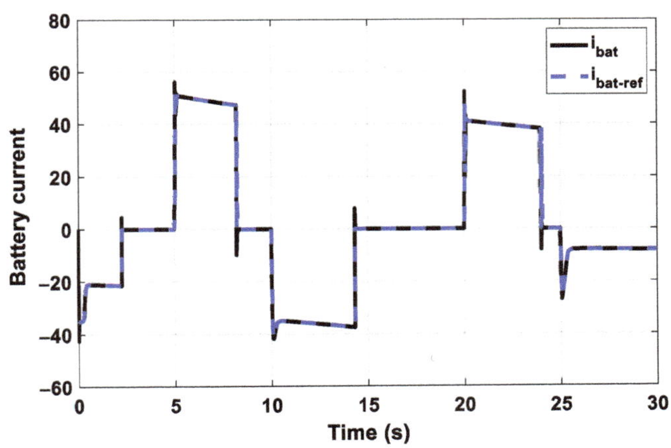

Figure 10.15 Charge and discharge current

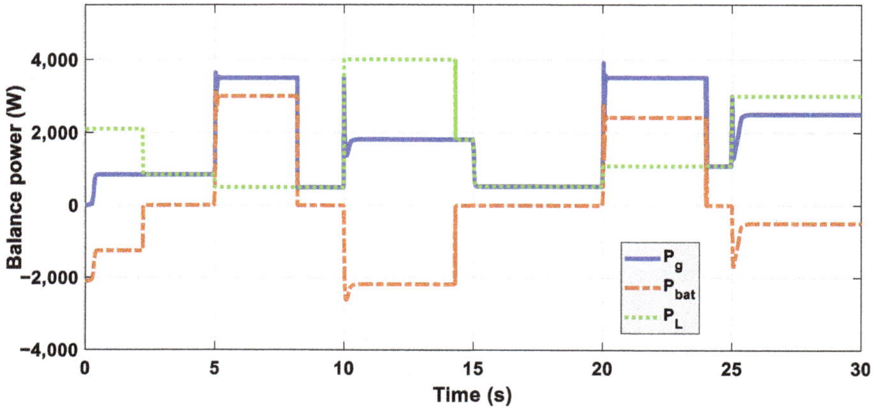

Figure 10.16 Balance power

Table 10.2 Synthesis of hardware resource utilization

		Available	Utilization	Utilization %
Resources	FF	106,400	140	0.13
	LUT	53,200	8,203	15.3
	LUTRAM	17,400	28	0.16

of the battery across the different operating modes. The battery's SOC increases during charging ($i_{battery}$>0) and decreases during discharging ($i_{battery}$<0). When the SOC reaches its maximum or minimum limit (SOC_{max}) or (SOC_{min}), the current becomes zero ($i_{battery}$=0), maintaining the SOC at its respective limit.

The proposed PMA ensures effective balance between power production and consumption, as demonstrated in Figure 10.16. As shown in Figure 10.16, at $t = 9$ s, the generated power surpasses the demanded power, and the SOC of the battery exceeds the maximum value SOC_{max}. As a result, Mode 2 is activated, which triggers the discharge of the battery ESS, as illustrated in Figure 10.13. In the time range [25, 30 s], the demanded power exceeds the generated power, and the SOC of the battery is more than 90%, which is the maximum threshold for the battery ESS. Consequently, Mode 3 is activated, as shown in Figure 10.13. In Mode 3, the system adjusts its operation to ensure efficient power distribution, balancing the energy supply and demand while maintaining the SOC within optimal limits.

10.6 Real-time implementation of the proposed PMA on the FPGA

Table 10.2 provides a detailed summary of hardware resource utilization on the FPGA Zed-Board for implementing the proposed PMA. The analysis reveals that

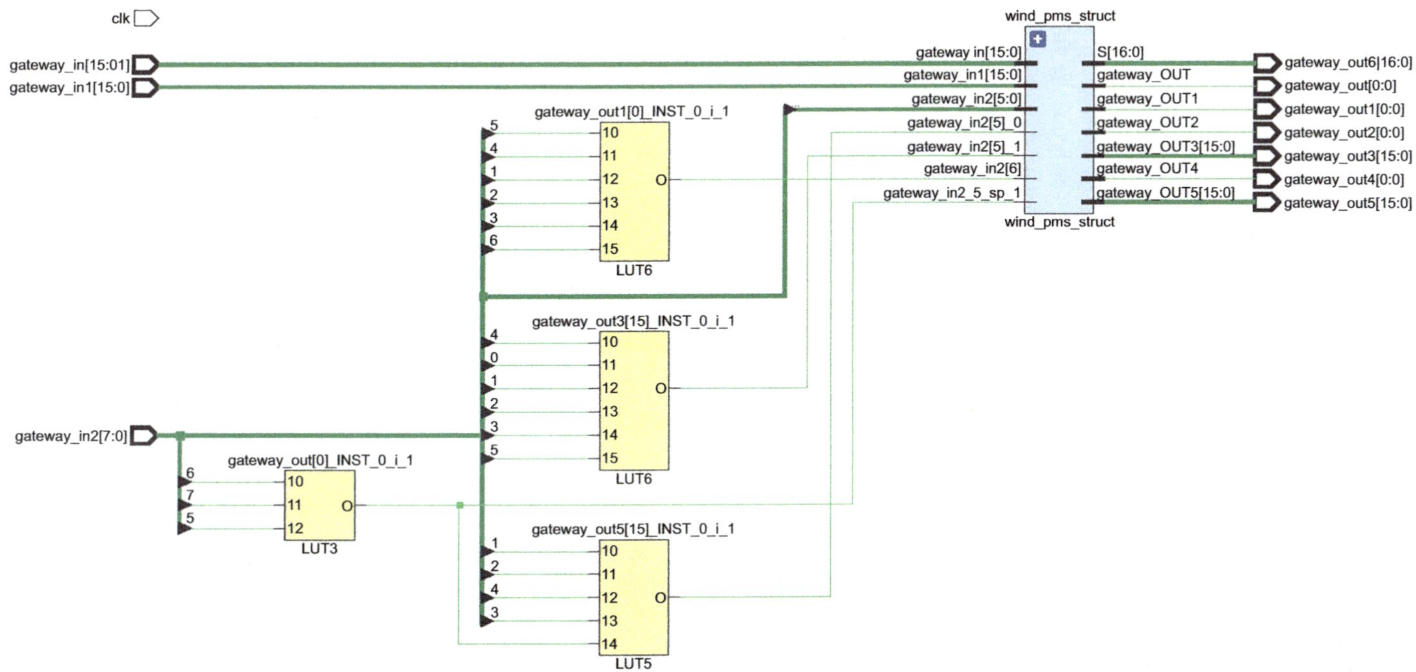

Figure 10.17 RTL schematic of the proposed PMA

the algorithm efficiently uses only a small fraction of the logic resources and occupies between 0.13% and 15.3% of the available memory. This indicates a highly optimized implementation that maximizes the use of FPGA resources while maintaining operational efficiency. The low resource consumption underscores the algorithm's suitability for real-time applications and its potential for integration into more complex systems without requiring extensive hardware modifications.

The RTL schematic, shown in Figure 10.17, is generated during the HDL synthesis phase of the design process. This schematic offers a visual representation of the design, optimized for testing, and uses generic symbols such as adders, multipliers, counters, and logical gates (AND, OR). These symbols are not specific to the Xilinx device but serve as placeholders for the components that will be implemented in hardware. This abstraction helps in verifying the functionality of the design before the actual device-specific implementation.

The XSG architectures for the proposed PMA have been validated through simulation. The simulation results ensure that the design operates as expected under various conditions. Following validation, the next phase involves the generation of VHDL code, which is used for hardware implementation. The synthesized results, as summarized in Table 10.2, provide a detailed overview of the resources utilized during the implementation of the PMA. These results are critical for assessing the efficiency of the design and ensuring that it meets the required performance and resource constraints for real-time operation on the FPGA Zed-Board. The efficient use of hardware resources, as indicated in Table 10.2, underscores the effectiveness of the PMA implementation and its suitability for deployment in practical systems.

10.7 Conclusion

In this chapter, we presented a real-time hardware implementation of a PMA on the FPGA Zed-Board, integrated with a battery ESS and a variable load. The PMA optimizes system performance and manages power disturbances caused by wind speed changes and varying loads. It ensures balance at the DC bus by matching the power from the wind turbine with the load demand and maintaining the battery's SOC between 30% and 90%, thus protecting the battery from overcharging or deep discharging. This approach effectively manages power flows in a WECS, providing stable power delivery and extending the battery's lifespan. For the hardware implementation, the XSG was utilized to implement the proposed PMA on the FPGA board. This method is very practical, as it does not require HDL knowledge, thereby reducing implementation time. Simulation results from MATLAB/Simulink using the XSG tool confirmed the PMS's ability to effectively manage production variations and system performance, validating the proposed approach.

References

[1] W. D. Brewer, R. Wengenmayr, and T. Bührke, *Renewable Energy: Sustainable Energy Concepts for the Energy Change*. New York: Wiley, 2013.

[2] R. Bali Swain, A. Karimu, and E. Gråd, "Sustainable development, renewable energy transformation and employment impact in the EU," *International Journal of Sustainable Development & World Ecology*, vol. 29, no. 8, pp. 695–708, 2022.

[3] S. El Bakali, H. Ouadi, and S. Gheouany, "Wind turbine and PV power prediction using a deterministic data-driven model with variational mode decomposition preprocessing," *Transactions of the Institute of Measurement and Control*, 2024, https://doi.org/10.1177/01423312241274044.

[4] M. Hannan, A. Q. Al-Shetwi, M. S. Mollik, *et al.*, "Wind energy conversions, controls, and applications: A review for sustainable technologies and directions," *Sustainability*, vol. 15, no. 5, p. 3986, 2023.

[5] H. Wang, G. Di Pietro, X. Wu, R. Lahdelma, V. Verda, and I. Haavisto, "Renewable and sustainable energy transitions for countries with different climates and renewable energy sources potentials," *Energies*, vol. 11, no. 12, p. 3523, 2018.

[6] J. L. Domínguez-García, O. Gomis-Bellmunt, F. D. Bianchi, and A. Sumper, "Power oscillation damping supported by wind power: A review," *Renewable and Sustainable Energy Reviews*, vol. 16, no. 7, pp. 4994–5006, 2012.

[7] S. M. Suboh, R. Sharma, W. Tushar, and N. B. Ahamad, "Enhancing active power dispatchability of WTG-BESS through integrated control considering wind forecast error," *IEEE Access*, vol. 12, pp. 2778–2792, 2024.

[8] F. Díaz-González, A. Sumper, O. Gomis-Bellmunt, and R. Villafáfila-Robles, "A review of energy storage technologies for wind power applications," *Renewable and Sustainable Energy Reviews*, vol. 16, no. 4, pp. 2154–2171, 2012.

[9] H. Zhao, Q. Wu, S. Hu, H. Xu, and C. N. Rasmussen, "Review of energy storage system for wind power integration support," *Applied Energy*, vol. 137, pp. 545–553, 2015.

[10] W. Liu, X. Sun, X. Yan, *et al.*, "Review of energy storage capacitor technology," *Batteries*, vol. 10, no. 8, p. 271, 2024.

[11] B. Meghni, D. Dib, M. Mordjaou, and A. Saadoun, "Effective supervisory controllers designed for optimal energy management in permanent magnet synchronous generator wind turbine with battery storage," *Wind Engineering*, vol. 40, no. 2, pp. 115–133, 2016.

[12] A. Elsanabary, K. Khairullah, N. F. Ab Aziz, and S. Mekhilef, "Power management and optimized control of hybrid PV-BESS-grid integrated fast EV charging stations," *Journal of Energy Storage*, vol. 90, p. 111802, 2024.

[13] M. Z. Daud, A. Mohamed, and M. Hannan, "An improved control method of battery energy storage system for hourly dispatch of photovoltaic power sources," *Energy Conversion and Management*, vol. 73, pp. 256–270, 2013.

[14] A. Brka, G. Kothapalli, and Y. M. Al-Abdeli, "Predictive power management strategies for stand-alone hydrogen systems: Lab-scale validation," *International Journal of Hydrogen Energy*, vol. 40, no. 32, pp. 9907–9916, 2015.

[15] L. Peng, L. Luo, J. Yang, and W. Li, "A wind power fluctuation smoothing control strategy for energy storage systems considering the state of charge," *Energies*, vol. 17, no. 13, p. 3132, 2024.

[16] M. Hermassi, S. Krim, Y. Krim, M. A. Hajjaji, A. Mtibaa, and M. F. Mimouni, "Xilinx-FPGA for real-time implementation of vector control strategies for a grid-connected variable-speed wind energy conversion system," in 2022 *5th International Conference on Advanced Systems and Emergent Technologies (IC_ASET)*, 2022: IEEE, pp. 49–54.

[17] M. Hermassi, S. Krim, Y. Kraiem, and M. A. Hajjaji, "Adaptive neuro fuzzy technology to enhance PID performances within VCA for grid-connected wind system under nonlinear behaviors: FPGA hardware implementation," *Computers and Electrical Engineering*, vol. 117, p. 109264, 2024.

[18] M. Hermassi, S. Krim, Y. Krim, M. A. Hajjaji, A. Mtibaa, and M. F. Mimouni, "Hardware FPGA implementation of an intelligent vector control technique of three-phase rectifier for wind turbine connected to the grid," in 2022 *IEEE 9th International Conference on Sciences of Electronics, Technologies of Information and Telecommunications (SETIT)*, 2022: IEEE, pp. 549–555.

[19] A. El Attafi, H. El Alami, B. Bossoufi, *et al.*, "Robust control of a wind energy conversion system: FPGA real-time implementation," *Heliyon*, vol. 10, no. 15, p. e35712, 2024.

[20] P. S. Panwar, M. Gupta, and A. Kumar, "PID embedded control and DC motor behaviour for FPGA-based simulation," in A. Kumar, S. Mondal, G. Verma, and P. Mani (eds.) *Embedded Devices and Internet of Things*. Boca Raton, FL: CRC Press, 2024, pp. 63–85.

[21] S. Krim and M. F. Mimouni, "Design of improved direct torque control based on a five level torque controller and a new Sugeno-Takagi fuzzy super-twisting controller applied to an induction machine," *Engineering Applications of Artificial Intelligence*, vol. 126, p. 106900, 2023.

[22] B. Murali Krishna, G. Rakesh Chowdary, C. Santhosh, S. K. Abdul Kalam, and K. P. Kishore Lakshman Naidu, "Implementation of Xilinx system generator based image processing algorithms through FPGA," *AIP Conference Proceedings*, vol. 2512, no. 1, p. 020011, 2024.

[23] M. Hermassi, S. Krim, Y. Kraiem, *et al.*, "Design of vector control strategies based on fuzzy gain scheduling PID controllers for a grid-connected wind energy conversion system: Hardware FPGA-in-the-Loop verification," *Electronics*, vol. 12, no. 6, p. 1419, 2023.

[24] Y. Krim, D. Abbes, S. Krim, and M. F. Mimouni, "Classical vector, first-order sliding-mode and high-order sliding-mode control for a grid-connected variable-speed wind energy conversion system: A comparative study," *Wind Engineering*, vol. 42, no. 1, pp. 16–37, 2018.

[25] F. Ebrahimi, A. Khayatiyan, and E. Farjah, "A novel optimizing power control strategy for centralized wind farm control system," *Renewable Energy*, vol. 86, pp. 399–408, 2016.

[26] Z. Cabrane, M. Ouassaid, and M. Maaroufi, "Battery and supercapacitor for photovoltaic energy storage: A fuzzy logic management," *IET Renewable Power Generation*, vol. 11, no. 8, pp. 1157–1165, 2017.

[27] B. Meghni, A. Saadoun, D. Dib, and Y. Amirat, "Effective MPPT technique and robust power control of the PMSG wind turbine," *IEEJ Transactions on Electrical and Electronic Engineering*, vol. 10, no. 6, pp. 619–627, 2015.

[28] M. Rekik, A. Abdelkafi, and L. Krichen, "A novel control strategy of a distributed generator operating in seven modes for ancillary services under grid faults," *International Journal of Electrical Power & Energy Systems*, vol. 47, pp. 100–108, 2013.

[29] A. Bidram, A. Davoudi, F. L. Lewis, and J. M. Guerrero, "Distributed cooperative secondary control of microgrids using feedback linearization," *IEEE Transactions on Power Systems*, vol. 28, no. 3, pp. 3462–3470, 2013.

[30] S. Krim, S. Gdaim, A. Mtibaa, and M. Faouzi Mimouni, "FPGA-based real-time implementation of a direct torque control with second-order sliding mode control and input–output feedback linearisation for an induction motor drive," *IET Electric Power Applications*, vol. 14, no. 3, pp. 480–491, 2020.

[31] C. Sisterna, M. Segura, M. Guzzo, G. Ensinck, and C. Gil, "FPGA implementation of an ultra-high speed ADC interface," in 2011 *VII Southern Conference on Programmable Logic (SPL)*, 2011: IEEE, pp. 161–166.

[32] B. Bossoufi, M. Karim, A. Lagrioui, M. Taoussi, and A. Derouich, "Observer backstepping control of DFIG-Generators for wind turbines variable-speed: FPGA-based implementation," *Renewable Energy*, vol. 81, pp. 903–917, 2015.

Chapter 11

Fuzzy logic for grid-connected doubly fed induction generator for wind energy

Abdeldjabbar Mohamed Kouadria[1,3], Selman Kouadria[2] and Mohamed Amine Bouzid[3]

Abstract

This study develops and designs a fuzzy controller technique to be used with grid-connected wind turbines and includes variable speed doubly fed induction generators. Fuzzy logic is a type of intelligent control system that is well regarded because of its robustness, which makes it an ideal and reliable choice for harvesting wind power for electricity generation. Fuzzy logic offers a good way to deal with linguistic uncertainty by modeling the vagueness and unreliability of information, thereby reducing the level of uncertainty in the system. Both control strategies were developed; we evaluated their outcomes to determine which was superior in reference tracking and robustness. The effectiveness of the suggested control technique is validated by theoretical analysis and MATLAB®/Simulink® simulations.

Keywords: Wind energy conversion system (WECS); doubly fed induction generator (DFIG); active and reactive power control; fuzzy logic controller (FLC)

11.1 Introduction

Concerns regarding environmental contamination from traditional energy sources have prompted rapid improvements in wind energy conversion technologies, a kind of renewable energy [1]. Wind power has recently increased in popularity and garnered significant scholarly interest [2,3]. For wind energy conversion systems (WECS), wind turbines with doubly fed induction generators (DFIGs) are becoming increasingly common because of their numerous advantages. These include the ability to operate at variable speeds, reduced converter size and ratings, and lower

[1]Department of Electrical Engineering, University of Abdelhamid Ibn Badis Mostaganem, Algeria
[2]Laboratoire de Génie Électrique et des Plasmas (LGEP), Ibn Khaldoun University, Algeria
[3]Laboratory Computer Engineering and Energy Engineering (L2GEGI), Ibn Khaldoun University, Algeria

power losses when compared to directly connected generators, making them more efficient and cost-effective for large-scale wind energy applications [4].

Variable-speed wind turbines are advantageous as they optimize power extraction at diverse wind velocities and reduce mechanical stress on WECS by accommodating fluctuations in wind power [5]. The utilization of the DFIG for wind power generation has lately garnered international interest due to its compact dimensions, elevated power density, efficiency, and torque-to-inertia ratio.

Induction generators are more reliable and less expensive than synchronous generators [6]. Their supplementary equipment and control systems are more basic. This study introduces a fuzzy control methodology for managing the power output of WECS using DFIGs.

This chapter is structured as follows: A review of the relevant research is presented in Section 11.2, while Section 11.3 outlines the entire system under investigation. Sections 11.4 and 11.5 address state modeling and the control strategy, respectively, while Section 11.6 concentrates on the design of the fuzzy logic approach. In Section 11.7, the results of the simulation that support the suggested DFIG control system are discussed. Finally, Section 11.8 concludes the study and suggests future directions for our research.

11.2 Literature review

Fuzzy logic control is one of the most efficient expert system methodologies employed in various control applications [7]. It circumvents the issues associated with the proportional–integral (PI) controller, hence enhancing the system's resilience. Constructing the rule-based system requires an understanding of the system's overall functioning and behavior, rather than an exact mathematical model, which accounts for its greater efficiency [8]. This chapter employs fuzzy logic controllers (FLCs) for the active and reactive power control routes' inner and outer control loops; using MATLAB/Simulink, we can execute simulations of the entire system model.

In [9], the integration of FLCs within variable speed wind turbines has been shown to yield superior power generation performance and enhanced efficiency. The significance of employing fuzzy controllers resides in their capacity for dealing with nonlinearity, reject disturbances, and augmenting system stability.

The reduction of active power fluctuations and the improvement of the generator's dynamic response are addressed in [10]; the direct power control (DPC) approach uses a FLC to adjust the bandwidth of the active power hysteresis controller.

In [11], the authors aimed to get a sinusoidal form of AC current with a reduced total harmonic distortion (THD) with unpredictable reactive power demand and unpredictable wind speeds. They accomplish this through combining two methods: modified hysteresis (MH) direct torque control and reactive power control-based fuzzy logic control.

In [12], a distinct technique was proposed, incorporating the use of multi-rotor wind energy systems based on DFIG and a novel fractional-order fuzzy control

(FOFC) algorithm. In order to achieve a very resilient method, this approach combines fuzzy control with fractional-order control.

In [13], three types of controllers were used to control the active and reactive power: a controller based on a neural network controller (NNC), a FLC, and the proportional–integral (PI). By substituting PI for an FLC and NNC, the drawback of traditional power controllers was overcome and performances were improved.

The combination of sliding mode and fuzzy logic have been discussed in [14]. Using a fuzzy sliding mode controller to regulate the conversion system's power allows for better performance than a traditional sliding mode controller. Reducing the chattering is the main objective of the fuzzy logic improvement.

A novel architecture based on fuzzy logic was used to control reactive power of a DFIG as presented in [15]. The I_{dr} signal is directly regulated by two fuzzy controllers. The method of control demonstrated improved steady-state behaviors in addition to rapid responses.

The authors in [16] proposed a strategy based on fuzzy logic control to replace both hysteresis regulators and the switching table. To address the limitations of the conventional DPC (C-DPC) method and achieve better control, this control relies on active and reactive power faults.

11.3 Description of studied system

Figure 11.1 illustrates the DFIG structure, where the stator is directly connected to the grid. The stator receives power directly from the AC mains, while the wound rotor is linked to the power electronics converter via slip rings, allowing it to exchange power with the grid. This configuration allows the DFIG to adjust its operational velocity in accordance with wind speed. The main approach is to link the grid's fixed frequency to the variable frequency asynchronous generator through a frequency converter. A rectifier's pulse width modulation (PWM) output processes the incoming direct current power and transforms it into alternating current (AC) through a PWM inverter.

An important benefit of DFIG is the reduction in converter size, which generally is between 25–30% of the machine's nominal power, dependent on the

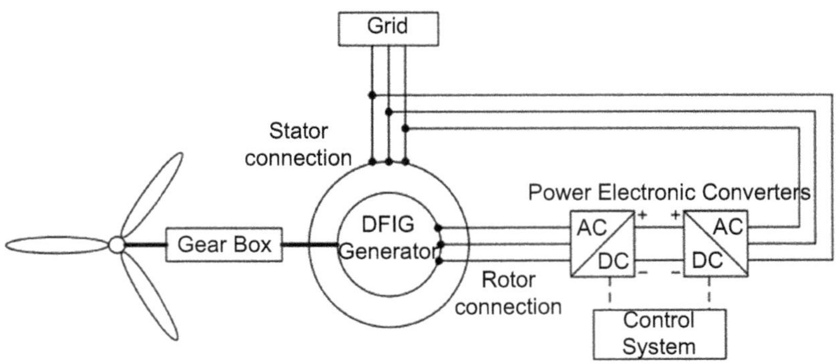

Figure 11.1 Wind turbine system constructed with a doubly fed induction generator

machine's slip [17]. This research illustrates that DFIGs can be enhanced by employing a FLC.

11.4 Modeling of studied system

11.4.1 Modeling of the wind turbine and gearbox

Based on [18,19], the mechanical power that can be extracted from the wind and used by the turbine shaft is given by:

$$T_{aer} = \frac{P_{aer}}{\Omega_t} = \frac{C_p(\lambda, \beta)\rho\pi R^2 v^3}{2\Omega_t} \tag{11.1}$$

where $C_p(\lambda, \beta)$ is the power coefficient, v represents the wind speed (m/s), R is the radius of the turbine disk, ρ is the air density, λ is the tip speed ratio, Ω_t is the turbine speed, and β is the pitch angle. The power coefficient of C_p is expressed in (11.1); it depends on the turbine characteristics [20].

$$C_p(\lambda, \beta) = 0.5176\left(\frac{116}{\lambda_i} - 0.4\,\beta - 5\right)e^{\frac{-21}{\lambda_i}} + 0.0068 \tag{11.2}$$

where λ and λ_i are given by

$$\begin{cases} \dfrac{1}{\lambda_i} = \dfrac{1}{\lambda + 0.08.\beta} - \dfrac{0.035}{\beta^3 + 1} \\ \lambda = \dfrac{\Omega_t R}{v} \end{cases} \tag{11.3}$$

11.4.2 Modeling and control of the DFIG

The DFIG is represented in the Park reference frame. The following equations relate to the voltage and flow of the DFIG [21,22]:

$$\begin{cases} V_{ds} = R_s.I_{ds} + \dfrac{d\Phi_{ds}}{dt} - \omega_s\Phi_{qs} \\ V_{qs} = R_s.I_{qs} + \dfrac{d\Phi_{qs}}{dt} + \omega_s\Phi_{ds} \\ V_{dr} = R_r.I_{dr} + \dfrac{d\Phi_{dr}}{dt} - (\omega_s - \omega)\Phi_{qr} \\ V_{qr} = R_r.I_{qr} + \dfrac{d\Phi_{qr}}{dt} + (\omega_s - \omega)\Phi_{dr} \end{cases} \tag{11.4}$$

The stator and rotor flux are represented as follows:

$$\begin{cases} \varphi_{ds} = L_s I_{ds} + M I_{dr} \\ \varphi_{qs} = L_s I_{qs} + M I_{qr} \\ \varphi_{dr} = L_r I_{dr} + M I_{ds} \\ \varphi_{qr} = L_r I_{qr} + M I_{qs} \end{cases} \tag{11.5}$$

Subscripts *d* and *q* indicate the *d*- and *q*-axes, respectively, while subscripts *s* and *r* pertain to the stator and rotor of the DFIG. The variables ωs and ωr (rad/s) represent the pulsations of the stator and rotor, respectively. The voltages V_{sd}, V_{sq}, V_{rd}, and V_{rq} correspond to the direct and quadrature components of the stator and rotor voltages, while I_{sd}, I_{sq}, I_{rd}, and I_{rq} represent the direct and quadrature components of the stator and rotor currents. The flux components are denoted by φ_{sd}, φ_{sq}, φ_{rd}, and φ_{rq} for the stator and rotor, respectively. R_s and R_r are the resistances of the stator and rotor, while L_s and L_r represent the leakage inductances for the stator and rotor. The term *M* signifies the magnetizing inductance, and *p* is the number of pole pairs.

The electromagnetic torque T_{em} can be expressed as

$$T_{em} = \frac{PM}{Ls}\left(\varphi_{qs} I_{dr} - \varphi_{ds} I_{qr}\right) \tag{11.6}$$

At the stator side, the active and reactive powers are specified as [23,24]:

$$\begin{cases} P_s = V_{ds}.I_{ds} + V_{qs}.I_{qs} \\ Q_s = V_{qs}.I_{ds} - V_{ds}.I_{qs} \end{cases} \tag{11.7}$$

By aligning the stator vector flux φ_{sd} with the *d*-axis and employing a biphasic *d–q* reference frame, constant electrical voltages and currents in steady-state operation can be achieved.

Hence, we can denote: $\varphi_{sd} = \varphi s$, $\varphi_{sq} = 0$.

The equation of the electromagnetic torque becomes then:

$$Tem = -\frac{PM}{Ls}\left(\varphi_{ds} I_{qr}\right) \tag{11.8}$$

The change in mechanical speed can be understood by applying the basic principles of dynamics.

$$J\frac{d\Omega_m}{dt} = T_m - T_m - f\Omega_m \tag{11.9}$$

where *J* is the system's moment of inertia, and *f* denotes the coefficient of friction.

11.5 Field-oriented control strategy applied to DFIG

To make the induction machine behave like a separately excited DC machine, the field-oriented control principle requires directing the field along one axis.

There are two types of power control for DFIG: direct control and indirect control.

11.5.1 Direct control of the doubly fed induction generator

This solution involves bypassing the coupling terms and implementing distinct regulators for active and reactive powers along each axis. The term "direct method" derives from the direct management of the machine's rotor voltages by the power regulators. A schematic of the direct control system of the DFIG is shown in Figure 11.2.

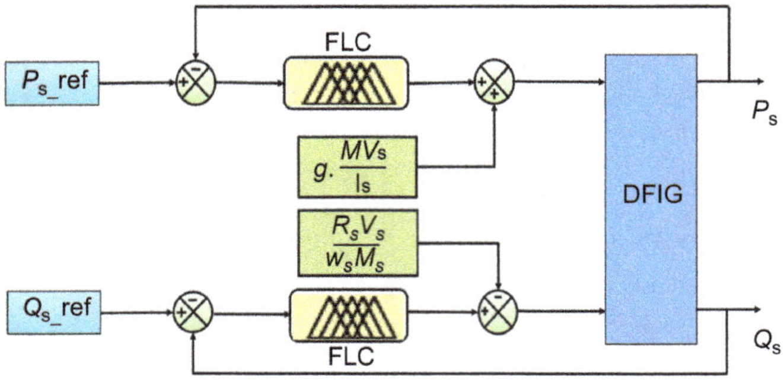

Figure 11.2 *Direct control of the doubly fed induction generator*

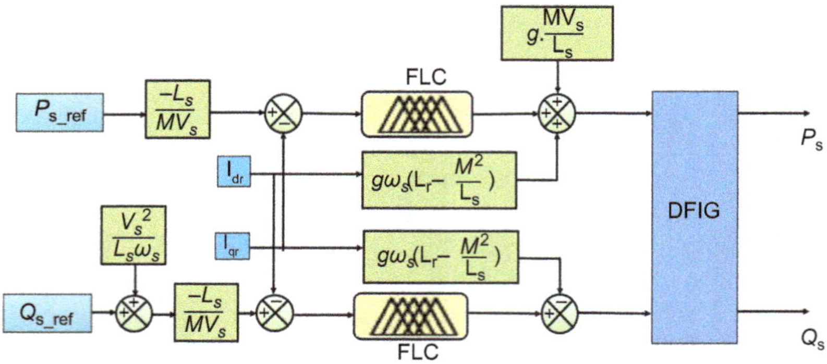

Figure 11.3 *Indirect control without power loop of the doubly fed induction generator*

11.5.2 Indirect control of the doubly fed induction generator

Indirect field-oriented control is a strategy commonly used for the control of electrical machines. It works by controlling the machine's torque and flux independently using feedback mechanisms and estimations of rotor flux position.

11.5.2.1 Indirect control without power loop

This method consists of regulating the rotor currents using a control loop (Figure 11.3). The reference values are imposed by the active and reactive powers set at the desired values. The regulators in this case are determined in the same way as for direct regulation.

11.5.2.2 Indirect control with power loop

To improve upon the previous method and eliminate static power errors, a second control loop for active and reactive power is incorporated into the rotor current control loop (Figure 11.4). This addition further enhances the system's performance.

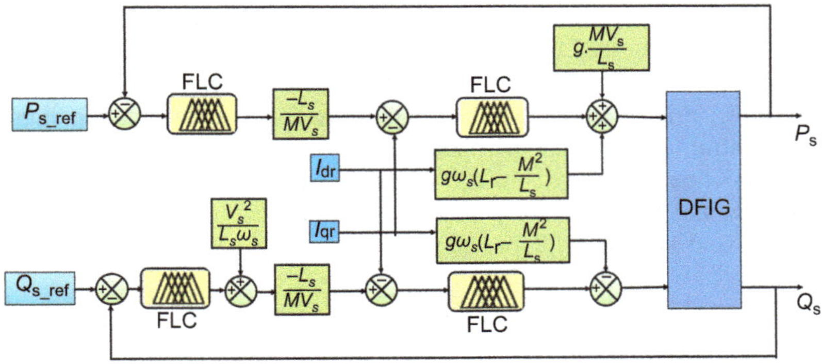

Figure 11.4 Indirect control with power loop of the doubly fed induction generator

The gain of the rotor current regulator is determined in the same way as in the previous method, and a second regulator is employed for power control. This eliminates static errors while maintaining the system's dynamics.

This method requires controlling the powers and rotor currents through a dual-loop system that considers and compensates for the coupling issues. The indirect technique regulates rotor currents to protect the DFIG generator by limiting the currents and improving the machine's flexibility.

11.6 Fuzzy logic controller

The FLC underpins the control system examined here for two primary reasons: first, it delivers excellent performance in managing the DFIG, essential for integrating the rotor's generated power into the grid; second, it is easy and convenient to implement.

FLCs have been effectively applied in motor control [25,26], variable-speed wind power generators [27,28], and several other domains [29,30]. Fuzzy logic control is characterized by various characteristics, details such as the total number of rules, the rules' antecedents and consequents, the defuzzification mechanism for output, the form and placement of each membership function, and the quantity of fuzzy variables. This presents numerous opportunities for developing high-performance controllers. Simultaneously, it integrates expert information into fuzzy rules and yields highly satisfactory performance, all without a detailed mathematical model of the system. It possesses inherent capabilities for managing noisy or imprecise data, enabling it to enhance its control functionality in operational contexts where linear control methods are inadequate, particularly during significant parameter fluctuations.

This research expands upon Mamdani's contributions by creating a FLC. This system operates as seen in Figure 11.5. The input membership procedures are employed to initially fuzzify the pristine inputs. Subsequently, the fuzzy inference engine processes the input fuzzy sets and applies the rules from the rule base to produce output fuzzy sets. Finally, the defuzzification process translates the output fuzzy sets into a precise numerical value that the system utilizes, employing the knowledge of the output membership functions.

In the proposed controller, two FLCs are established: both of which are devoted to controlling the DFIG's reactive power (Q_s) and active power (P_s). Two inputs and a single output fuzzy variable constitute each controller's structure.

Table 11.1 illustrates the phases of fuzzification and defuzzification employed by fuzzy control to synthesize the fuzzy controller for active and reactive powers. These rules define the approach used for system control [31,32].

The input and output variables' membership functions are shown in Figure 11.6. They are defined as follows: *N* for Negative, *EZ* for Equal Zero, and *P* for Positive.

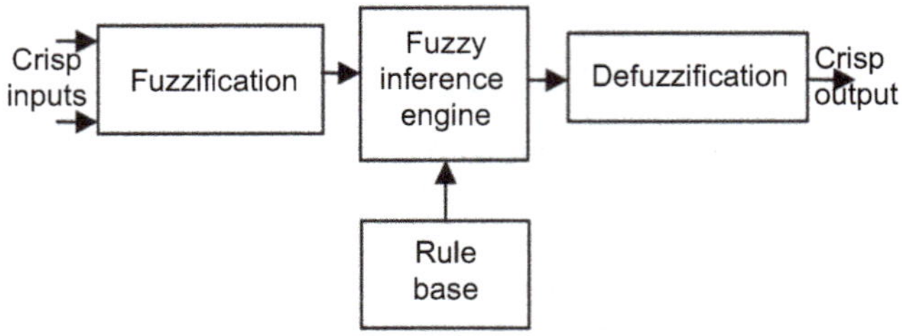

Figure 11.5 Fuzzy system structure

Table 11.1 Rule bases of fuzzy controller

E De	N	EZ	P
N	N	N	EZ
Z	N	EZ	P
P	Z	P	P

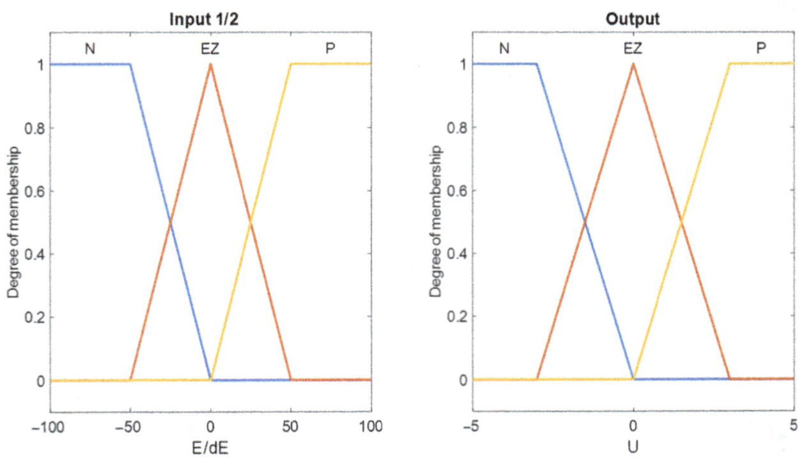

Figure 11.6 Membership functions for proposed fuzzy controller

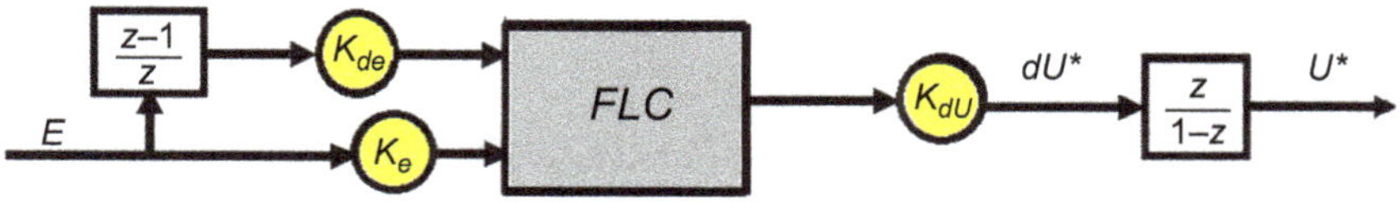

Figure 11.7 Diagram of fuzzy logic controller

11.6.1 Regulator conception

The structure of the current FLC is given in Figure 11.7.

To guarantee that the active power P_s and reactive power Q_s align with their reference values, the controller observes the error signal's pattern and its derivative within the active or reactive power control loops. It subsequently modifies the output U appropriately.

Each fuzzy controller utilizes the error signal E and its derivative dE/dt as inputs. The scaling factors K_e and K_{de} are employed to normalize these two signals. The command signal is generated by multiplying the output scale factor K_{dU} with du/dt, followed by integrating the result to obtain the output control signal U.

11.7 Simulation results and discussions

To evaluate the efficiency of the proposed control techniques, an authentic wind speed profile was applied to test controllers against disturbances that affects the real power output.

To ensure a unit power factor, the MPPT method is employed to generate the reference for active power ($P_{s\text{-}ref}$) while setting the reference for reactive power at almost zero ($Q_{s\text{-}ref} = 0$). Our primary objective is to achieve optimal power output and efficiency.

11.7.1 Direct control

In direct control, the coupling terms are neglected, allowing for independent regulation of active and reactive powers by a controller on each axis. V_{qr}, the quadratic component of rotor voltage, will regulate active power, while V_{dr}, the direct component, will control reactive power.

The direct vector control configuration of the DFIG is illustrated above, where each axis is managed by a single regulator.

The results, which are illustrated in Figures 11.8 and 11.9, indicate disturbances in both active and reactive powers, accordingly decreasing the effectiveness of the technique in applications with variable speeds.

11.7.2 Indirect control

Indirect control involves considering the coupling terms and compensating for them by establishing a two-loop system to regulate currents and powers. In this strategy,

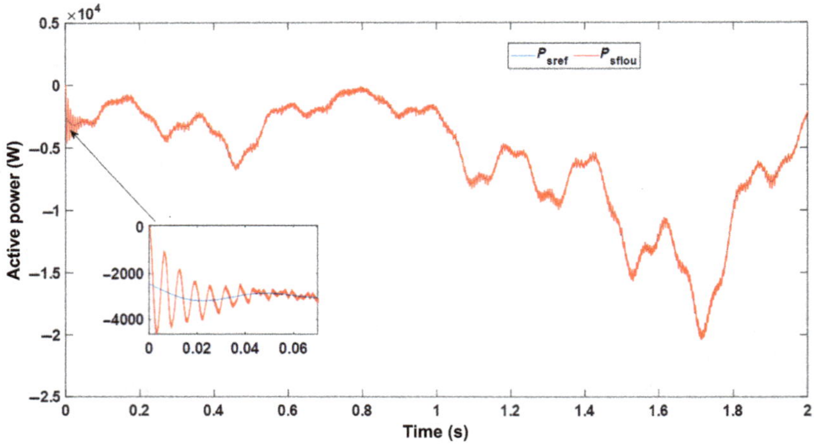

Figure 11.8 Active power control

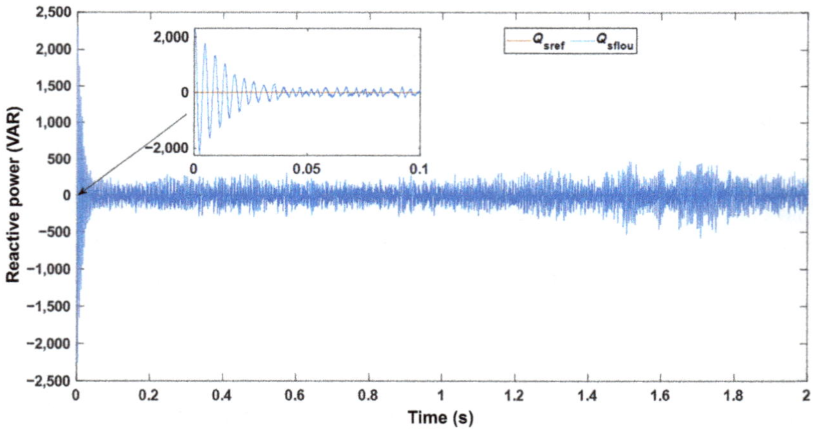

Figure 11.9 Reactive power control

the active (or reactive) power regulator generates the current reference I_{qr} (or I_{dr}). Finally, the current regulators generate the rotor voltage references V_{qr} (or V_{dr}).

11.7.2.1 Indirect control without power control loop

Figures 11.10 and 11.11 show how the suggested method smoothly follows the reference signal of a dynamic system's active and reactive power. Figures 11.12 and 11.13 show the currents.

The results demonstrate the variation in active power based on the operating speed of the DFIG. The power fluctuates significantly. In systems without a power

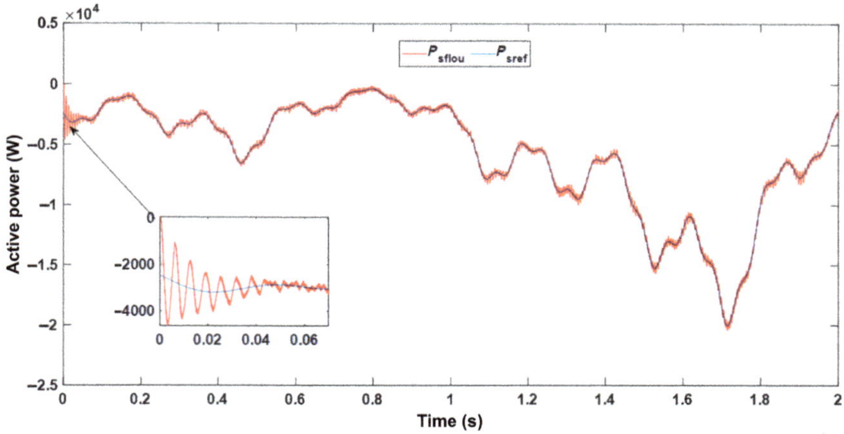

Figure 11.10 Active power control

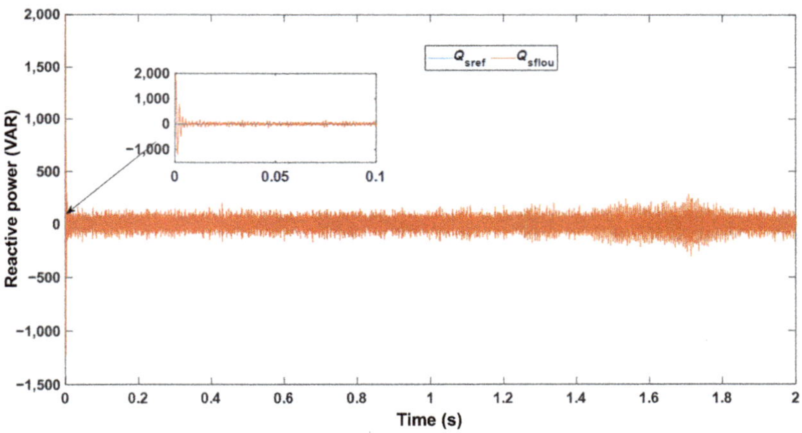

Figure 11.11 Reactive power control

control loop, the focus is on controlling the current to regulate the torque and speed. As a result, the active power is not directly regulated.

11.7.2.2 Indirect control with power control loop

The generator accurately keeps to the power references for both active and reactive power, as illustrated in Figures 11.14 and 11.15. The direct rotor current I_{dr} and the quadrature rotor current I_{qr} are the two factors that define the stator active power P_s and reactive power Q_s, respectively.

In Figures 11.16 and 11.17, it is clear that the powers decouple from each other, ensuring that the unity power factor is maintained. Here, the results obtained

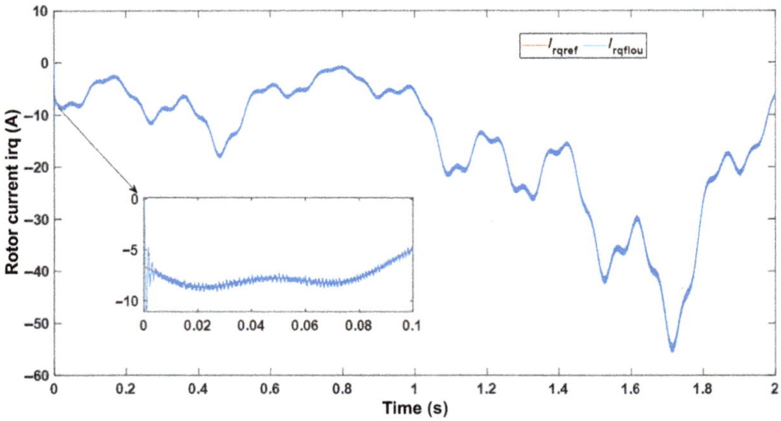

Figure 11.12 Rotor current i_{rq}

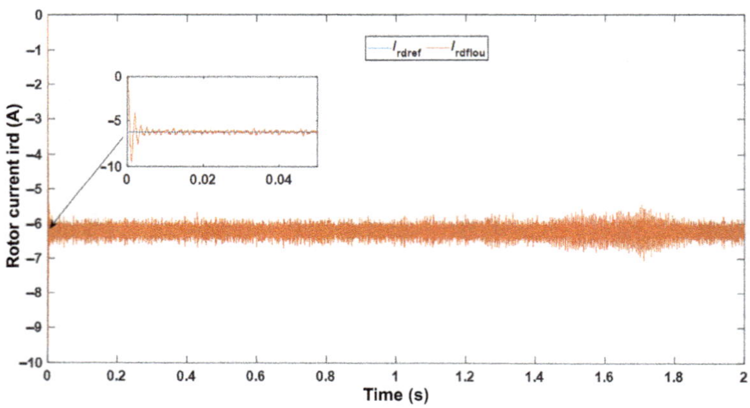

Figure 11.13 Rotor current i_{rd}

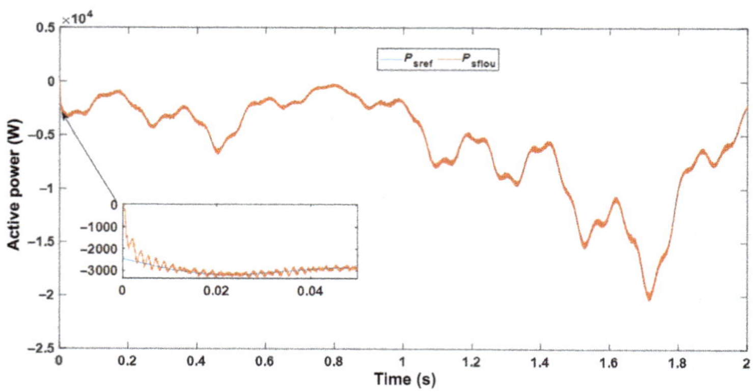

Figure 11.14 Active power control

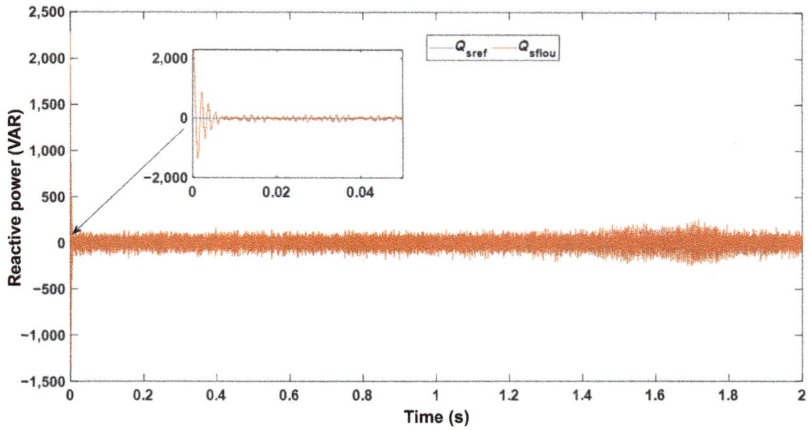

Figure 11.15 Reactive power control

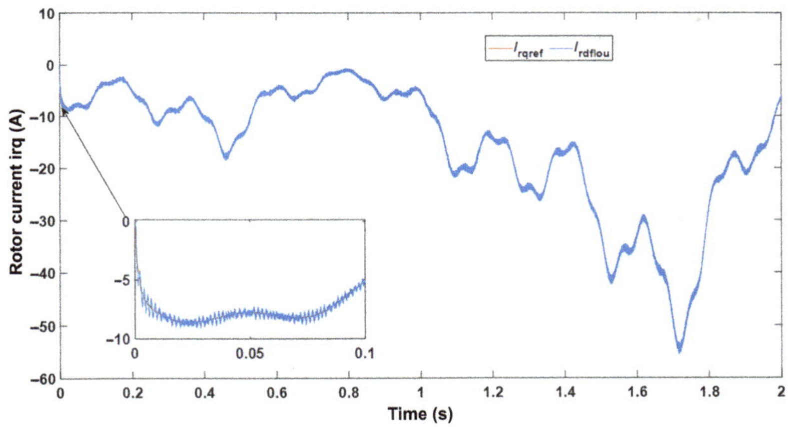

Figure 11.16 Rotor current i_{rq}

using an indirect fuzzy control strategy with power loop control demonstrate significant enhancements in both dynamic and static performances when compared with those obtained without the power control loop.

11.7.3 Comparison

It is clearly shown in Figure 11.18 the amplitude reduction of the oscillations and almost zero error reference tracking in the case of indirect control with power control loop. It is verified that the indirect control with a power control loop method shows better performance than the two previous methods.

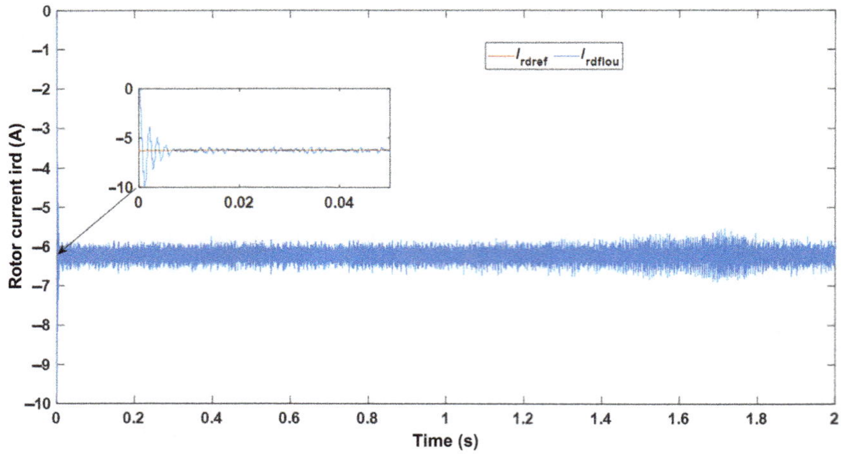

Figure 11.17 Rotor current i_{rd}

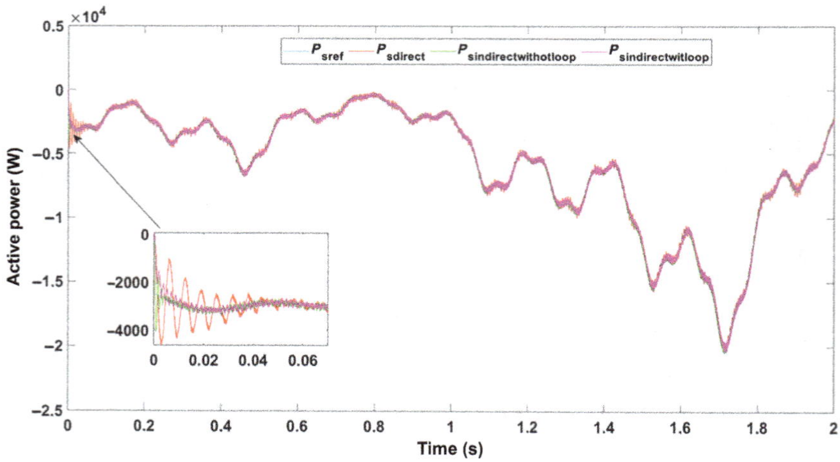

Figure 11.18 Active power control

Under the indirect control scheme with a power control loop, the oscillation amplitudes are lower and transpire in a much shorter duration, resulting in decreased static error when contrasted with other techniques.

11.8 Conclusion

A complete structure for generating electrical energy is illustrated using DFIG. The direct control approach for active and reactive power constitutes the initial system among three fuzzy logic-based control systems. There are two ways for indirectly controlling active and reactive power: one excludes power management, while the other employs a power control loop.

We present the simulation results to demonstrate the efficiency of the proposed control method. The FLC has exceptional performance in regulating both active and reactive power.

The indirect control method with the power control loop improved performances of the system and makes it superior. Responses of the controller show a very interesting result toward reference tracking, sensitivity to perturbation.

Appendix A

See Table A.1 for the overview of wind turbine and DFIG parameters.

Table A.1 Wind turbine and DFIG parameter overview [31]

	Parameter	**Symbol**	**Value**
Turbine	Rated power	P	10 kW
	Rotor diameter	D	3 m
	Number of blades		3
	Gearbox	G	5.4
	Inertial	J	0.042 $kg{\cdot}m^2$
	Viscous coefficient	f	0.017 $N{\cdot}m{\cdot}s^{-1}$
DFIG	Rated power	P	7.5 kW
	Rated stator voltage	V_s	220 V
	Rated stator frequency	F	50 Hz
	Number of pole pairs	pn	2
	Rotor winding resistance	R_r	1.8 Ω
	Stator winding inductance	L_s	0.20202 H
	Rotor winding inductance	L_r	0.20384 H
	Magnetizing inductance	M	0.195 H
	Inertia of system	J	0.2 $kg{\cdot}m^2$
	Viscous coefficient	f	1×10^{-3} $N{\cdot}m{\cdot}s^{-1}$

References

[1] T.A. Hamed and A. Alshare, "Environmental impact of solar and wind energy – A review," *Journal of Sustainable Development of Energy, Water and Environment Systems*, vol. 10, br. 2, str. 1–23, 2022. [Online]. https://doi.org/10.13044/j.sdewes.d9.0387.

[2] A. Rahman, O. Farrok, and M.M. Haque, "Environmental impact of renewable energy source based electrical power plants: Solar, wind, hydroelectric, biomass, geothermal, tidal, ocean, and osmotic," *Renewable and Sustainable Energy Reviews*, vol. 161, p. 112279, 2022. [Online]. https://doi.org/10.1016/j.rser.2022.112279.

[3] F. Manzano-Agugliaro, A. Alcayde, F.G. Montoya, A. Zapata-Sierra, and C. Gil, “Scientific production of renewable energies worldwide: An overview,” *Renewable and Sustainable Energy Reviews*, vol. 18, pp. 134–143, 2013. https://doi.org/10.1016/j.rser.2012.10.020.

[4] M. Cheng and Y. Zhu, “The state of the art of wind energy conversion systems and technologies: A review,” *Energy Conversion and Management*, vol. 88, pp. 332–347, 2014. https://doi.org/10.1016/j.enconman.2014.08.037.

[5] D. K. Bhutto, J. Ahmed Ansari, S. S. Hussain Bukhari, and F. Akhtar Chachar, “Wind energy conversion systems (WECS) generators: A review,” *2019 2nd International Conference on Computing, Mathematics and Engineering Technologies (iCoMET)*, Sukkur, Pakistan, 2019, pp. 1–6, 10.1109/ICOMET.2019.8673429.

[6] F. M. Hughes, O. Anaya-Lara, N. Jenkins, and G. Strbac, “Control of DFIG-based wind generation for power network support,” *IEEE Transactions on Power Systems*, vol. 20, no. 4, pp. 1958–1966, 2005, 10.1109/TPWRS.2005.857275.

[7] C. W. de Silva, *Intelligent Control: Fuzzy Logic Applications* (1st ed.). Boca Raton, FL: CRC Press; 1995. https://doi.org/10.1201/9780203750513.

[8] Z. Kovacic and S. Bogdan, *Fuzzy Controller Design: Theory and Applications* (1st ed.). Boca Raton, FL: CRC Press; 2005. https://doi.org/10.1201/9781420026504.

[9] M. Almaged, A. Mahmood, and Y. Alnema, “Design of an integral fuzzy logic controller for a variable-speed wind turbine model,” *Journal of Robotics and Control (JRC)*, vol. 4, no. 6, pp. 762–768, 2023. https://doi.org/10.18196/jrc.v4i6.20194.

[10] S. Khateri-abri, S. Tohidi, and N. Rostami, “Improved direct power control of DFIG wind turbine by using a fuzzy logic controller,” *2019 10th International Power Electronics, Drive Systems and Technologies Conference (PEDSTC)*, Shiraz, Iran, 2019, pp. 458–463, 10.1109/PEDSTC.2019.8697581.

[11] S. Tamalouzt, Y. Belkhier, Y. Sahri, N. Ullah, R. N. Shaw, and M. Bajaj, “New direct reactive power control based fuzzy and modulated hysteresis method for micro-grid applications under real wind speed,” *Energy Sources, Part A: Recovery, Utilization, and Environmental Effects*, vol. 44, no. 2, pp. 4862–4887, 2022. https://doi.org/10.1080/15567036.2022.2081741.

[12] H. Benbouhenni, N. Bizon, M. I. Mosaad, I. Colak, A. B. Djilali, and H. Gasmi, “Enhancement of the power quality of DFIG-based dual-rotor wind turbine systems using fractional order fuzzy controller,” *Expert Systems with Applications*, vol. 238, Part A, p. 121695, 2024, https://doi.org/10.1016/j.eswa.2023.121695.

[13] S. Aoun, A. Boukadoum, and L. Yousfi, “Advanced power control of a variable speed wind turbine based on a doubly fed induction generator using field-oriented control with fuzzy and neural controllers,” *International Journal of Dynamics and Control*, vol. 12, pp. 2398–2411, 2024. https://doi.org/10.1007/s40435-023-01345-9.

[14] K. Boulâam and A. Boukhelifa, "Fuzzy sliding mode control of DFIG power for a wind conversion system," 2014 *16th International Power Electronics and Motion Control Conference and Exposition*, Antalya, Turkey, 2014, pp. 353–358, 10.1109/EPEPEMC.2014.6980518.

[15] N. Aghanoori, M. Mohseni, and M. A. S. Masoum, "Fuzzy approach for reactive power control of DFIG-based wind turbines," 2011 *IEEE PES Innovative Smart Grid Technologies*, Perth, WA, Australia, 2011, pp. 1–6, 10.1109/ISGT-Asia.2011.6167124.

[16] K. F. Sayeh, S. Tamalouzt, and Y. Sahri, "Improvement of power quality in WT-DFIG systems using novel direct power control based on fuzzy logic control under randomness conditions," *International Journal of Modelling and Simulation*, vol. 45, no. 1, pp. 55–67, 2025. https://doi.org/10.1080/02286203.2023.2270757.

[17] G. D. Marques and M. F. Iacchetti, "DFIG topologies for DC networks: A review on control and design features," *IEEE Transactions on Power Electronics*, vol. 34, no. 2, pp. 1299–1316, 2019, 10.1109/TPEL.2018.2829546.

[18] M. A. Kouadria, T. Allaoui, and M. Denai, "Power quality improvement in isolated wind-diesel power system," *WSEAS Transactions on Systems and Control*, vol. 16, pp. 490–498, 2017.

[19] A. M. Kouadria, T. Allaoui, M. Denaï, and G. Pissanidis, "Power quality enhancement in off-grid hybrid renewable energy systems using type-2 fuzzy control of shunt active filter." In: Y. Bi, S. Kapoor, and R. Bhatia (eds.), *Intelligent Systems and Applications*. Springer, Cham, 2016, pp. 345–360. https://doi.org/10.1007/978-3-319-33386-1_17.

[20] R. Chen, D. Qin, and C. Liu, "Dynamic modelling and dynamic characteristics of wind turbine transmission gearbox-generator system electromechanical rigid-flexible coupling," *Alexandria Engineering Journal*, vol. 65, pp. 307–325, 2023. https://doi.org/10.1016/j.aej.2022.10.036.

[21] Y. Yang, Z. Guo, Q. Song, Y. Zhang, and Q. Li, "Effect of blade pitch angle on the aerodynamic characteristics of a straight-bladed vertical axis wind turbine based on experiments and simulations," *Energies*, vol. 11, no. 6, pp.1514, 2018. https://doi.org/10.3390/en11061514.

[22] M. A. Kouadria, S. Kouadria, and H. Chaib, "Sliding mode control for doubly fed induction generators system-based a wind turbine," *Studies in Engineering and Exact Sciences*, vol. 5, no. 2, p. e7199, 2024. https://doi.org/10.54021/seesv5n2-139.

[23] X. Liu and X. Kong, "Nonlinear model predictive control for DFIG-based wind power generation," *IEEE Transactions on Automation Science and Engineering*, vol. 11, no. 4, pp. 1046–1055, 2014, 10.1109/TASE.2013.2284066.

[24] S. Li, T. A. Haskew, K. A. Williams, and R. P. Swatloski, "Control of DFIG wind turbine with direct-current vector control configuration," *IEEE Transactions on Sustainable Energy*, vol. 3, no. 1, pp. 1–11, 2012. 10.1109/TSTE.2011.2167001.

[25] L. Xu and P. Cartwright, "Direct active and reactive power control of DFIG for wind energy generation," *IEEE Transactions on Energy Conversion*, vol. 21, no. 3, pp. 750–758, 2006. 10.1109/TEC.2006.875472.

[26] P. Guillemin, "Fuzzy logic applied to motor control," *IEEE Transactions on Industry Applications*, vol. 32, no. 1, pp. 51–56, 1996. 10.1109/28.485812.

[27] F. Betin, D. Pinchon, and G.-A. Capolino, "Fuzzy logic applied to speed control of a stepping motor drive," *IEEE Transactions on Industrial Electronics*, vol. 47, no. 3, pp. 610–622, 2000. 10.1109/41.847902.

[28] T. L. Van, T. H. Nguyen, and D.-C. Lee, "Advanced pitch angle control based on fuzzy logic for variable-speed wind turbine systems," *IEEE Transactions on Energy Conversion*, vol. 30, no. 2, pp. 578–587, 2015. 10.1109/TEC.2014.2379293.

[29] A. G. Abo-Khalil, D.-C. Lee, and J.-K. Seok, "Variable speed wind power generation system based on fuzzy logic control for maximum output power tracking," *2004 IEEE 35th Annual Power Electronics Specialists Conference (IEEE Cat. No.04CH37551)*, Aachen, Germany, 2004, pp. 2039–2043, Vol. 3, 10.1109/PESC.2004.1355431.

[30] C. T. S. Dagang, G. Kenne, and F. A. Muluh, "Fuzzy logic direct torque/power control for a self-excited induction generator driven by a variable wind speed turbine," *International Journal of Dynamics and Control*, vol. 9, pp. 1210–1222, 2021. https://doi.org/10.1007/s40435-020-00709-9.

[31] A. Kasbi, and A. Rahali, "Performance optimization of doubly-fed induction generator (DFIG) equipped variable-speed wind energy turbines by using three-level converter with adaptive fuzzy PI control system," *Materials Today: Proceedings*, vol. 47, Part 10, pp. 2648–2656, 2021. https://doi.org/10.1016/j.matpr.2021.05.406.

[32] S. Kouadria, E. M. Berkouk, Y. Messlem, and M. Denai, "Improved control strategy of DFIG-based wind turbines using direct torque and direct power control techniques," *Journal of Renewable and Sustainable Energy*, vol. 10, no. 4, p. 043306, 2018.

Chapter 12

Energy management with classical Boolean method for grid-connected hybrid systems with wind, PV and battery

Anas Hassari[1,2], *Kamal Hirech*[1,2] *and Mohammed Rhiat*[1,2]

Abstract

This chapter presents a study on an energy management system using a classical Boolean method within a grid-connected hybrid system that includes a wind turbine, photovoltaic panels, and battery storage. The study details the system's configuration and the algorithms implemented to optimize energy flow between the renewable sources, the storage system, and the grid, based on load demand and battery state of charge, ensuring consistent energy availability for the load. The results illustrate how this system manages switch control in response to changing conditions and the potential energy production from renewables over a day, based on meteorological data—specifically, solar radiation and wind speed from Morocco. This location is highlighted for its high solar and wind potential and the country's commitment to increasing renewable energy contributions while reducing fossil fuel dependence.

Keywords: Grid-connected hybrid systems; microgrid; solar PV; wind farm; batteries; renewable energy; energy management system block; switches

12.1 Introduction

Electrical systems in most countries around the world are undergoing significant changes in a context characterized by the rising costs of fossil fuels, the need to meet the growing energy demand, and the reduction of greenhouse gas emissions. In recent years, renewable energy share has increased considerably across many countries, including those with abundant fossil fuel resources.

Morocco, as a country relying on fossil fuel imports, has set the goal of obtaining 52% of its electricity from renewable sources by 2030. According to the Ministry of Energy Transition and Sustainable Development, renewable sources

[1]PESIP Team – LGEM Laboratory, Mohamed First University, Morocco
[2]Higher School of Education and Training, Mohamed First University, Morocco

represent 38% of the installed capacity in 2022, primarily from hydroelectric power (16.7%), wind power (13.48%), and solar energy (7.82%). In the future, Morocco plans to invest heavily on wind power and solar energy, given their availability and the impact of the drought on hydroelectric generation [1].

To address these issues, the share of decentralized production based on renewable energies has increased significantly in recent years, especially in Moroccan rural areas to contribute to its development and energy autonomy [2].

The systems used in decentralized production must maximize the use of renewable energy sources, minimize the cost of energy production, and reduce the equivalent CO_2 emissions within the microgrid. Furthermore, they must ensure the availability of electricity and optimal energy flow exchange with the grid in cases of surplus or shortage of production [3]. Hybrid systems based on renewable energies are generally employed to achieve these objectives.

Many criteria must be satisfied within the microgrid. The most important are the ones covering socioeconomic, environmental, and technical aspects. There are two classifications of these criteria: quantitative and qualitative. Quantitative criteria include the optimized levelized cost of energy, the system reliability, carbon dioxide emission, and the efficiency, while the qualitative ones feature the flexibility for future expansion, ease of installation and operation, and social acceptability [4].

Microgrids combining PV panels, wind turbines, batteries, maximum power point tracking (MPPT) algorithms [5–7], and energy management system (EMS) are commonly employed in Morocco as they effectively meet the previously mentioned criteria and align with the country's promotion of wind and solar energy [8,9].

In some studies, diesel generators are used instead of wind turbines [10] to support renewable production. However, others have proven that the microgrid under study is feasible, reliable, and efficient [11] in sunny windy areas which are numerous in Morocco [12,13]. Moreover, the system has the advantage of no CO_2 emissions compared to the last one.

In some cases, the power generated from renewable energies exceed load demand and the battery is full. It is necessary to use a system that allows the use of this excess by injecting it to the grid to avoid wastage. In other cases, renewable energy produced is insufficient to cover the load and the state of charge (SOC) of the battery is very low. Conventional grid must compensate the energy shortage.

Due to the unpredictable behavior of renewable resources, the EMS offers the best solution to these challenges, as it allows for the optimal exchange of energy flow with the conventional grid and avoid power outage [14]. Moreover, it limits the impact of variation in input parameters on the system's stability, safety, and efficiency.

Different approaches are applied to EMS in a microgrid: classical method, meta-heuristic, stochastic, model predictive control, artificial intelligence, and multi-agent systems [15]. Among these, the classical Boolean method is particularly popular, especially for simpler microgrid configurations [16].

This study focuses on a microgrid featuring PV, wind turbine, energy storage system (ESS), and EMS based on classical Boolean method. It also covers MPPT algorithm (P&O) for PV system and wind turbine to extract maximum power and the algorithms implemented within EMS to efficiently manage energy across the entire system. Simulation results are presented to show the response of all components of the system especially EMS, based on wind speed, solar irradiance, temperature, the SOC of the battery, and load profile during a day. Conclusion will be made on the advantages and the drawbacks of the system, after analyzing these results and comparing them with others studies in the same field.

12.2 Description of the studied system

The microgrid to be studied as shown in Figure 12.1 encompasses:

- PV installation which includes seven series-connected PV modules with a maximum power 280 W each, boost converter, and MPPT controller (P&O).

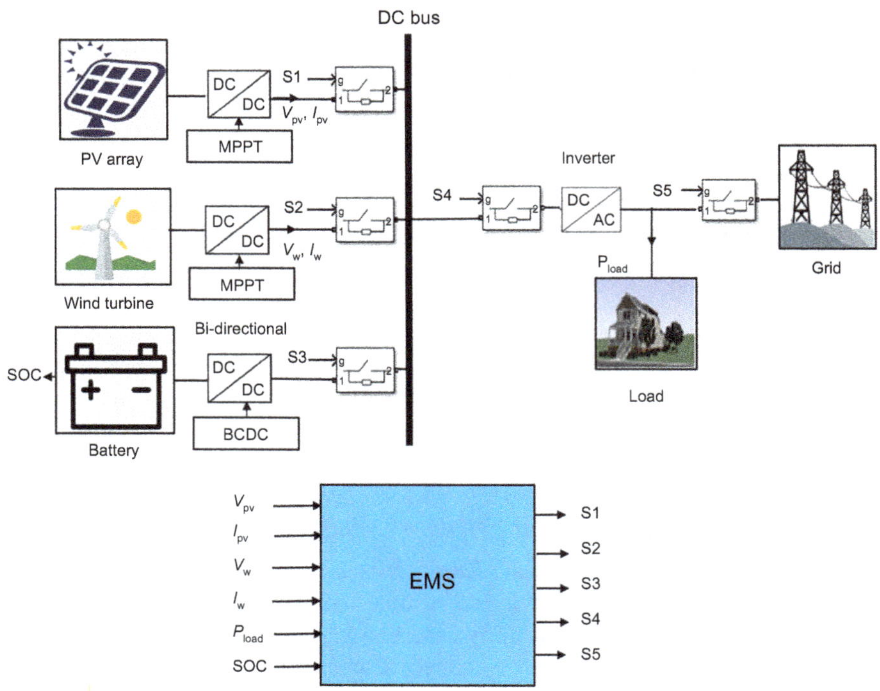

Figure 12.1 Hybrid system combining PV, wind turbine, battery, and energy management system

- Wind turbine installation which includes Wt of 1,500 W nominal mechanical power, PMSM, rectifier, boost converter, and MPPT controller (P&O).
- EES which includes battery, bidirectional converter, charging and discharging controller (BCDC).
- DC bus.
- Inverter.
- AC load with a RMS value voltage of 220 V.
- EMS.
- Automated switches that are controlled by EMS.

The wind turbine installation and PV array are connected to DC bus through DC/DC converter. The battery is also linked to this bus through a bidirectional converter. The load can be either powered by the DC bus or the grid. The excess of the energy can be injected into the grid. If the load requires more energy, the grid will ensure power supply.

12.3 Energy management system

The EMS plays a crucial within the microgrid. Here are some of the main functions it fulfills:

- **Energy production management**: It coordinates energy production from PV and wind turbine. Moreover, it ensures that each source is optimally utilized based on weather conditions, load requirements, and network constraints.
- **Network stabilization**: It contributes to maintaining network stability by regulating voltage, frequency, and other electrical parameters.
- **Demand management**: It can implement demand management strategies to optimize energy use and reduce peak loads. This may include techniques such as load control, load smoothing, and dynamic pricing to encourage more efficient energy consumption.
- **Continuity of power supply**: Ensuring continuous, stable, reliable power supply to the load and avoiding power cut.
- **Energy storage system management**: It controls these storage systems to optimize their use and extend their lifespan.

12.3.1 Energy management system description

An EMS based on classical Boolean methods uses binary logic (true/false or 1/0) to manage the state of switches, devices, and other components within an electrical network. Boolean control is particularly useful in scenarios requiring simple, binary decisions, such as on/off states or connect/disconnect commands, making it a highly reliable approach in structured EMS tasks. Boolean logic forms the basis of rule-based control, where predefined conditions determine the operation of switches and other elements in the system.

EMS as shown in Figure 12.1 has six inputs:

- PV installation output voltage (V_{pv}),
- PV installation output current (I_{pv}),
- Wt installation output voltage (V_w),
- Wt installation output current (I_w),
- Battery state of charge (SOC), and
- Load power (P_{load}).

EMS has also five outputs:

- S1: The switch connecting PV installation with DC bus.
- S2: The switch connecting Wt installation with DC bus.
- S3: The switch connecting ESS with DC bus.
- S4: The switch connecting the load with DC bus.
- S5: The switch connecting the grid with the load.

The renewable power (P_{ren}) can be calculated using the formula below:

$$\boldsymbol{P_{ren}} = \eta_{\boldsymbol{inv}} \cdot (\boldsymbol{P_{PV}} + \boldsymbol{P_W}) = \eta_{\boldsymbol{inv}} \cdot (\boldsymbol{V_{PV}} \cdot \boldsymbol{I_{PV}} + \boldsymbol{V_W} \cdot \boldsymbol{I_w})$$

where

η_{inv}: the efficiency of the inverter
$\boldsymbol{P_{pv}}$: the power generated by PV installation
$\boldsymbol{P_w}$: the power generated by Wt installation

Before coordinating energy flows, EMS must isolate the parts of the microgrid affected by anomalies:

- If the output voltage of PV installation (V_{pv}) is equal to zero that means there is a short circuit in PV installation. It must be disconnected from the microgrid.
- The same for Wt installation

EMS must coordinate energy flows as follows:

- *If* $P_{load} > P_{ren}$: Renewable power generated doesn't meet load demand. There are two cases:
  - *If* $SOC < 10\%$: It is necessary to disconnect the load from the microgrid to avoid deep discharge of the battery. Power supply will be ensured only by the grid.
  - *If* $SOC \geq 10\%$: The battery will compensate the lack of energy.
- *If* $P_{load} \leq P_{ren}$: Renewable power generated exceed load demand. There are two cases:
  - *If* $SOC < 90\%$: The excess of energy will charge the battery.
  - *If* $SOC \geq 90\%$: The battery must be disconnected to avoid over-charge. The excess of energy will be injected into the grid.

EMS block will control the switches based on the conditions mentioned before. The algorithm implemented within EMS will perform this function.

12.3.2 Energy management system algorithm

In Figure 12.2, the EMS block has the role not only of controlling energy exchanges between the different components, on the one hand, but also of regulating the storage charge/discharge process on the other hand (protection against overcharges and deep discharges), to preserve the physicochemical properties of batteries and extend their lifespan. In addition to that, it disconnects the parts affected by anomalies.

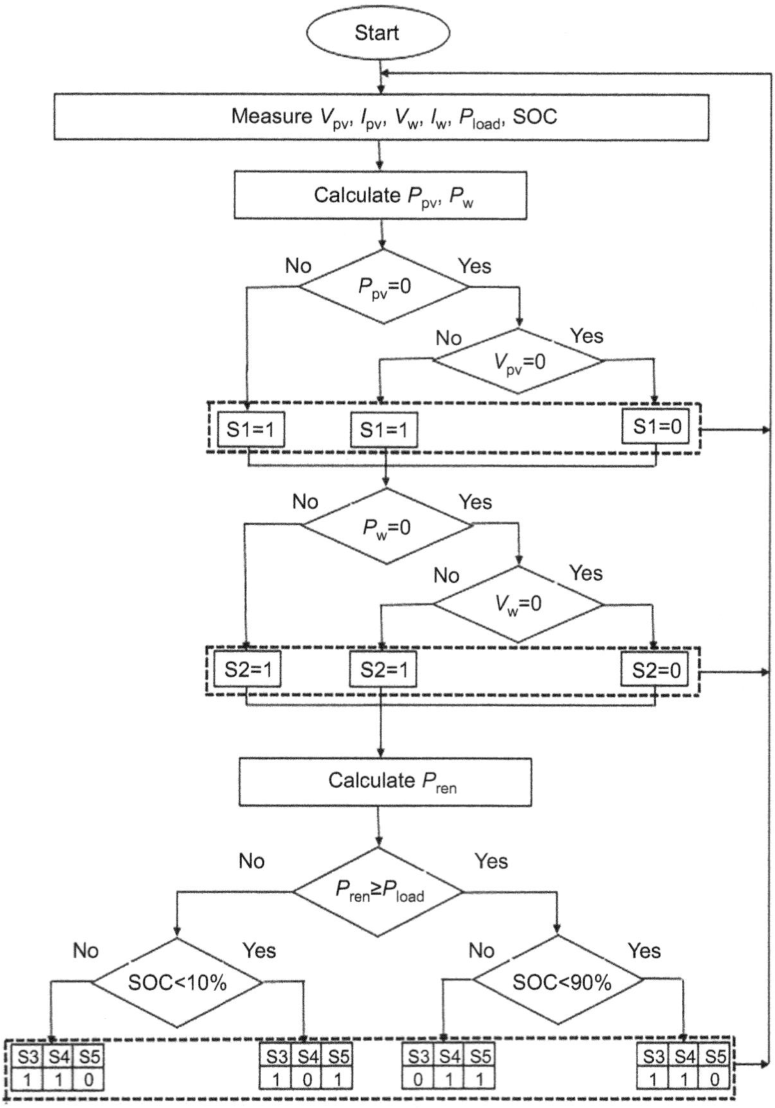

Figure 12.2 EMS algorithm flowchart for the studied system

The management algorithm is based on six scenarios:

- The two first scenarios depend if there are anomalies within PV or Wt installation. In this case, they must be isolated from the microgrid before coordinating energy flow.
- Supply to the consumer and storage of energy coming from the PV and Wt installation if the energy produced is higher than load demand. The battery is connected to DC bus because its SOC is still below the maximum threshold.
- Injection of the excess of energy produced into the grid if the SOC has reached its maximum. In this case, the battery is disconnected from DC bus to avoid overcharge.
- Supply the consumer via the battery and renewable energy if renewable power generated doesn't meet load demand. In this case, the SOC of the battery is still higher than its minimal threshold.
- The renewable energy production doesn't meet load demand and the battery SOC has reached its minimum. All the energy produced will charge the battery and the grid will ensure alone power supply to the load.

12.4 Simulation and discussion

12.4.1 Test conditions for the studied system

In this part, we will study the behavior of the system during a day according to the following test conditions:

- PV, Wt installation and load test conditions are shown in Figure 12.3.
- The initial SOC of the battery is equal to 90%.
- There are no anomalies within PV or Wt installation.

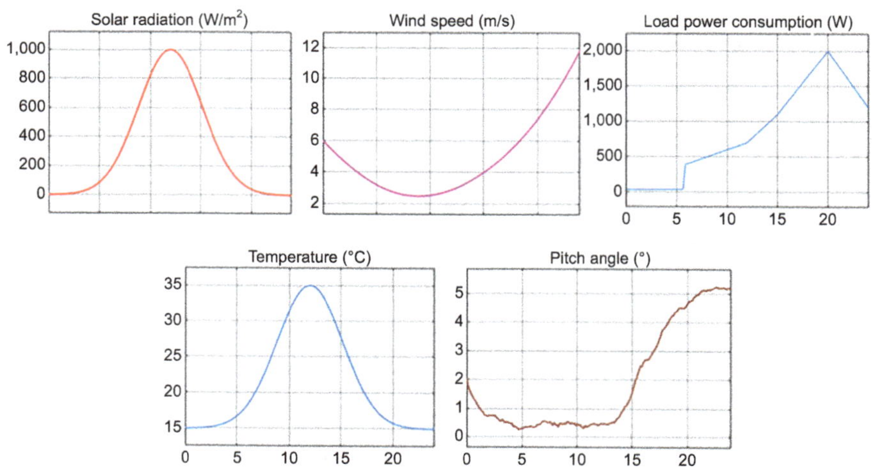

Figure 12.3 Test conditions for PV, Wt, and the load

Solar irradiance reaches its highest point at noon and is minimal at night, while temperature fluctuates between 15 °C and 35 °C throughout the day. Wind speed peaks at midnight and decreases to its lowest level around 10 am, a pattern also followed by the pitch angle. Load power consumption is lowest between midnight and 5:45 am, gradually increasing afterward to reach its peak at 8 pm, with a maximum load capacity of 2,000 W.

12.4.2 Simulation results

After running the simulation, we obtain the following graphs illustrating the renewable power generated, load demand, battery state of charge (SOC), and switch signals (Figure 12.4). Since there are no anomalies within PV or Wt installation, S1 and S2 signal keep the same value 1. Only three switches signals are shown (S3, S4, S5).

From midnight to 12:30 am, the renewable energy generated is slightly less than the load demand, with the battery compensating for this minimal shortfall, keeping the SOC near 90%. During this time, switches 3 and 4 are closed, while switch 5 remains open. From 12:30 am to 5:45 am, renewable power generation surpasses load consumption, quickly bringing the battery SOC to 90%, after which it disconnects from the microgrid, and excess energy is directed to the grid. In this period, switches 4 and 5 are closed, while switch 3 is open.

Between 5:45 and 7:25 am, load demand exceeds generated power, causing the battery to discharge. However, from 7:25 to 9:30 am, power generation again surpasses load demand, and the battery's SOC increases until it reaches 90% at 9:30 am. During this interval, switches 3 and 4 are closed, and switch 5 remains open. From 9:30 am to 3:00 pm, renewable energy continues to exceed demand, the battery remains disconnected, and excess energy is once more directed into the grid. Switches 4 and 5 are closed, with switch 3 open.

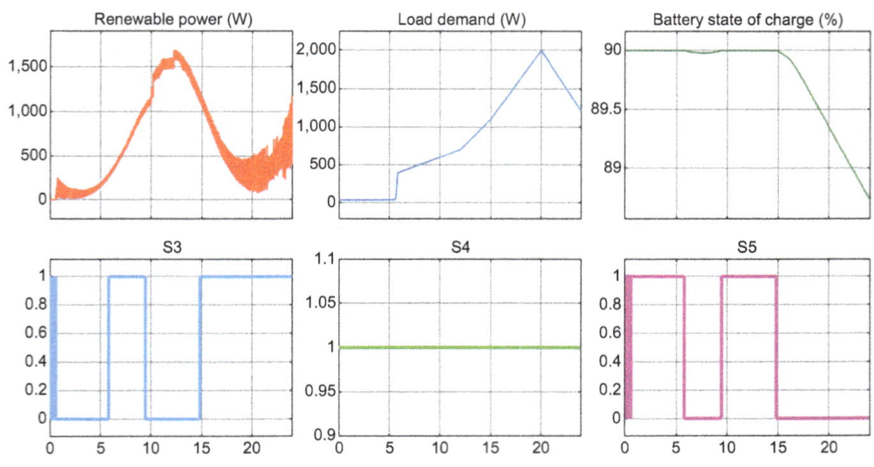

Figure 12.4 Simulation results for renewable power, load power, and battery SOC and switch signals

Finally, from 3:00 pm to midnight, load demand exceeds renewable generation, and the battery compensates for this shortage. Throughout this period, switches 3 and 4 are closed, and switch 5 remains open.

12.5 Conclusion

Based on the results obtained, we conclude that a hybrid system is highly effective for providing a continuous supply of electrical energy. Solar radiation and temperature are critical factors influencing PV panel output, while wind speed significantly impacts wind turbine productivity. MPPT controllers within the microgrid play an essential role in optimizing system performance and efficiency. The EMS enables the hybrid system to meet load demands, maintain microgrid stability, extend component lifespan, and manage optimal energy exchange with the grid. Additionally, the EMS ensures microgrid protection by detecting and addressing system anomalies.

Boolean control, being straightforward to design, implement, and troubleshoot, is well suited for applications requiring simple, reliable operations with consistent outputs. Its capability for real-time switch operations with minimal computation makes it ideal for fast-response applications such as fault isolation and load shedding. However, Boolean methods are less effective for complex decision-making involving non-binary variables or multiple interdependent objectives, and their rule-based nature limits adaptability in dynamic environments such as grids with high renewable penetration. Moreover, once established, Boolean rules can be inflexible, requiring time-consuming reprogramming for large systems when adjustments are necessary.

References

[1] Lahbous, M., Touzani, L., and Ousidahou, A. (2023). LA SECHERESSE ET SON IMPACT SUR L'ECONOMIE MAROCAINE. *La Revue D'économie et D'environnement*, *3*(2), 20–42.

[2] Ben-Meir, Y., Opfer, K., and Et Akhezzane, I. (2021). Mobilizing decentralized, participatory energy transition in Morocco. *Georgetown Journal of International Affairs*, *22*(1), 80–89.

[3] Thomas, D., Deblecker, O., Genikomsakis, K., and Ioakimidis, C. S. (2017). Smart house operation under PV and load demand uncertainty considering EV and storage utilization. In *IECON 2017-43rd Annual Conference of the IEEE Industrial Electronics Society* (pp. 3644–3649). Piscataway, NJ: IEEE.

[4] Ocon, J. D., Cruz, S. M. M., Castro, M. T., Aviso, K. B., Tan, R. G. R., and Promentilla, M. A. B. (2018). Optimal multi-criteria selection of hybrid energy systems for off-grid electrification. *Chemical Engineering Transactions*, *70*, 367.

[5] Idrissi, Y. E. A., Assalaou, K., Elmahni, L., and Aitiaz, E. (2022). New improved MPPT based on artificial neural network and PI controller for

photovoltaic applications. *International Journal of Power Electronics and Drive Systems*, *13*(3), 1791–1801.

[6] Soetedjo, A., Lomi, A., and Mulayanto, W. P. (2011). Modeling of wind energy system with MPPT control. In *Proceedings of the 2011 International Conference on Electrical Engineering and Informatics* (pp. 1–6). Piscataway, NJ: IEEE.

[7] Adaliou, A. H., Lamreoua, A., Isknan, I., *et al.* (2023). High-performance MPPT based on developed fast convergence for PV system-experimental validation. In *International Conference on Digital Technologies and Applications* (pp. 694–703). Cham: Springer Nature Switzerland.

[8] Šimelytė, A. (2020). Promotion of renewable energy in Morocco. In M. Tvaronavičienė and B. Ślusarczyk (eds.), *Energy Transformation Towards Sustainability* (pp. 249–287). Amsterdam: Elsevier.

[9] Rhiat, M., Karrouchi, M., Hassari, A., *et al.* (2023). Design and simulation of a low-cost solar irradiance meter for PV applications. *E3S Web of Conferences*, *469*, 00076.

[10] Djilali, A. B., Yahdou, A., Bounadja, E., Benbouhenni, H., Zellouma, D., and Colak, I. (2024). Energy management of the hybrid power system based on improved intelligent perturb and observe control using battery storage systems. *Energy Reports*, *11*, 1611–1626.

[11] Chrifi-Alaoui, L., Drid, S., Ouriagli, M., and Mehdi, D. (2023). Overview of photovoltaic and wind electrical power hybrid systems. *Energies*, *16*(12), 4778.

[12] Bakkari, M., Bossoufi, B., El Kafazi, I., Bouderbala, M., and Karim, M. (2024). A review of wind energy potential in Morocco: New challenges and perspectives. *Wind Engineering*, *48*(1), 101–120.

[13] Tazi, G., Jbaihi, O., Ghennioui, A., Merrouni, A. A., and Bakkali, M. (2018). Estimating the renewable energy potential in Morocco: Solar energy as a case study. *IOP Conference Series: Earth and Environmental Science*, *161*(1), 012015.

[14] Kermani, A., Jamshidi, A. M., Mahdavi, Z., *et al.* (2023). Energy management system for smart grid in the presence of energy storage and photovoltaic systems. *International Journal of Photoenergy*, *2023*(1), 5749756.

[15] Eyimaya, S. E., and Altin, N. (2024). Review of energy management systems in microgrids. *Applied Sciences*, *14*(3), 1249.

[16] Abdulgader, M., Yang, C., and Kaur, D. (2016). "Efficient energy management system based on GA and classical Boolean approach." *International Journal of Computational Intelligence and Applications*, *15*(3), 1650012.

Chapter 13

Control of photovoltaic-wind energy systems using MPC and PSO

Chaymae Boubii[1,2], Ismail El Kafazi[2] and Rachid Bannari[1]

Abstract

The integration of renewable energy sources, particularly photovoltaic (PV) and wind energy, presents significant challenges due to their inherent variability and nonlinear dynamics. This chapter introduces a novel hybrid energy system that combines PV and wind power, managed by an advanced control strategy that integrates model predictive control (MPC) with particle swarm optimization (PSO). The MPC&PSO approach is designed to optimize power output and ensure the stability of the system under dynamic environmental conditions.

Through detailed simulations conducted in MATLAB/Simulink, the proposed control strategy demonstrated superior performance in maintaining high efficiency and stability across a range of operating scenarios. The stability of the system was rigorously analyzed using Lyapunov's theorem, confirming its ability to operate reliably even in the face of rapid fluctuations in solar irradiance and wind speed.

The findings suggest that the MPC&PSO hybrid control strategy offers a robust solution for managing the complexities of renewable energy integration, providing both theoretical and practical contributions to the development of sustainable energy systems. Future research directions include the application of this control strategy to other renewable energy systems and real-world implementation for further validation.

Keywords: Hybrid system; MPC; PSO; Lyapunov

[1]Engineering Sciences Laboratory, Écoles Nationales des Sciences Appliquées (ENSA), Ibn Tofail University, Kenitra, Morocco

[2]Laboratory SMARTILAB, Moroccan School Engineering Sciences, École Marocaine des Sciences de l'Ingénieur (EMSI), Rabat, Morocco

13.1 Introduction

Over the past few years, the major impetus of this energy transition has come from addressing two parallel challenges: meeting growing demand for power and stemming spiraling carbon emissions. With diminishing fossil fuel reserves and increasing environmental concerns, renewable energy sources such as wind and solar have both become viable alternatives. Although these sources of energy are relatively common and renewable, they come with an inherent variability and unpredictability that represents formidable challenges for both grid stability and efficient operation management [1–7].

Solar photovoltaic (PV) systems and wind turbines are two of the things that come to mind when thinking about renewables; this is because PVs convert sunlight into electricity and in turn give a shift of energy converted by force from moving air (kinetic) during high-speed winds seasons. However, since these energy sources are intermittent, their power output fluctuates, which makes grid stabilization and reliable supply challenging [8–10]. Schematically, advanced control strategies must be developed to handle real-time coordination and dynamic interactions of hybrid energy systems.

This chapter addresses these constraints using a novel hybrid energy system blending PV and wind power based on advanced control strategies. The operation of the proposed system is designed while considering energy production and grid stability simultaneously through model predictive control (MPC) aided with particle swarm optimization (PSO) [11–16]. MPC can predict future states of the system and, considering these predictions, determines actuator values to infer control input; this is opposed with PSO augmenting optimization through fast optimal controller parameters search [11–16]. It is also demonstrated that the system can provide a stable generation of power and efficient energy integration, even under variability characteristics typically associated with wind and solar resources based on rigorous Lyapunov analysis to validate its stability.

This welcomes the contribution to the development of renewable energy technologies, given this new perspective on hybrid energy system management. Combining MPC with PSO significantly advances a control solution for renewable energy systems by providing resistance against the uncertainty, variation, and unpredictability of these civilian size grids. Thus, it includes good response to existing gaps in renewable energy field and also establishes a platform for future advancement forms of theories on sustainable functions [16].

13.2 Background/theoretical framework

13.2.1 The growing demand for renewable energy

In order to meet growing hunger energy and address the environmental consequences of burning fossil fuels, nations around the world are transforming their energy landscape. Thus, solar and wind power sources have become more recognized as successful alternative power resources because they are highly available that yields in renewable energy source [17]. A few decades ago, coal was the dominant source of electricity generation, while renewable resources had a minimal share. Today, wind farms and solar power plants are increasingly integrated into transmission grids, reducing GHG emissions. However, their unpredictable nature presents challenges for grid integration [2–4].

13.2.2 Challenges in renewable energy integration

Solar PV systems, along with wind turbines, are principal technologies of the renewable energy industry. Where PV installations convert solar energy into useable electricity, wind turbines generate electric power from kinetic energy in the wind. Both solar PV and wind power generation are weather-dependent, and therefore intermittent. Such intermittent power output causes the grid to enter an unstable status and it is not easy for a reliable continuous supply of electric power, which threatens the economic development of society. The cost reduction trajectory for grid-scale renewable energy integration is no longer straightforward [5–7]. The increasing penetration of renewable energy technologies has underscored the deficiencies in traditional fossil-fuel based power systems to handle these fluctuations, reiterating the necessity for state-of-the-art control strategies.

13.2.3 Hybrid renewable energy systems

Hybrid energy systems comprising PV and wind power are suggested as a feasible solution to stabilize the variability effects in renewable energy generation. Because solar power is generally available during the day, while wind power may be more abundant at night or in different weather conditions (depending on how they complement each other), a hybrid system can offer greater stability and consistency over time [8–10]. But integrating these systems does add complexity in how we manage and control those systems, so now you need more sophisticated strategies to get the best out of the investment.

13.2.4 Advanced control strategies: MPC and PSO

Many advanced control strategies have been applied to solve the difficulties of hybrid renewable energy systems, for example MPC and PSO. MPC is a control strategy that forecasts the future behavior of a system and alters its manipulated variables to meet desired objectives. This is especially useful in dealing with the multi-timescale, probabilistic, and dynamic nature of most renewable energy systems [18–20]. Unlike PSO, which is an optimization algorithm inspired by the social behavior of birds and fish, MPC predicts future system behavior and adjusts control inputs accordingly. PSO is used to optimize control parameters iteratively by evaluating different potential solutions and selecting the best one based on predefined quality metrics [21–23].

The integration of MPC and PSO for managing hybrid PV-wind systems provides a conservative method to optimize the output power with system stability.

MPC provides a MPC frame, while PSO enhances the optimization process by efficiently solving for optimal values and enables searching the best control parameters. This hybrid method efficiently manages renewable energy sources despite their inherent nonlinearities and uncertainties.

13.2.5 Stability analysis using Lyapunov's theorem

The integration of hybrid renewable energy systems in power grid necessitates stable generation [24]. The theory of stability analysis in control systems is well grounded based on Lyapunov's theorem. Applying Lyapunov's method, the stability of a system that experiences fluctuation in wind and solar power generation is guaranteed [16]. This body of theory underpins the building of robust hybrid energy systems for grid stability.

13.3 Methodology

13.3.1 System configuration

The proposed system is a hybrid energy system that consists of both PV and wind energy systems in order to add up on the efficiency level when compared with individual implementations. At the heart of the system are solar panels, a wind turbine, power converters, and grid connection, orchestrated to cooperate in an optimized manner for high energy production with secure operation within a modern electrical network.

The solar panels installed in a series (or parallel) circuit are called PV. Most solar PV systems convert sunlight into electricity or direct current (DC) using photovoltaic technology. The DC–DC boost converter increases the voltage from the solar panels to a level suitable for grid integration [25]. Within the PV system, maximum power point tracking (MPPT) technology is employed to adjust the operating point of panels as needed in order for them to output their maximum power under different environmental conditions.

This configuration uses a doubly fed induction generator (DFIG) that is connected to the rotor of wind turbine-based generator. The DFIG is used for wind-power-generating systems since the DFIG system can attain high efficiency across a broad range of wind speeds, owing to its variable-speed control. This flexible behavior is obtained using two power converters: the rotor side converter (RSC) and the grid side converter (GSC). The RSC regulates rotor currents to control both active and reactive power, while the GSC maintains the DC-link voltage, and ensures synchronization with the grid frequency [26].

The PV and wind systems are integrated through a common DC link that combines the outputs of the PV boost converter with those from DFIG power converters. A common bus ensures seamless, bi-directional flow of energy between AC and DC sides as well the grid resulting in optimal utilization of power flows. The common grid-connected invertor synchronizes the AC power generated by both PV systems (after inversion) and wind system, with its output back-injected into the grid. This inverter not only changes the power from DC source at shared

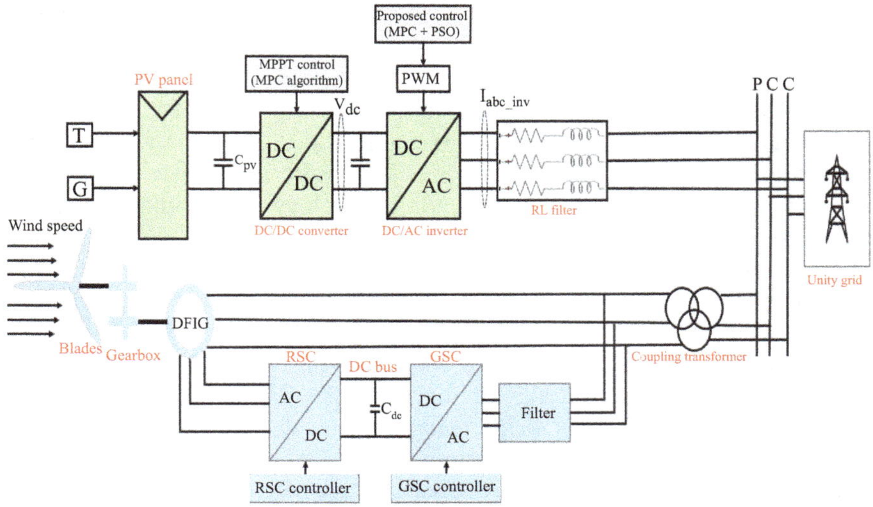

Figure 13.1 Hybrid system

bus into suitable current for grid use, i.e., AC, but also plays a vital part into maintaining high standards of power quality as well as system stability with hybrid and regional network functionalities, according to [20].

Figure 13.1 shows the schematic of the hybrid system, including PV and wind energy systems in combination with associated power converters connected to grid [7]. In the above figure, one can interpret how this energy is flowing from renewable sources to DC/AC inverters followed by having RL filters and then coupling transformers before tangential entry into unity grid through point of common coupling. Figure 13.1 also focuses on utilization of advanced control strategies, i.e., MPC for MPPT control of PV system and the combined hybridization approach, the MPC&PSO as an overall controlled methodology.

Besides the hardware, it offers advance control algorithms to coordinate all complex activities between PV and wind with grid. These algorithms will optimize power outputs for both renewable sources while guaranteeing that the energy supplied to the grid meets all relevant stability and quality specifications.

In summary, the format of choice for a hybrid system is to beat down general nature challenge between solar and wind. The system will use the two sources together to create a more dependable and consistent flow of power, which means it won't be as affected by the natural variability in each.

13.3.2 Control strategy

A hybrid PV and wind energy system is much more efficient to work using some advanced control strategies such as the management of renewable energies which have a higher portion nonlinear dynamics and variability. To accomplish this, the proposed system utilizes MPC as well as PSO, acting for optimization of performance with stability.

A robust control strategy to manage hybrid system in dynamic conditions based on MPC is the heart of this approach. The benefits of MPC for such a situation are inevitability monitoring its future states and adapting the inputs accordingly. This predictive capability is of particular importance to renewable energy systems, where environmental conditions such as solar irradiance and wind speed can evolve at a rapid pace [27].

In Figure 13.2, the MPC controller optimizes for power output from both PV and wind in a hybrid scenario. The main control function for the PV system, MPC is a part of the MPPT algorithm that dynamically adjusts solar panel operating point to harvest an optimal amount of energy under changing environmental conditions. MPC-based MPPT: The MPC algorithm predicts the optimal voltage and current levels, ensuring that the PV system operates at its maximum power point under varying sunlight conditions. The algorithm of the MPC control for MPPT in a PV system is shown in Figure 13.2, which presents each step from input initialization to cost function minimization [27].

MPC is used to control the operation of DFIG for a wind energy system. The control is achieved by controlling rotor currents through the RSC, thus varying active and reactive power output of DFIG. The GSC controls the DC-link voltage and regulates power injected into/taken off the grid synchronized with respect to

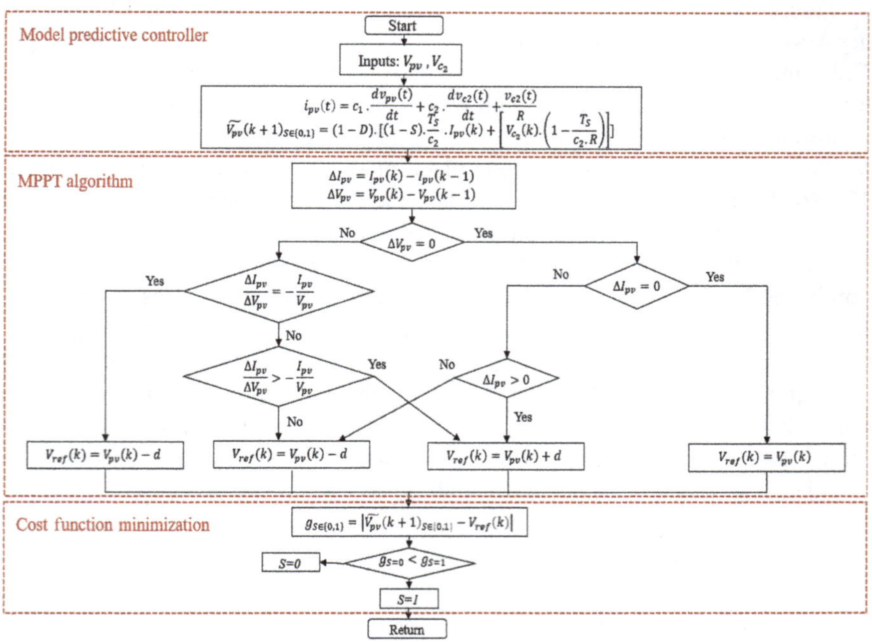

Figure 13.2 MPC algorithm

grid frequency. MPC ensures the DFIG operates efficiently over a wide range of wind speeds by predicting how both the future behavior of the wind turbine and associated power converters will interact with it [16].

PSO is embedded within the MPC framework to enhance the optimization process by fine-tuning controller gains and handling nonlinearities. PSO is a derivative-free algorithm for solving unconstrained optimization problems based on swarm intelligence. It iteratively improves candidate solutions in a population of existing "swarm" or potential solutions. This study used the structure of PSO, as shown in Figure 13.3, starting from parameter initialization and finishing at optimal control values selection [28].

In this chapter, we apply PSO for fine-tuning the cost function in the MPC framework using a hybrid system. The cost function often contains elements of power delivered, the stability of system operation and energy efficiency. PSO continuously changes the control parameters to reduce this cost function so that it is at its lowest where better and optimal-system operation happens. For instance, PSO is applied to optimize the power converters by adjusting duty cycle, and also the reference signals for inverters; these are key factors that affect system performance.

It balances the prediction accuracy with an optimization-based efficiency, which makes a combination of them strongly suitable for control. MPC provides an accurate control action through the prediction of system future behavior and controlling at real time, while PSO further improves this procedure by global optimization of all control parameters. The proposed hybrid control strategy has many advantages in minimizing the effects of uncertainties and nonlinearities that are present in renewable energy systems, which makes a PV-wind power system work perfectly on both efficiency improvement and stable operation.

At the end, MPC and PSO used in the control strategy together provide high efficiency, maximize power output of hybrid system easily to be able achieved better stability with changeable environment. Not only does this advanced control

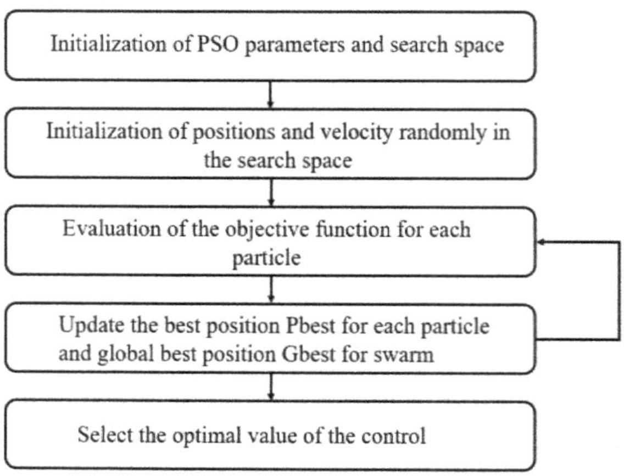

Figure 13.3 PSO steps

strategy enhance the operation of all these systems, it is also acts as a pioneer to vitalize and green up renewable energy infiltration inside grid environment.

13.3.3 Stability analysis using Lyapunov's theorem

Ensuring the stability of a hybrid PV and wind energy system is essential for reliable operation in grid-connected applications. Stability analysis is necessary to ensure that the system can operate in a stable way regardless of solar irradiance and wind speed variations which are constant features of sequential control systems. This chapter is based on Lyapunov's theorem to effectively prove that these control strategies are stable.

13.3.3.1 Application of Lyapunov's theorem

In mathematical form, Lyapunov's theorem serves as an important tool used to verify the stability of dynamic systems. Lyapunov's theorem states that a system is stable if there exists a scalar function (Lyapunov function) that is positive definite and has a negative definite time derivative. This technique is commonly used in control theory to demonstrate the stability of nonlinear systems, including hybrid energy system discussed here.

13.3.3.2 Definition of the Lyapunov function for the hybrid system

Lyapunov function is formulated for the hybrid system based on its component energy storage and used to apply Lyapunov's theorem. Lyapunov function is a scalar function of the system state variables (i.e., voltage and current in PV or wind subsystems). The function is designed to be positive definite, meaning that $V(x) > 0$ for all non-zero states of the system, and $V(x) = 0$ only when the system is at equilibrium.

For the PV subsystem, the Lyapunov function $V_{pv}(x)$ is based on the energy stored in the capacitor of the boost converter:

$$V_{pv}(x) = \frac{1}{2}.c_2.V_{c2}^2 \tag{13.1}$$

where c_2 is the capacitance of the capacitor and V_{c2} is the voltage at its terminals.

For the wind subsystem, the Lyapunov function $V_{wind}(x)$ is defined in terms of the kinetic energy of the rotor and the magnetic energy stored in the rotor inductance:

$$V_{wind}(x) = \frac{1}{2} \cdot J \cdot \omega_r^2 + \frac{1}{2} \cdot L_r \cdot I_{rd}^2 + \frac{1}{2} \cdot L_r \cdot I_{rq}^2 \tag{13.2}$$

where J is the rotor moment of inertia, ω_r is the rotor angular velocity, and L_r is the rotor inductance. The terms I_{rd} and I_{rq} represent the rotor currents in the d and q axes, respectively.

The Lyapunov function for the complete hybrid system is then:

$$V(x) = V_{pv}(x) + V_{wind}(x) \tag{13.3}$$

$$V(x) = \frac{1}{2} \cdot c_2 \cdot V_{c2}^2 + \frac{1}{2} \cdot J \cdot \omega_r^2 + \frac{1}{2} \cdot L_r \cdot I_{rd}^2 + \frac{1}{2} \cdot L_r \cdot I_{rq}^2 \tag{13.4}$$

13.3.3.3 Positivity of the Lyapunov function

The first condition of the Lyapunov theorem requires that the function $V(x)$ be positive for all values of the state of the system, except at a point where it is zero (usually the origin, corresponding to the equilibrium state).

In our case, $V(x)$ is the sum of quadratic terms representing stored energies, which are always positive or zero. Thus, $V(x) > 0$ for any state x other than the steady state (where $V_{c2} = 0$, $\omega_r = 0$, $I_{rd} = 0$, $I_{rq} = 0$).

13.3.3.4 Calculation of the time derivative of the Lyapunov function

Stability is guaranteed if the time derivative of $V(x)$, noted $\dot{V}(x)$, is negative or zero for all x. So, let's calculate this derivative for the hybrid system.

The time derivative of $V_{pv}(x)$ is given by

$$\dot{V_{PV}}(x) = c_2.V_{c2}.\dot{V_{c2}} \tag{13.5}$$

From the previous analysis of the PV subsystem, we know $\dot{V_{PV}}(x) \leq 0$, which means that the energy stored in the capacitor can only increase or remain constant, but does not decrease.

Similarly, the time derivative of $V_{wind}(x)$ is

$$V_{\dot{wind}}(x) = J.\ \omega_r\ .\ \dot{\omega}_r + L_r\ .\ I_{rd}\ .\ \dot{I_{rd}} + L_r\ .\ I_{rq}\ .\ \dot{I_{rq}} \tag{13.6}$$

Previous analyses have shown $V_{\dot{wind}}(x) \leq 0$, which ensures that the kinetic energy of the rotor and the magnetic energy in the rotor inductors can only increase or remain constant.

Thus, the total time derivative of the Lyapunov function for the hybrid system is

$$\dot{V}(x) = \dot{V_{PV}}(x) + V_{\dot{wind}}(x) \leq 0 \tag{13.7}$$

13.3.3.5 Conclusion of the hybrid system stability

On the hybrid system, one can show that this Lyapunov function φ is positive and negative τ with respect to the time. This, according to the Lyapunov theorem, means that hybrid system is stable. This is crucial to maintain the system in a stable and efficient manner even under different environmental conditions or consumptions.

Moreover, these results ensure the reliability of predictive control with PSO for hybrid skimmed-milks system (MPC&PSO) used in this chapter and will present further theoretical validation to its eventual field application.

In hybrid PV-wind system, Lyapunov function has to be designed considering the coupling of two systems, i.e. (PV and wind) as well interaction from control inputs based on MPC&PSO techniques. Based on the convergence analysis, system stability with nonlinear compensators is guaranteed over a wide range of operational regions using this control strategy. That ranges from changes in environmental factors like sudden shifts in solar irradiance and wind speed that are frequent under real applications.

Therefore, the employment of Lyapunov's theorem in this work proves a solid ground for stability assurance of hybrid PV-wind system. The validity of the proposed method is proved by rigorous theoretical analysis on stability, which strengthens its reliability and efficiency to some extent for implementation in modern power grids with hybrid systems developed here.

13.3.4 Simulation setup

MATLAB®/Simulink® is one of the most popular simulation tools available, as it provides a wide array of features for modeling and simulating dynamic systems effectively. Simulation environment was set up to faithfully mimic the actual working conditions of hybrid system such as interaction between PV subsystem and wind subsystem, power converters, control algorithms.

System modeling: The hybrid system was simulated in MATLAB/Simulink using models of both the PV and wind energy subsystems.

The model is based on the single-diode equivalent circuit, to capture the nonlinear behavior of a solar panel. The dynamical equations that geometrically describe the generator itself together with those of RSC and GSC are considered to model wind energy system, in particular DFIG. The incorporation of these models in a common DC bus enabled the simulation under several environmental conditions for a complete hybrid system.

Implementation of the control algorithm: MPC and PSO algorithms were implemented in a Simulink environment. An MPC strategy for the simultaneous optimization of MPPT in PV system and DFIG operation control is proposed. The control parameters of MPC framework were optimized using this algorithm so that the system can work at its best-known efficiency using PSO algorithm.

Real-time simulations were used to evaluate the performance of the control algorithms for a range of scenarios.

Set of simulations: We ran simulations with different settings to assess the robustness, and figure out what benefits this could bring. Different operational conditions were simulated by varying key parameters like solar irradiance, wind speed, and grid voltage. Irradiance levels for the PV system varied between 200 W/m^2 to 1,000 W/m^2 (Figure 13.4), corresponding from cloudy to clear sky conditions (Figure 13.5). Wind speeds range from 5 m/s to 10 m/s which is within the typical operational wind speed for a wind turbine, as shown in Figure 13.5 The grid voltage was set to the nominal value, with slight deviations being introduced in order to mimic some of the most common disturbances on distribution network.

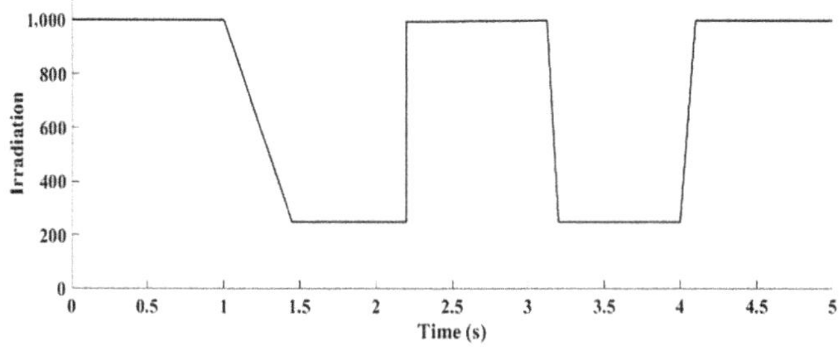

Figure 13.4 Variable irradiations

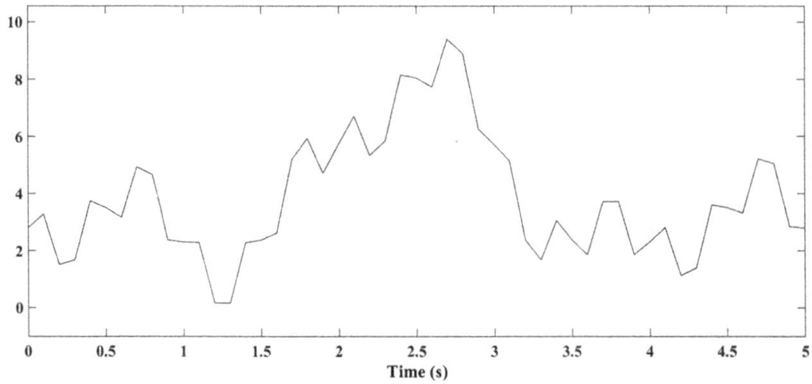

Figure 13.5 Wind speed

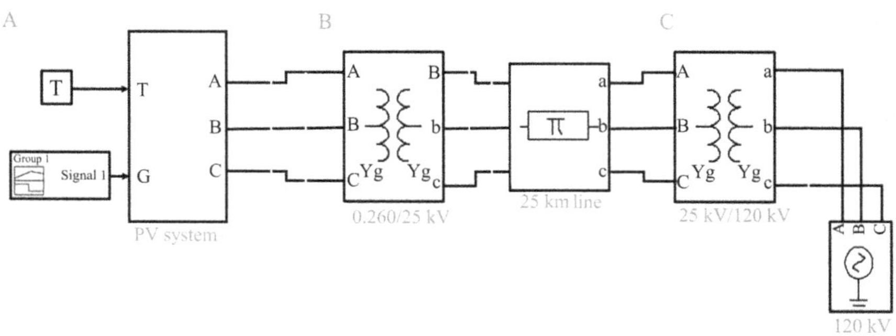

Figure 13.6 PV system

13.3.4.1 Simulation results

Figure 13.6 shows the schematic of the PV) system. The PV system in Figure 13.6 consists of a few basic materials, which is properly converting sun energy into electric electricity and directly reflow data together with grid. How the system works: the PV array (section A) is where solar energy shoots down and appears as electricity. The DC output of the PV array is then processed by a DC–AC inverter, which converts it into AC, ready for grid connection.

The higher incoming AC power is then stepped up from a lower voltage level of the PV system to an intermediate higher (section B) using step-up transformer. The case of voltage then going from 0.26 kV to 25 kV. This intermediate voltage level is crucial in helping transmit signals over longer distances.

This power goes through transmission line (part C), 25 km distance, to minimize the losses of power during its transmitted. The transmission line is modeled to simulate how it would behave with real-world transmission, impedance, and losses.

At the receiving end, a transformer steps up the voltage from 25 kV to 120 kV before feeding it into the high-voltage grid. The last product at 120 kV is then sent to the grid for accommodating end-consumers.

The general architecture of the wind energy system shown in Figure 13.7 is similar to that used in PV systems but this one harvest wind energy instead. The system starts with a wind turbine (part A), which changes the kinetic energy of the wind into mechanical power. The generator transforms this mechanical energy into electrical energy, delivering alternating current to the internal components of the system (with a voltage level appropriate for that in use within such systems).

The output AC power from the wind turbine is initially at a lower voltage level, which is stepped up from 0.575 kV to 25 kV by a step-up transformer (section B). This voltage level is typically chosen to match transmission requirements and decrease losses at long distances.

Similarly, the power is sent back over a 25 km transmission line (section C) in parallel to PV system. The transmission line has been optimized to minimize power losses and increase the amount of efficient conveyed electricity at a grid connection point. A further transformer transmits 25–120 kV which is higher voltage and compatible with a high-voltage grid connection. The final output is now rated at 120 kV. The line then ties into the grid and can be distributed out as needed for consumption.

Again, both the PV and wind systems have been designed to work independently although they share a common topological structure that makes them straightforward

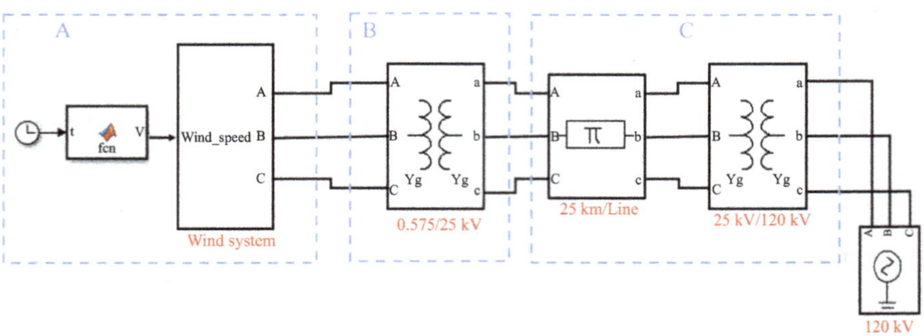

Figure 13.7 Wind system

in their integration with the grid. At different levels, step-up transformers are used, making it easy for the energy produced at these renewable sources to be transmitted across long distances with least losses. The 120-kV final voltage level also allows the proper high-voltage grid connection for both systems, which is crucial when it comes to transmitting renewable energy over long distances.

It is important to evaluate the performance of these systems in different wind and solar irradiance operating conditions that require simulations. Analyzing this behavior in a simulated environment allows to improve their respective design and operation so that they help effectively with grid stability and reliability.

13.3.4.2 Results: PV system

Figure 13.8 presents part A output of the PV system over 5 s period. The waveform at this time has high frequency oscillate over the fundamentals, because of

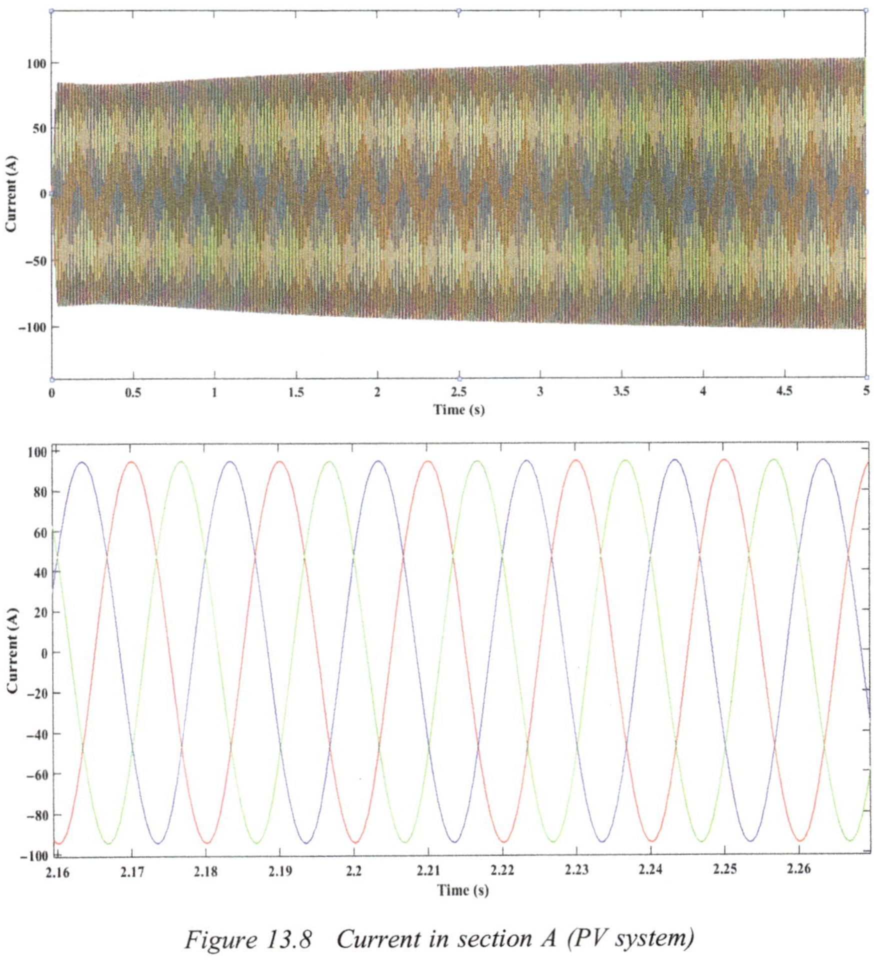

Figure 13.8 Current in section A (PV system)

switching harmonics that are generally added by power electronics (e.g., inverters or converters in a system).

Despite these fluctuations, the current flow generally remains consistent but for small variations induced by a load alteration or changes in operating conditions. This means that the PV system is operating properly and outputting a stable current at this time.

The illustrated active power (P_s) and reactive power (Q_s) generated by section A of the PV system is shown in Figure 13.9. The active power (red) begins to gradually rise over the course of the simulation, reflecting how well a system can convert solar energy into power visually. As the reactor end controls active power, we can see in green that reactive power remains very low through output. This is desirable for grid-connected systems since it keeps the reactive power exchange with the grid to a minimum which reduces losses and increases overall efficiency.

Finally, Figure 13.10 gives the waveforms that are developed by section A of voltage and current of the PV system. The signal (red) represents waveform and the signal (blue) deals with current, both sinusoidal just as we would expect from a basic inductive circuit which means a phase lagging of current and voltage. The regularity of these sine waves confirms that the PV system is producing clean, steady power free from distortion (which will be explained in detail later), which is necessary for it to synchronize with the grid without issues and also avoid losses. The lower graph in Figure 13.10 also reveals the stable voltage and current output over a 5-s window, underscoring that this is an essential part of what makes it possible for us to show constant performance at steady state.

This is the voltage output from section B of the PV system for 5 s as shown in Figure 13.11. These high-frequency oscillations overlay on top of the main waveforms in voltage indicating switching harmonics produced by power electronic devices (inverters, converters) used within the system. These vibrations are common to the systems which employ pulse-width modulation for voltage regulation.

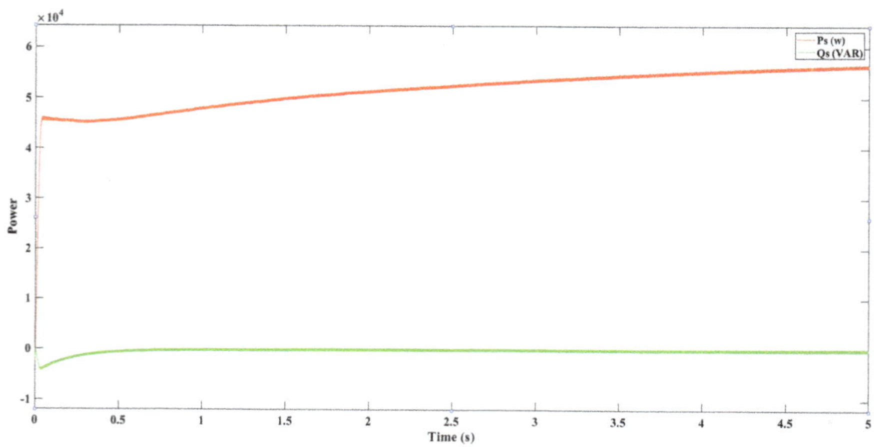

Figure 13.9 Active and reactive power in section A (PV system)

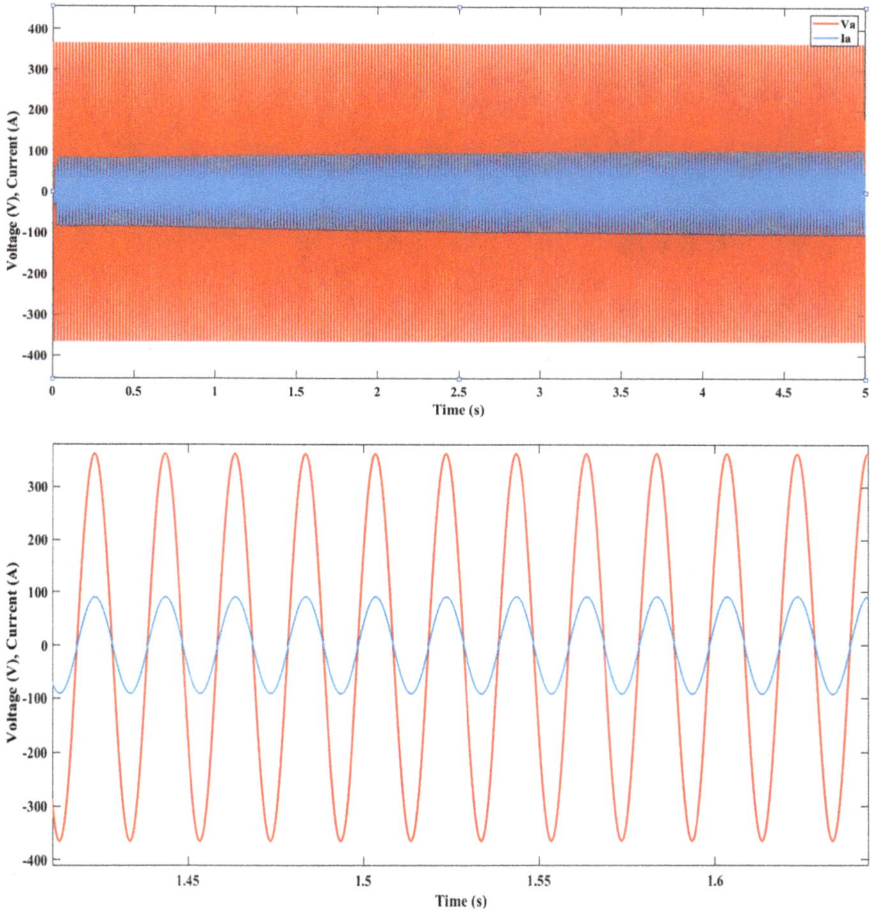

Figure 13.10 Voltage and current in section A (PV system)

The high-frequency components are due to ripple but the output voltage waveform otherwise remains stable indicating that under normal conditions, such as load variations, this system may also be able to maintain a constant V_{out} state.

Figure 13.12 shows the current output from section B of the PV system, depicting the current in each phase (I_a, I_b, I_c) over the same 5-s period. Similar to the voltage waveform, the current also exhibits high-frequency oscillations, which are consistent across all three phases. The balanced nature of these waveforms across the phases indicates that the system is operating symmetrically, which is crucial for minimizing losses and ensuring efficient power transfer. The presence of consistent oscillations further suggests that the current output is stable, with no significant deviations or distortions that could impact the performance of the PV system.

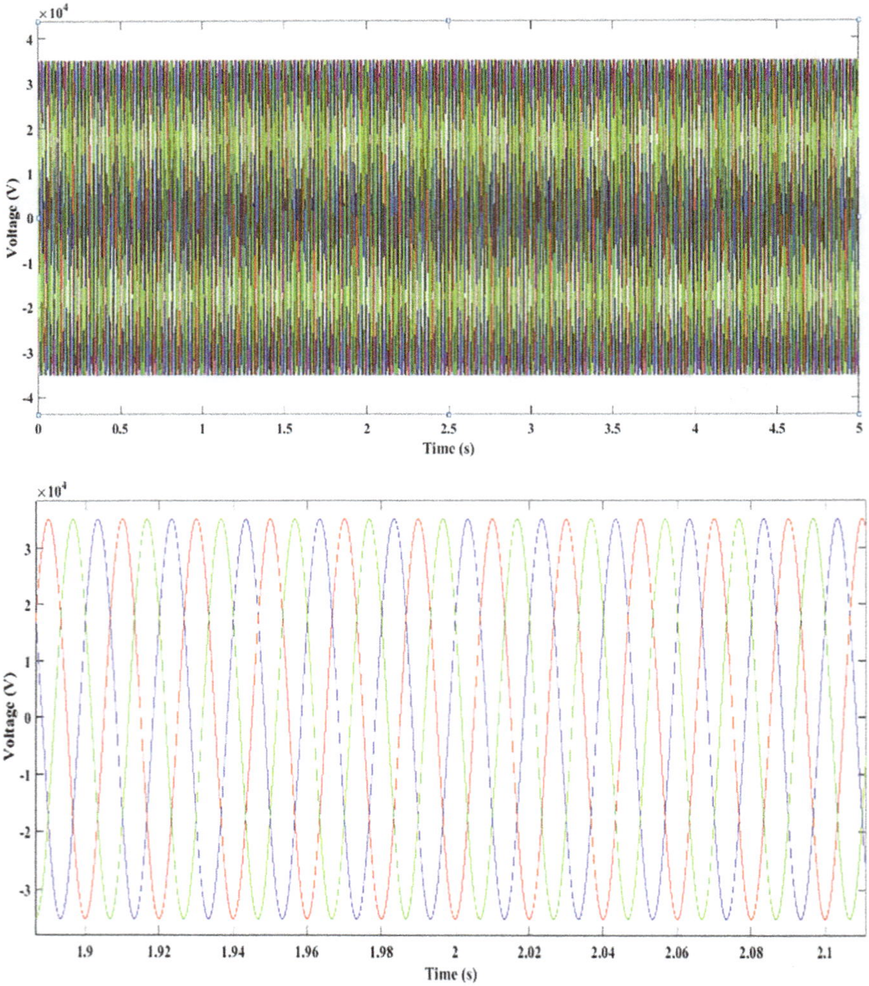

Figure 13.11 Voltage in section B (PV system)

Active power (P_s) and reactive power (Q_s) generated by section B of the PV system are shown in Figure 13.3. Over the course of the simulation duration, we can notice a steady rise in active power (red) which corroborates with how well our system stabilizes itself and is able to transform solar energy into electricity. The output power gradually increases as the system stabilizes. The reactive power (green) stays near zero for the majority of the simulation, meaning that it's running at close to unity power factor. This in turn is advantageous as the transfer of reactive power to and from grid will be less which means lower losses/greater efficiency overall.

The monotonous growth of the active power output demonstrates that the system successfully converts energy to another kind, and nearly zero reactive

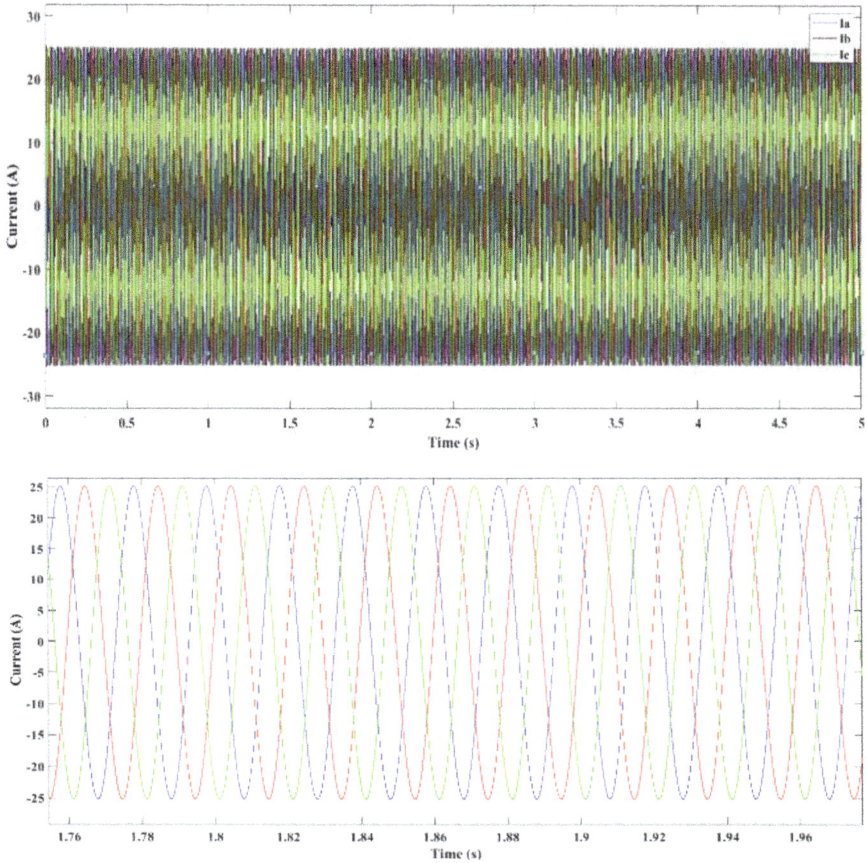

Figure 13.12 Current in section B (PV system)

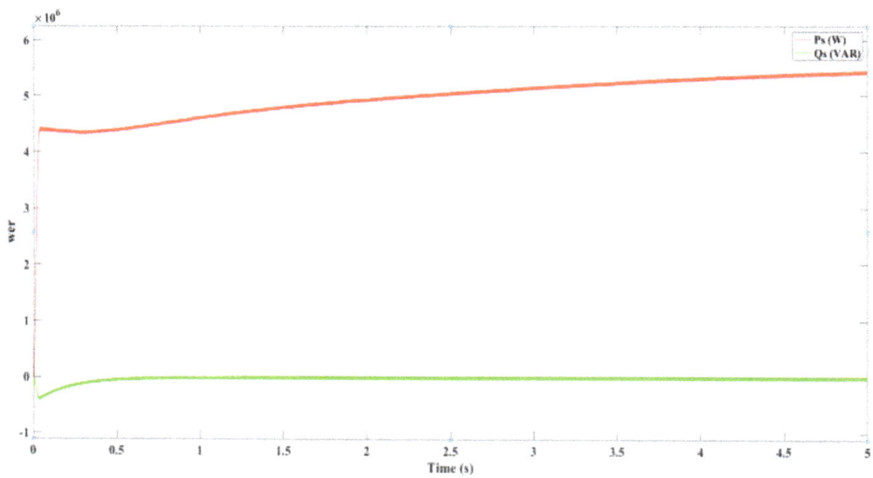

Figure 13.13 Active and reactive power in section B (PV system)

power level signifies high-quality electric energy production. Voltage output of the PV system (5 s in section C) is shown in Figure 13.14; oscillations in the voltage waveform that are high-frequency oscillations are tailored to systems with power electronic converters as inverter solutions. The converter switching associated with harmonics will probably contribute to the oscillations observed in this waveform. Nevertheless, except for these high-frequency components the voltage is stable and constant on all three phases (V_a, V_b, V_c), which means that this system correctly regulates its output. This also means that the system is balanced between all three phases, a good sign for reducing voltage drops and improving grid capacity.

Figure 13.15 shows the current output of section C of these PV systems, displaying over a cycle period (5 s) with phase currents (I_a, I_b, I_c) in it. Similarly, the waveform consists of high-frequency oscillations as well and this is expected due to a switch-mode power electronic system. And they are showing similar oscillations in all

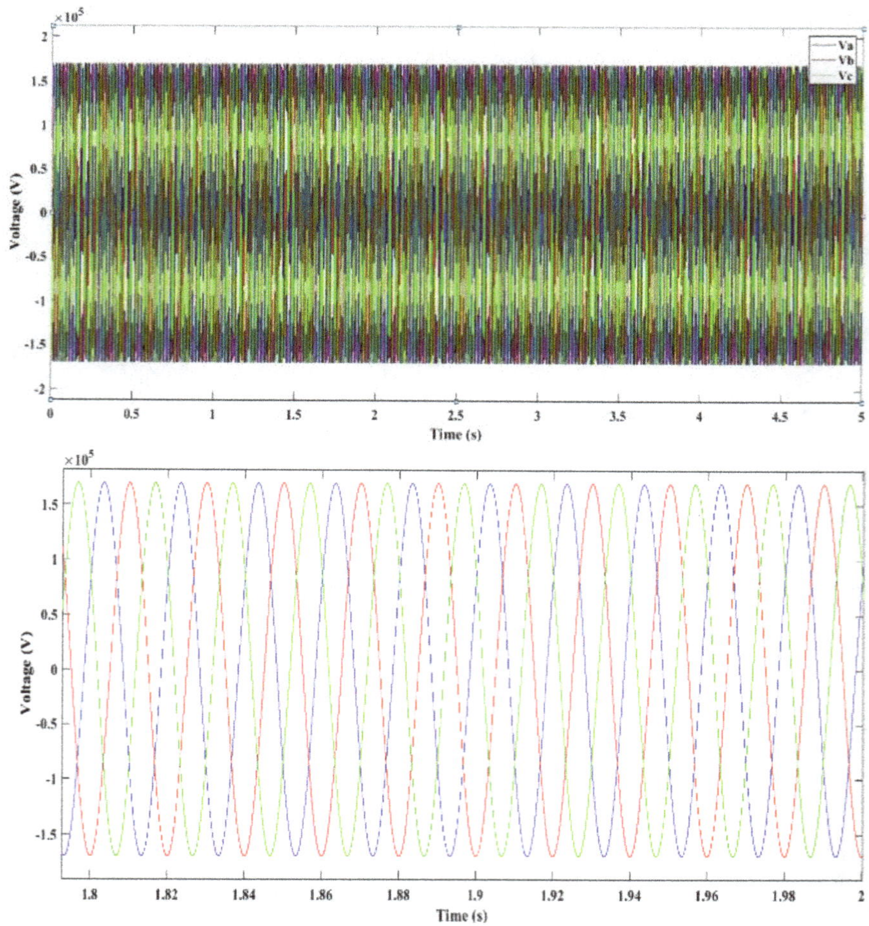

Figure 13.14 Voltage in section C (PV system)

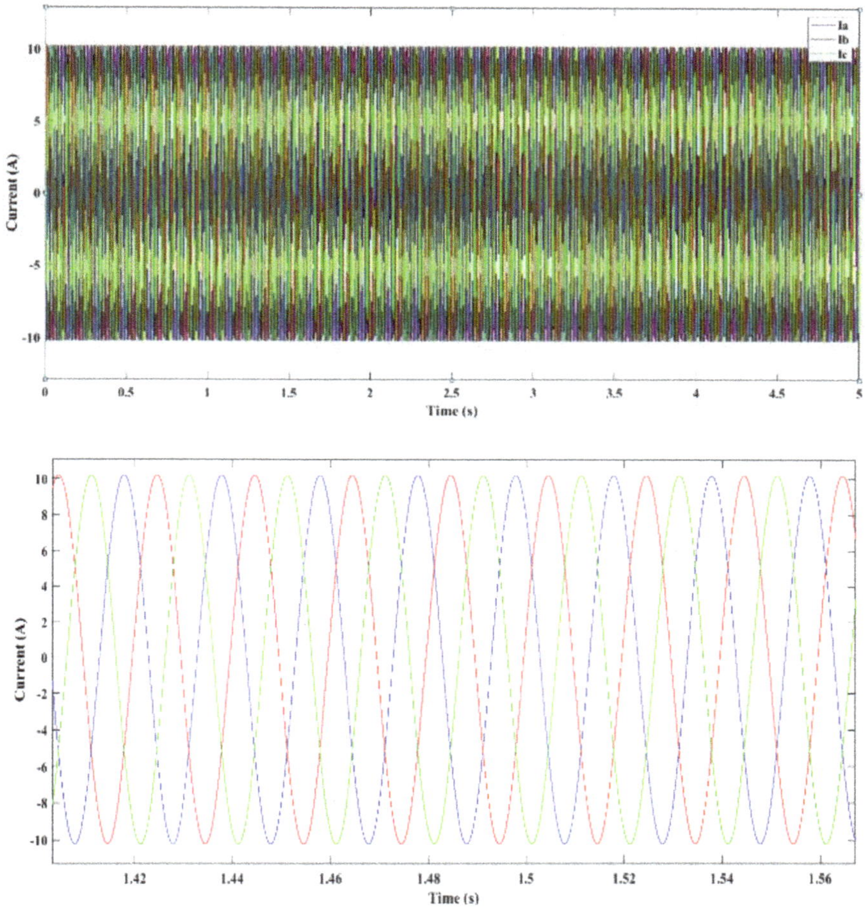

Figure 13.15 Current in section C (PV system)

phases as well; this is how you know the current for each phase distribution evenly. The current waveforms across the phases should be symmetrical and stable, indicating that the system is operating as intended with a high degree of distortion or imbalance.

The active power (*P*) and reactive power (*Q*) produced by section C of the PV system are shown in Figure 13.16 Active power (red) represents the irradiation and is gradually increasing over time throughout our simulation period, as we can see from this chart that the system stabilizes to harvest more solar energy converting it into electrical energy. The reactive power (green in color) throughout the simulation is close to zero, implying that the unit maintains a nearly unity power factor operation. This is beneficial in reducing the reactive power exchange with grid and thus minimizes losses, thereby improving the efficiency of energy delivery. The increase of active power keeps a record for optimal conversion characteristic with the system; the minimal reactive power makes sure that high quality and clean energy is given forth by it.

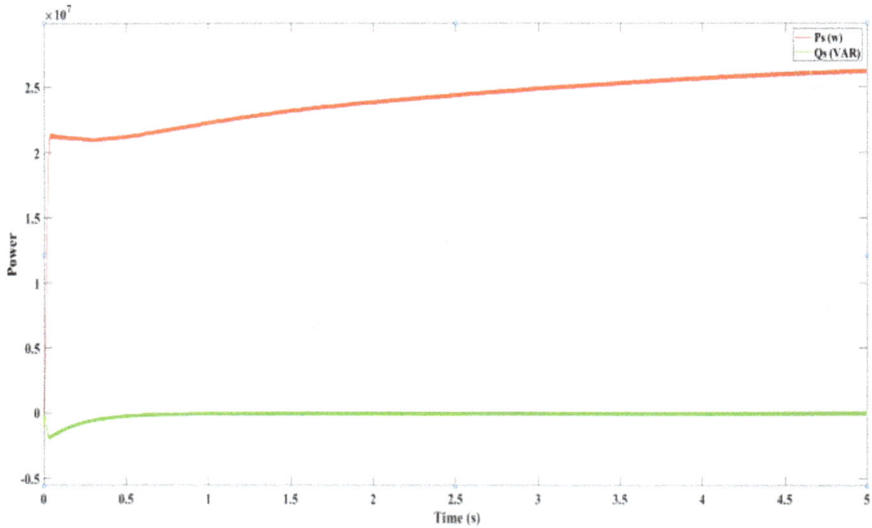

Figure 13.16 Active and reactive power in section C (PV system)

13.3.4.3 Results of wind system

In Figure 13.17, the output current from section A of the wind energy system is shown for a 5-s reign. The current waveform across the three phases (I_a, I_b, I_c) has relatively consistent, high-frequency oscillations that represent variations in switching actions within the system power electronic converters. Waveform is unbalanced but the same waveform in all phase may indicate that the system is balanced and running properly. The presence of such a steady current wave means that the wind turbine is generating electrical power without any significant distortion or phase unbalance.

Figure 13.18 shows the voltage output from section A of the wind system, measured across the three phases (V_a, V_b, V_c) over a period (5 s). On the voltage waveform, high-frequency oscillations will also be seen in a fashion similar to that of the present waveform. Diesel generators produce variable AC power, so the inverter must repeatedly switch on and off to transform this over time and remain grid-compatible. The voltage waveforms were stable and balanced in all phases, indicating that the system maintains a constant output of voltage which is very important for efficient power transmission with, as far as possible, not big losses.

Figure 13.19 shows the active power (P_s) and reactive power (Q_s) from section A of the wind system; the red line represents active power which starts to settle more smoothly over time, signaling effective conversion of wind energy into electrical power in the system. The problem now is the reactive power (green) that you can see has a very large disturbance with oscillations particularly at 2.5 s. This suggests that there was a transient event or instability in the system that temporarily affected the reactive power output. Despite this, the system appears to recover, as indicated by the stabilization of the reactive power toward the end of the

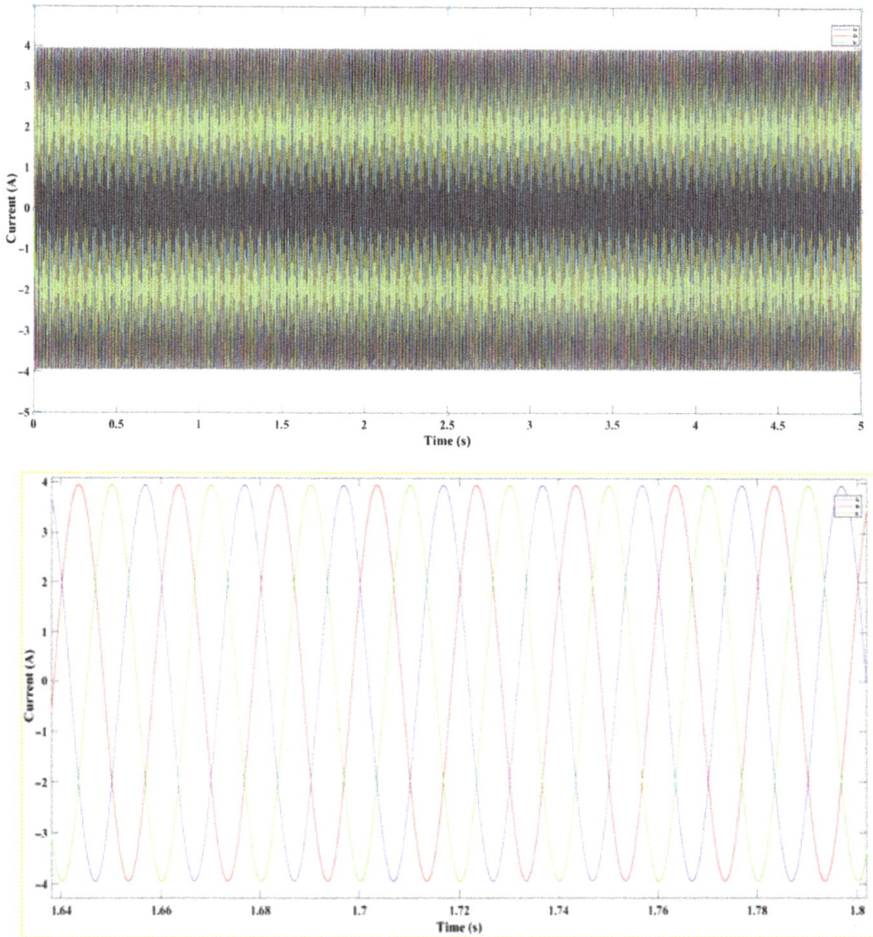

Figure 13.17 Current in section A (wind system)

simulation. The overall low level of reactive power suggests that the system operates near unity power factor most of the time, which is ideal for minimizing losses and ensuring efficient power delivery to the grid.

Figure 13.20 shows a set of the actual output data from section B using test wind energy system for the time span of period: 0–5 s. The output current across the three phases (I_a, I_b, I_c) is shown. The current waveform in all the three phases is shown above and it exhibits high-frequency oscillations that are a common feature of systems with power electronic converters. The aforementioned oscillations are identical in all of the three phases, indicating a generally healthy operation for that phase. The amplitudes of the current in all three phases seem consistent and reveal a well-functioning system, although one rooted in wind, which can fluctuate depicting this steady output.

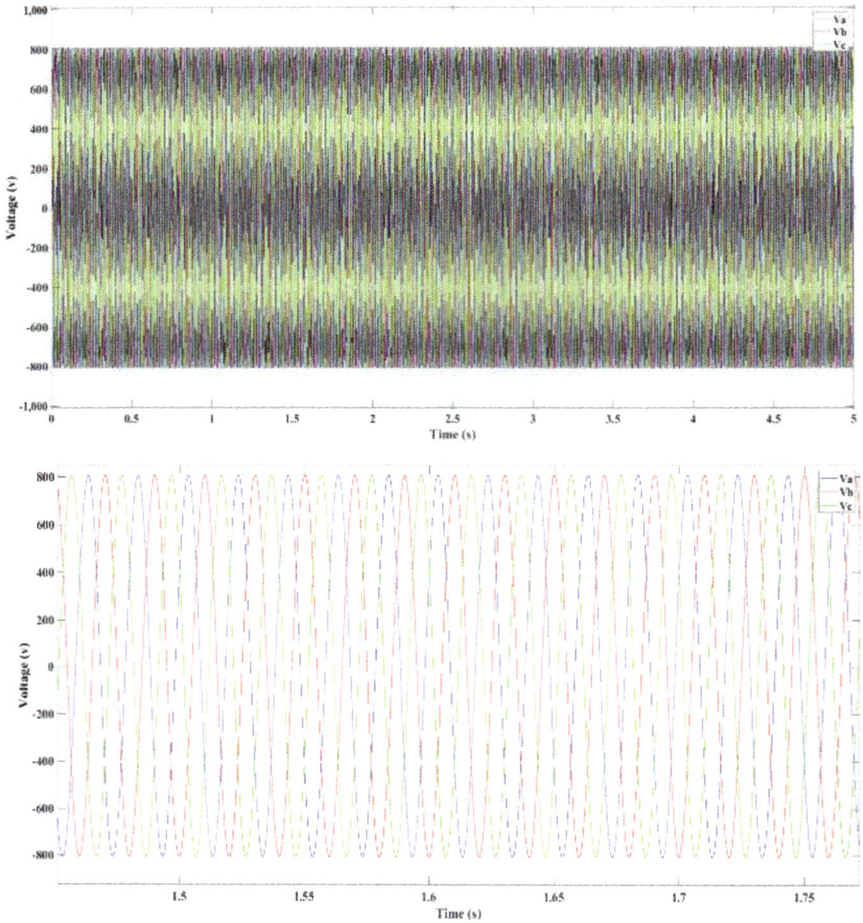

Figure 13.18 Voltage in section A (wind system)

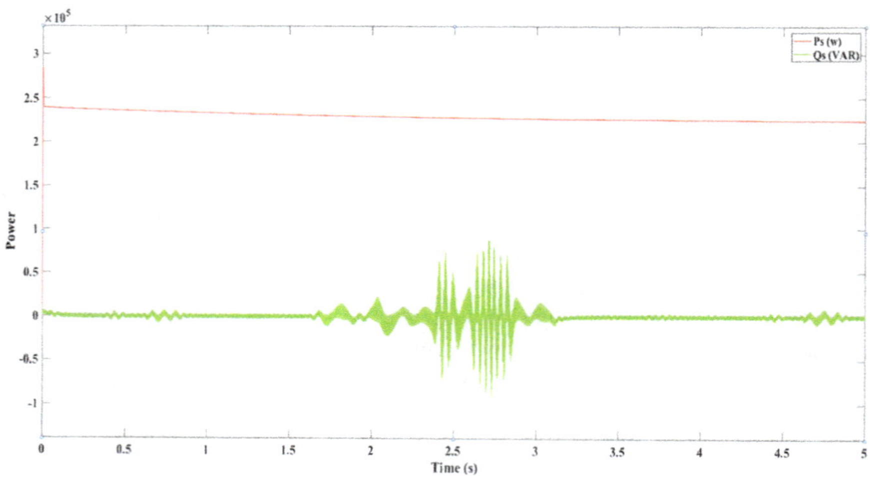

Figure 13.19 Active and reactive power in section A (wind system)

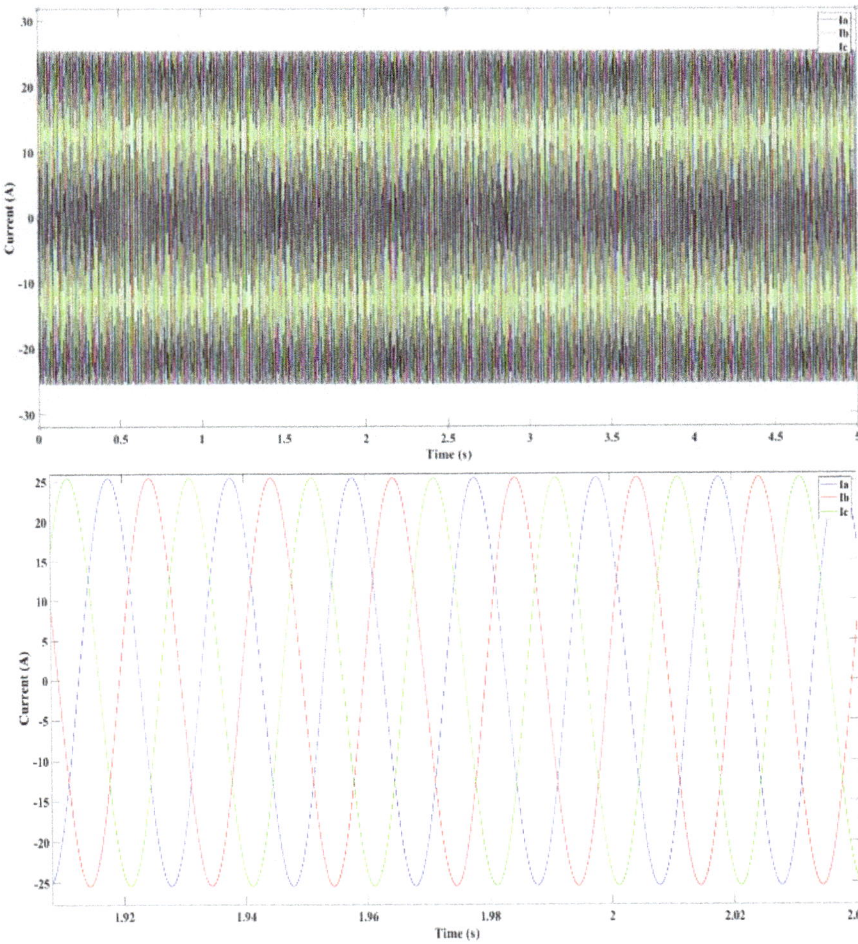

Figure 13.20 Current in section B (wind system)

In Figure 13.21, one can view the voltage output for section B of the wind system in line with phase at t=0–5 s. Like the current waveform, pertaining to modern inverters, switching operations within secure and accurate bounds caused high-frequency oscillations in voltage as shown below. Nevertheless, the voltage is stable and balanced in all three phases, important for power quality and efficient feeding back to grid. The stable voltage values between all three phases tell us that the wind system is now producing a constant and robust source of output according to its state.

In this regard, the active power (P) with its reactive counterpart is depicted for section B of the wind system in Figure 13.22 The red line represents the active power, and shows a slow decrease during the simulation period (potential reduction in wind speed or perhaps system conditions). Compared to real power (red), the

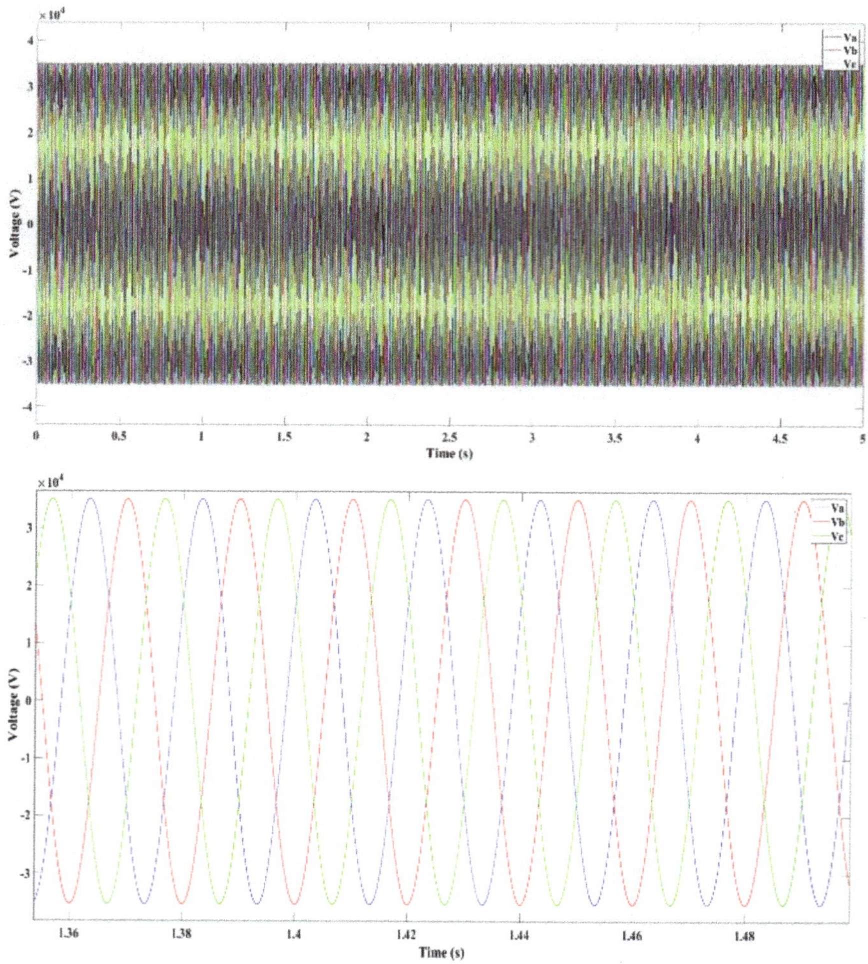

Figure 13.21 Voltage in section B (wind system)

reactive power (displayed in green) remains low near zero during simulations, showing that desirable system operation takes advantage of a good nearly-unity power factor. This is beneficial also to keep the reactive power exchange with the grid as low as possible, hence lowering losses and increasing overall efficiency. A drop in active power could indicate lower wind availability, but reactive power remains very stable during changing modes ensuring high-power quality.

The output voltage of the wind energy system which is part C from the wave component is shown in Figure 13.23 over a period of several seconds. These high-frequency oscillations in the voltage waveforms across three phases (V_a, V_b, V_c) are common in systems with power electronic converters. The oscillations are balanced and consistent across the three phases, indicating a

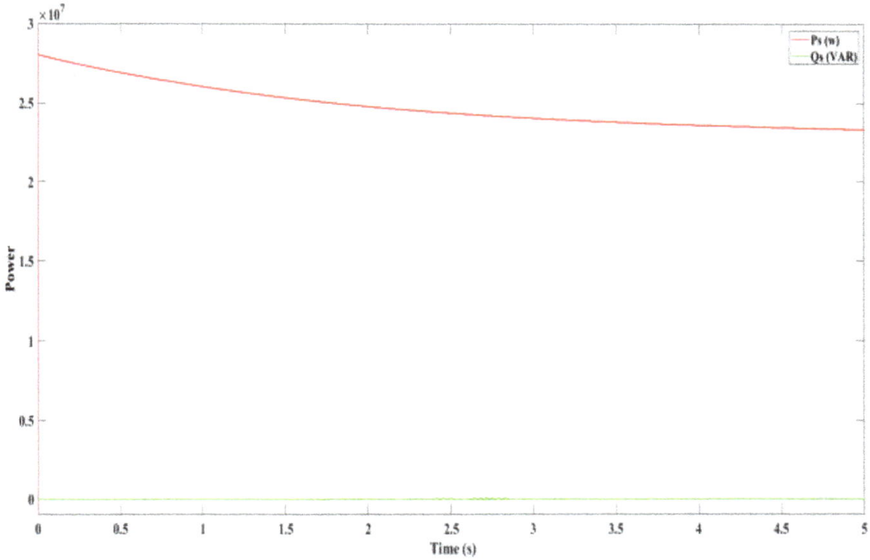

Figure 13.22 Active and reactive power in section B (wind system)

stable voltage in the system. The fact that the waveforms are symmetric indicates proper phase balance, which is critical for efficiently delivering power and minimizing losses in a system.

Figure 13.24 presents the current output of section C from the wind system: (I_a, I_b, I_c) represents current across three phases within a 5-s timeframe, respectively. The current waveform also has high-frequency oscillations because of the switching actions in all converters. The similarity in present waveforms throughout all periods suggests that the wind system is executing well with a balanced load circumstance. This balance is necessary to keep the system running smoothly and provide steady and dependable power into the grid.

Figure 13.25 shows the active power (P) and reactive power (Q) generated by section C of the wind system. The active power, represented in red, shows a slight decline over the simulation period, which might suggest a decrease in wind speed or an adjustment in the system's operational parameters. Despite this, the reactive power, shown in green, remains close to zero throughout the simulation, indicating that the system is operating near unity power factor. The near-zero reactive power is advantageous because it implies that the system is not generating or absorbing significant reactive power, thereby reducing losses and improving overall system efficiency.

13.3.4.4 Results of hybrid system

In Figure 13.26, part A of the hybrid system generated active power (P_s) and reactive power (Q_s) over a period of 5 s. This is illustrated in red as the active

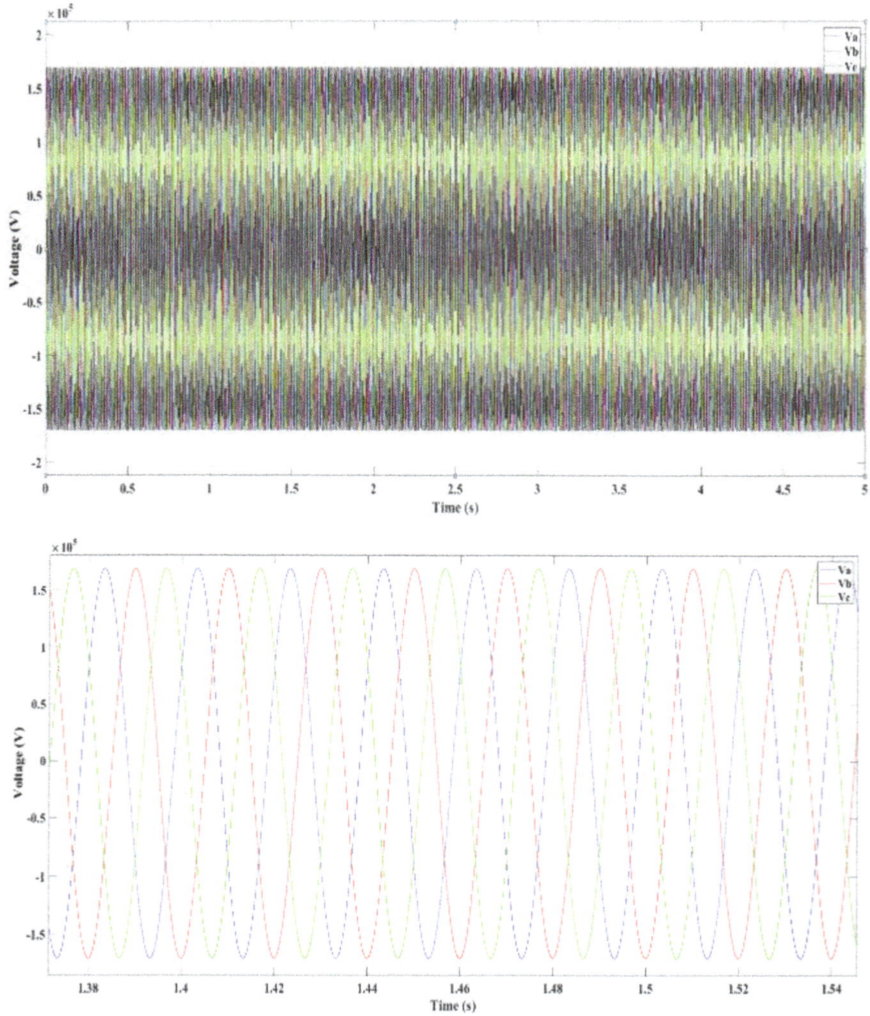

Figure 13.23 Voltage in section C (wind system)

power (it builds up slowly through the simulation, representing a system that can switch easily between energy forms) grows with time indicating how well it dampens and then converts mechanical to electrical work. The green line represents reactive power and for this signal there is a temporary disturbance of around 2.5 s with wavy properties. This implies a transient, or temporary short-term unbalance in the generation capacity of reactive power. However, the system quickly recovers, with reactive power returning to near-zero levels, which indicates that the system is operating close to a unity power factor. This is beneficial for minimizing losses and enhancing the overall system efficiency.

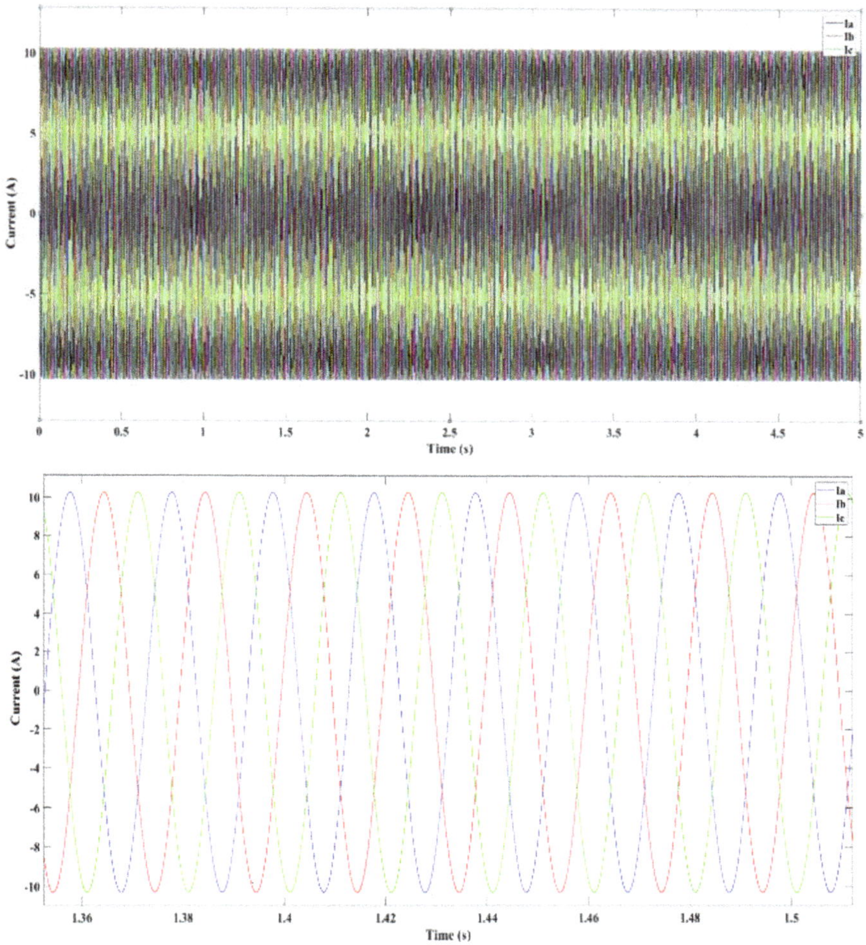

Figure 13.24 Current in section C (wind system)

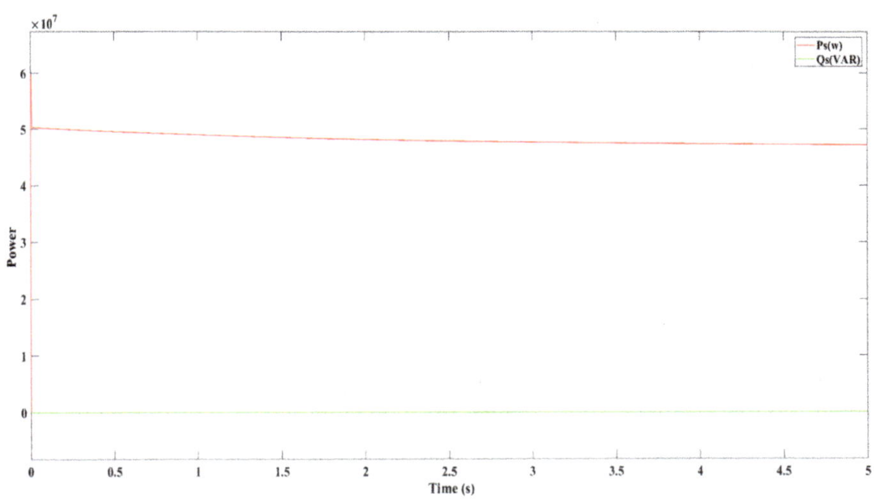

Figure 13.25 Active and reactive power in section C (wind system)

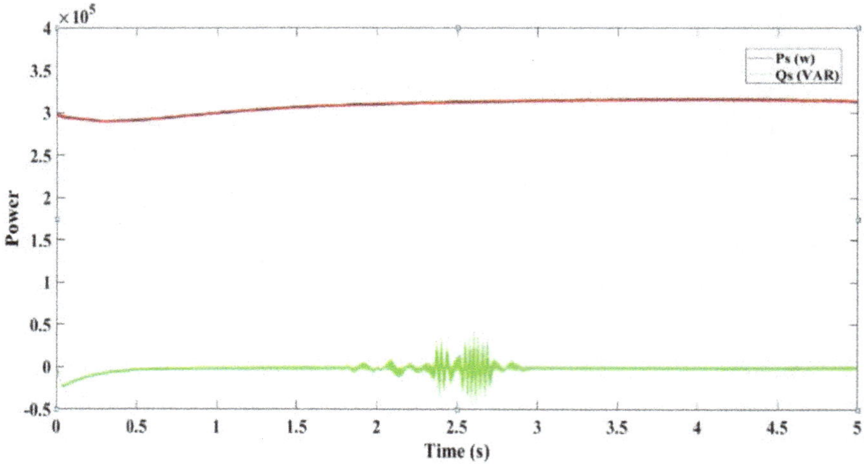

Figure 13.26 Active and reactive power in section A (hybrid system)

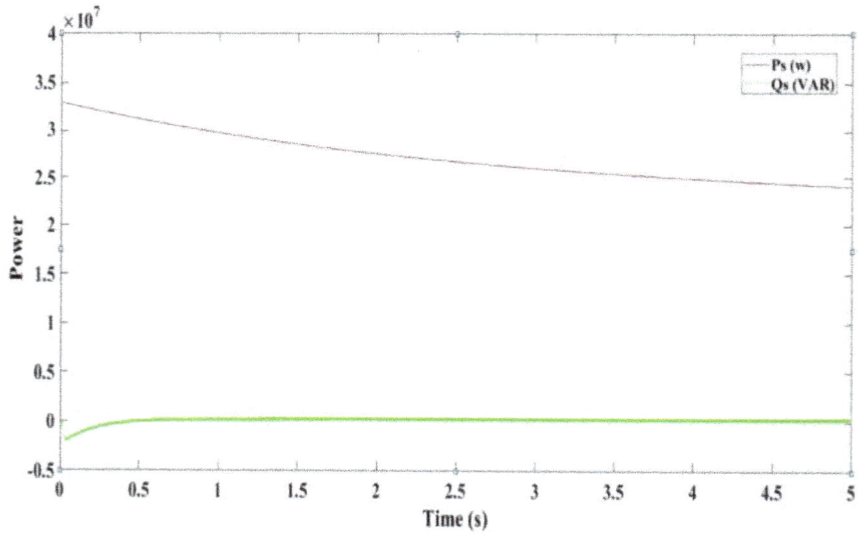

Figure 13.27 Active and reactive power in section B (hybrid system)

The active and reactive power of the hybrid system in section B is shown in Figure 13.27. During a simulation, the active power falls just a little, suggesting less energy is available (e.g., solar irradiance or wind speed). The reactive power is maintained close to zero during the entire simulation implying that, even as active load reduces, the system continues to have nearly unity of operation. The consistent reactive power shows that the system in part B is actively feeding in a high quality

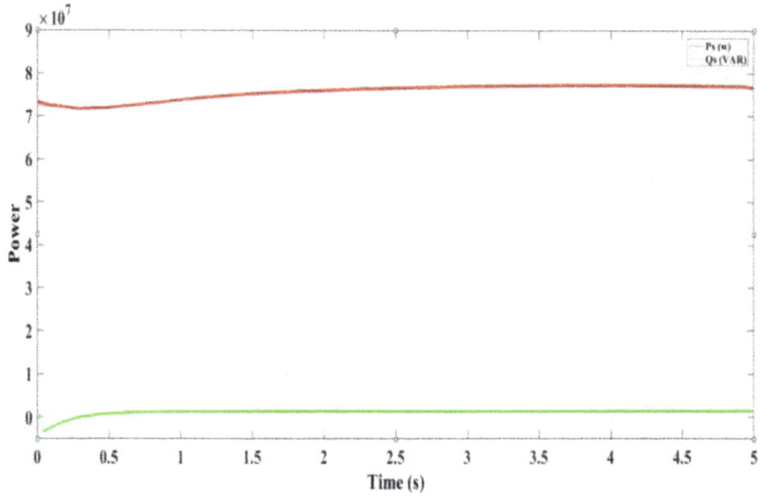

Figure 13.28 Active and reactive power in section C (hybrid system)

of active power and ensures its essential part to hold grid stability by maintaining the minimum bounds.

The active and reactive power outputs for Part C of the hybrid system are shown in Figure 13.28. It continues to rise slowly; the active power rises gradually as its ability of system stabilization being more and more effective in efficiently capturing useful incoming energy. Since reactive power stays minimal and near zero, the system operates on a high-power factor configuration that lowers Qs flow and increases efficiency. Part C demonstrates why a well-designed power system which follows the error slope in both the active and reactive planes is able to hold on stability across parameter variations.

The simulations showed that the hybrid system can provide stable and efficient performance across various operating conditions. The MPC and PSO algorithm regulated the power output appropriately so that it controlled the optimal point of operation for rapid changes in solar irradiance, as well as wind speed.

In general, the simulation configuration had established an integrated performance verification system for hybrid energy storage. Using the system modeling with high level of detail and more advanced control algorithms, we could test its real-world performance and stability virtually through a large number of iterations.

13.4 Conclusion

In this chapter, a new MPC&PSO-based control strategy was proposed for the management of hybrid PV and wind energy systems. The integration of these control strategies has been proven successful to combat both the variability and nonlinear dynamics present in power systems with renewable generations. Based on detailed simulations, supported by a rigorous stability analysis based on the Lyapunov theorem, the proposed system has been proven to maintain stable operation and high efficiency under broad conditions.

The principal novelty of this work is to present not only a robust control scheme that maximizes the power output of the hybrid system but also guarantee its operation in wide ranges against dynamic environmental conditions. This MPC algorithm with the support of PSO is a robust tool for tuning control parameters in real time so that our system can adapt readily to varying solar and wind conditions. This flexibility is important as more renewables, which by nature of their source have inherently low-capacity allowances, come into the system.

Simulation results demonstrate the feasibility of the suggested control strategy and indicate significant improvements on power quality, system response, and general stability. This helped provide a strong basis for such findings and together proved the system remains stable among adverse conditions, as per Lyapunov's theorem (used within stability analysis).

Nevertheless, this study serves as a step forward in research area and many areas still remain to be explored. Possible future investigations need to consider the implementation of this control strategy in a wider variety or renewable energy systems (whether with storage or other generation) such as those used for buildings, agriculture, and remote areas. Moreover, more sophisticated optimization algorithms would improve the performance of the MPC&PSO method. Overall, actual real-world implementation and testing of the presented system would be a major next step needed to realize its full potential in practice.

In a nutshell, this chapter shows that the hybrid PV-wind integration with MPC and PSO is practical and promising. The current research helps address this conundrum of optimality and stability in renewable energy integration, moving us closer to the design of sustainable and resilient future energy systems.

References

[1] Hamoodi, S.A., Hamoodi, A.N., and Mohammed, R.A.N. Design and simulation of smart grid based on solar photovoltaic and wind turbine plants. *Journal Européen des Systèmes Automatisés* 2024, vol. 57, no. 4, pp. 953–961. https://doi.org/10.18280/jesa.570403.

[2] Ishaque, K. and Salam, Z. A review of maximum power point tracking techniques of PV system for uniform insolation and partial shading condition. *Renew. Sustain. Energy Rev.* 2013, vol. 19, pp. 475–488.

[3] Tolba, M., Rezk, H., Diab, A.A.Z., and Al-Dhaifallah, M. A novel robust methodology based salp swarm algorithm for allocation and capacity of renewable distributed generators on distribution grids. *Energies* 2018, vol. 11, p. 2556.

[4] Tolba, M.A., Rezk, H., Tulsky, V., Diab, A.A.Z., Abdelaziz, A.Y., and Vanin, A. Impact of optimum allocation of renewable distributed generations on distribution networks based on different optimization algorithms. *Energies* 2018, vol. 11, p. 245.

[5] Cheng, Z., Zhou, H., and Yang, H. "Research on MPPT control of PV system based on PSO algorithm." In: 2010 *Chinese Control and Decision Conference*, 2010: Piscataway, NJ: IEEE, pp. 887–892.

[6] Singh, G.K. "Solar power generation by PV (photovoltaic) technology: A review," *Energy* 2013, vol. 53, pp. 1–13.

[7] Baloch, M.H., Chauhdary, S.T., Ishak, D., *et al.* Hybrid energy sources status of Pakistan: An optimal technical proposal to solve the power crises issues. *Energy Strategy Rev.* 2019, vol. 24, pp. 132–153.

[8] Baloch, M., Abro, S., Sarwar Kaloi, G., *et al.* A research on electricity generation from wind corridors of Pakistan (two provinces): A technical proposal for remote zones. *Sustainability* 2017, vol. 9, p. 1611.

[9] Lun, S., Du, C., Sang, J., Guo, T., Wang, S., and Yang, G. An improved explicit I–V model of a solar cell based on symbolic function and manufacturer's datasheet. *Sol. Energy* 2014, vol. 110, pp. 603–614.

[10] Bai, J., Liu, S., Hao, Y., Zhang, Z., Jiang, M., and Zhang, Y. Development of a new compound method to extract the five parameters of PV modules. *Energy Convers. Manag.* 2014, vol. 79, pp. 294–303.

[11] Orosz, T., Rassõlkin, A., Kallaste, A., *et al.* Robust design optimization and emerging technologies for electrical machines: Challenges and open problems. *Appl. Sci.* 2020, vol. 10, p. 6653. https://doi.org/10.3390/app10196653.

[12] Youcef, B. Contribution à l'étude et à la Commande Robuste d'un Aérogénérateur Asynchrone à Double Alimentation. *Ph.D. thesis, Université Mohamed Khider – Biskra*; 2014.

[13] Guda, S.R. Modeling and power management of a hybrid wind-microturbine power generation. Master's thesis, Université de Bozeman, *Monata*; 2005.

[14] Hirose, T. and Matsuo, H. Standalone hybrid wind-solar power generation system applying dump power control without dump load. *IEEE Trans. Ind. Electron.* 2012, vol. 59, pp. 988–997.

[15] Hossain, M.K. and Ali, M. Transient stability augmentation of PV/DFIG/SG-based hybrid power system by parallel-resonance bridge fault current limiter. *Electr. Power Syst. Res.* 2016, vol. 130, pp. 89–102.

[16] Boubii, C., Kafazi, I.E., Bannari, R., *et al.* Synergizing wind and solar power: An advanced control system for grid stability. *Sustainability* 2024, vol. 16, p. 815. https://doi.org/10.3390/su16020815.

[17] Asif, M., Sahin, G., and Khalid, M. (Eds.). *Handbook of Energy and Environment in the 21st Century: Technology and Policy Dynamics* (1st ed.). Boca Raton, FL: CRC Press; 2024. https://doi.org/10.1201/9781032715438.

[18] Vidhya, K., Arfan, G., and Erping, Z. Modelling and simulation of maximum power point tracking algorithms and review of MPPT techniques for PV applications. In: *5th International Conference on Electronic Devices, Systems and Applications (ICEDSA)*, 6–8 Dec 2016, 2016: pp 1–4.

[19] Nandurkar, S.R., and Rajeev, M. Modeling simulation and design of photovoltaic array with MPPT control techniques. *Int. J. Appl. Power. Eng. (IJAPE)* 2014, vol. 3, no. 1, pp. 41–50.

[20] Babaa, S.E., Armstrong, M., and Pickert, V. Overview of maximum power point tracking control methods for PV systems. *J. Power Energy Eng.* 2014, vol. 2, pp. 59–72.

[21] Faranda, R., and Leva, S. Energy comparison of MPPT techniques for PV systems. *WSEAS Trans. Power Syst.* 2008, vol. 3, no. 6, pp. 446–455.

[22] Singh, G., Lentijo, S., and Sundaram, K. The impact of the converter on the reliability of a wind turbine generator. In: *Proceedings of the ASME 2019 Power Conference*, Salt Lake City, UT, 15–18 July, V001T06A016. New York, NY: ASME; 2019a.

[23] Singh, G., Matuonto, M., and Sundaram. K. Impact of imbalanced wind turbine generator cooling on reliability. In: *2019 10th International Renewable Energy Congress (IREC)*, 26–28 March, pp. 1–6. New York, NY: IEEE; 2019c. https://doi.org/10.1109/IREC.2019.8754545.

[24] Levine, W.S. (Ed.). *The Control Systems Handbook: Control System Advanced Methods* (2nd ed.). Boca Raton, FL: CRC Press; 2011.

[25] Talla, V., Kellogg, B., Gollakota, S., and Smith, J.R. Battery-free cellphone. *Proc. ACM Interact. Mob. Wearable Ubiquitous Technol.* 2017, vol. no. 2, Article 25 (June 2017), p. 20. https://doi.org/10.1145/3090090.

[26] Pena, R.S., Asher, G.M., and Clare J.C. "A doubly fed induction generator using back to back PWM converters supplying an isolated load from a variable speed wind turbine." *Proc. Inst. Electr. Eng.—Power Appl.* 1996, vol. 143, no. 5, pp. 380–387.

[27] Boubii, C., Kafazi, I.E., Bannari, R., and Bhiri, B.E. A comparison between MPC, perturb and observe, and incremental conductance method. In: *Proceedings of the 2023 5th Global Power, Energy and Communication Conference (GPECOM)*, Nevsehir, Türkiye, June 14–16, 2023, pp. 198–202.

[28] Boubii, C., El Kafazi, I., Bannari, R., *et al.* Integrated control and optimization for grid-connected photovoltaic systems: A model-predictive and PSO approach. *Energies* 2023, vol. 16, p. 7390. https://doi.org/10.3390/en16217390.

Chapter 14

Integration of electrical vehicle into smart grid, opportunities and challenges

Adamou Amadou Adamou[1] and Chakib Alaoui[1]

Abstract

As nations confront escalating air pollution and rising prices of fossil fuels, the acceptance of electric vehicles (EVs) is rapidly increasing. However, integrating electric vehicles into the electric utility remains a challenge since most of the grids worldwide cannot take the sudden increase of power demand from the newly acquired EVs. This challenge can potentially be transformed into an opportunity for the power utility and the EV users thanks to the implementation of vehicle-to-grid (V2G) concepts. V2G concepts would allow the massive embracing of EVs while performing the needed ancillary services to sustain the electric utility. It is possible to exploit the battery package of an EV, that is in parking for 95% on average, for frequency regulation, peak shaving, load shifting, voltage recovery, renewable energy resources adoption, and other needed services. However, the embracing of the EV to sustain the utility does not come easy; it requires the implementation of smart grid infrastructure, an increase of the supply of energy, and the implementation of V2G concepts on the vehicles.

Keywords: Electric vehicles; smart grid; V2G; ancillary services; renewable energy resources

14.1 Introduction

The upgrading of electric utility is essential in today's increasing demand for electricity, air pollution, and depleting accessibility to fossil fuel. These grids are being re-engineered into smart, adaptive, and sustainable networks. These systems are designed to comply with the complex demands of a rapidly evolving energy landscape, which is characterized by the need to integrate renewable sources and the imperative to reduce carbon emissions [1]. Electric vehicles (EVs) are fulfilling a key role in this transformation; they impact both the transportation and the energy sectors. Their incorporation into smart grid represents a dynamic and multifaceted

[1]Euromed Polytechnic School (EPS), Euromed University of Fes, Morocco

process that promises to redefine the way electricity is consumed and managed. Through mechanisms such as vehicle-to-grid (V2G), vehicle-to-building (V2B), and the generalized vehicle-to-X (V2X), EVs are poised to become not merely consumers of electricity but also providers. The "prosumers" contribute to grid stability, energy efficiency, and the facilitation of renewables integration [2]. This chapter offers a comprehensive examination of the part of EVs within the smart grid ecosystem. It explores the potential of EVs to enhance grid resilience, enable the transition to a low-carbon economy, and support the creation of a more sustainable and efficient energy system [3].

Historically, the traditional electric utility has been a robust, and yet a rigid structure designed to deliver electricity in a one-way flow from producers to consumers. However, this system is increasingly challenged under the growing weight of today's energy demands and the necessity to integrate renewable sources. The integration of renewable sources presents their own challenges, including network inadequacy and the intermittency of resources like solar and wind. In response to these challenges, the smart grid combines the most recent techniques and technologies. Unlike the traditional utility, the smart grid offers real-time data communication, self-healing capabilities, and enhanced efficiency [4]. These features not only improve the reliability of electricity supply but also support the seamless transition to adopting energy resources that are more distributed, resilient, secure, and clean. The smart grid's capability to adopt renewable sources effectively into the utility is vital for the energy transition. It allows for a bidirectional flow of electricity, where consumers can also become producers—prosumers—contributing to energy generation through solar and other renewable technologies [5]. This modern grid system leverages the gains of the digital transformation, including the Internet of Things and big data. By enabling these advancements, smart grid plays a pivotal role in meeting the increasing demand for renewables and ensuring a sustainable future.

Parallel to the progress of the electric utility is the transformative journey of the transportation domain. EVs are rapidly becoming the standard-bearers for a less-polluting and more sustainable mode of transport. They are not mere replacements for their combustion-engine counterparts but are poised to become popular elements within the broader energy ecosystem. Today's EVs are furnished with sophisticated systems, allowing a self-diagnostics and interaction with the environment [6–8].

The technologies of V2G and V2X are at the front of this integration. V2G technology empowers EVs to function as mobile energy reservoirs, capable of feeding surplus electricity back into the utility. This aids in managing peak loads and fortifies grid stability during times of fluctuating energy supply. Moreover, the technology ensures the efficient recharge of the vehicles. It allows the electricity to be drawn during off-peak periods or during renewable production excess. This optimizes utility operations and reduces the dependence on fossil sources. The manuscript is arranged as follows: segment 2 presents the fundamentals of the smart grid technologies and an overview of the ancillary services. Segment 3 presents some fundamental concepts of the V2G, its services to the utility, and the challenges to its implementation. This manuscript ends with a conclusion in segment 4.

14.2 Smart grid

14.2.1 Fundamentals of smart grid technology

While some upgrades were made to satisfy the constantly rising demand, the utility still operates the way it did more than 120 years ago—electricity flows over the grid from central power plants to consumers, while its reliability is safeguarded by preserving surplus capacity [9]. Communication is very restrained; the customers must call the grid operators in the occurrence of faults. Fossil fuels is a prevailing basis of electricity. So the vision of a smart grid was born. An electric utility is equipped with the most recent monitoring, communications, control, and estimation technologies to improve its reliability, efficiency, carbon print, and most importantly in our context, successful adoption of EVs. Table 14.1 shows a brief comparison among the customary electric utility and smart grid [10].

Smart grid represents a congregation of solutions aimed at realizing these objectives through the empowerment of its customers by giving them the possibility to generate energy with segmental renewable energy devices with the possibility to sell surplus energy to the electric utility. This would improve the capacity of the transmission lines and distribution systems. For this operation to succeed, customers (prosumers) should be provided with real-time pricing information, facilitating higher levels of adoption and exploitation of renewable energy, and more importantly in this essay's context, the proper integration of a massive quantity of EVs into the utility. An illustration of smart grid is shown in Figure 14.1.

Moreover, the traditional utility has demonstrated its limitations regarding its capacity for two-way communication. This is needed for an efficient and a rapid management of the grid, where it is possible to automatically plug/unplug renewables into the grid, while controlling the charging/discharging status of EVs, and adopting new technologies like controllable appliances, and advanced smart

Table 14.1 Brief comparison between the customary grid and smart grid [10]

Traditional power grid	Smart grid
Electromechanical (relays, transformers, meters, etc.)	Digital/microprocessor (smart relays, static converters, smart meter, etc.)
One-way communication (if local)	Global/integrated two-way communication
Centralized generation (major thermal station, nuclear facility, etc.)	Distributed generation systems (renewables, EVs, battery energy storage system (BESS), etc.)
Limited protection, monitoring, and control (Local SCADA systems, etc.)	Wide-area monitoring system (WAMS), adaptive protection, etc.
Blind (customers must call grid operators in the case of a blackout)	Self-monitoring
Manual restoration	Automated, self-healing
Check equipment manually	Monitor equipment remotely
Limited control system contingencies	Pervasive control system
Estimated reliability	Predictive reliability

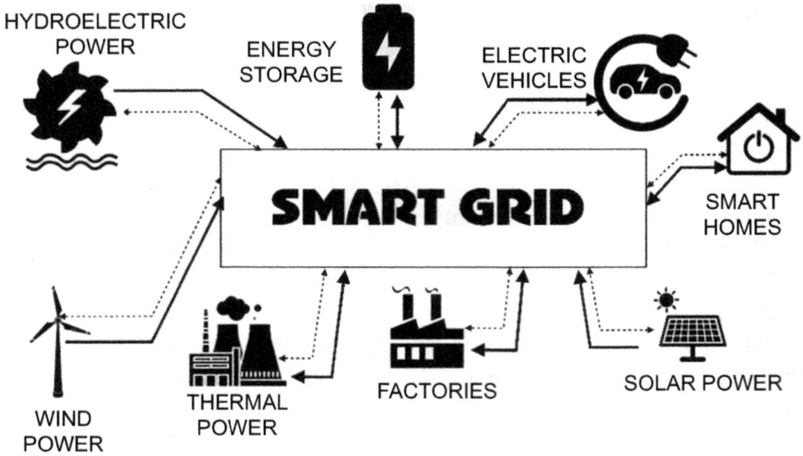

Figure 14.1 Smart grid technologies

meters, managing storage systems like batteries for plug-in hybrid EVs. Two-way communication capabilities are being adopted to open the possibility for the adoption of smart grid technics and to allow "smart" charging of EVs [11].

The EV charging infrastructure will play a key role to the massive adoption of these vehicles. However, the charging of EVs is very sporadic and random in nature, and the large-scale recharging of EVs via the utility has brought unparalleled trials to the reliable and stable operation of the electric grid. Such a penetration and random recharge of EVs would cause voltage stresses, grid instability, and blackouts [12]. Therefore, coordinated charging of multiple EVs are required not only to bound the additional stress on the electric grid, but can also offer the needed ancillary services for the upgrading of the grid's resilience and efficiency.

Introducing vehicle-to-grid (V2G) concept represents a solution for a coordinated recharge of EV fleets. In V2G way, vehicles can deliver energy to the grid and needed services like spinning reserve or others [13]. It was shown in several studies that uncontrolled EV charging increase the system losses and may threaten the utility stability and reliability [14–16].

14.2.2 Ancillary services

Ancillary services (AS) represent a number of products and services that, while separated from energy production, improve the reliability and security of a utility. The Federal Energy Regulatory Commission has established AS as *The services necessary to support the transmission of electric power from seller to purchaser given the obligations of control areas and transmitting utilities within those control areas to maintain reliable operations of the interconnected transmission system.* More precisely, FERC branded six AS as follows:

1. Reactive power and voltage control
2. Loss compensation

3. Scheduling and dispatch
4. Load following
5. System protection
6. Energy imbalance.

14.2.2.1 Scheduling and dispatch

Although scheduling and dispatch are two separate services, they are usually joined together for practical reasons. Scheduling discusses the before-the-fact assignment of resources to meet anticipated loads. Scheduling involves different periods, like a few minutes to some weeks ahead. Dispatch represents a real-time control of all generation and transmission resources that are currently available to the grid operator. The latter must have control over generating units and operate them in a fashion that minimizes fuel consumption and other operational expenditures. In addition, the system's operator must minimize equipment damage and service interruptions.

EVs can be exploited in the service of scheduling and dispatch if the system operator knows their availability, state of charge, and geographical location.

14.2.2.2 Load following

Load following aims to achieve an equilibrium between loads and sources. This equilibrium is measured by a frequency that is maintained at 60 Hz (or 50 Hz). Practically, each control area must generate enough capacity for its loads. It must possess enough provision for contingencies to maintain a constant frequency. The NERC's A1 criteria demand a balance within a control area at least once every 10 min. The NERC's A2 criteria require that the control area's energy imbalance be within a specified limit (0.2–0.4% of peak demand) every 10 min. These two NERC's criteria gave birth to additional ancillary services that support them, including frequency control and spinning reserve.

Ancillary services aim to increase the reliability of transmission and distribution system components and most importantly to assure the essential equilibrium between demand and generation. Traditionally, Ancillary services are offered through incremental metering, billing, and by controlling generators consumption to minimize fuel and other operating costs [17].

14.2.2.3 Frequency regulation

Frequency regulation (FR) safeguards the parity of electricity demand and supply in a constant manner, typically over periods from few seconds to few minutes. When supply surpasses demand, the frequency increases. This constitutes a spontaneous change in active power output as a consequence of a frequency change. The frequency must be kept within legal and operational boundaries.

Traditionally, the speed controller (also called the governor) of each rotating generator acts to upsurge the generation power to recover the system's balance. Traditionally, a frequency controller has a hierarchical structure with primary, secondary, and tertiary control as shown in Figure 14.2. Primary control is continuous, automatic control placed locally at the generators. It is usually based on proportional control, and it corrects the power imbalance between produced and consumed power

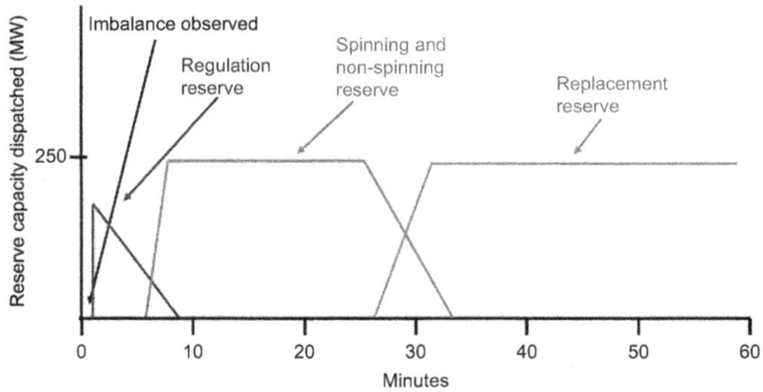

Figure 14.2 Frequency controller structure: primary, secondary, and tertiary control [18]

within 15–30 s. If the primary control does not restore the frequency to its set point, the secondary control is activated. Secondary control is a slower, centralized, and automatic controller which releases primary control. It is activated automatically after primary within 100–200 s. Tertiary control is an even slower (30 min), centralized controller, which again releases the secondary control. The succession from the primary to secondary, and then to the tertiary control is accompanied by a smooth recovery to the electric utility frequency set point as shown in Figure 14.2.

When it comes to renewable sources connected to the utility such as wind, solar, biogas, hydroelectric, each generation has a dedicated and proper "reserve" power to accomplish frequency regulation when active. It is indispensable to note that it is easier to resolve the case of over-frequency, which requires a power decrease. Nevertheless, this is not the case of under-frequency, which would require a power increase, because of the unpredictability of the renewable sources. Energy storage systems (ESS) might be employed for the case of under-frequency. Battery energy storage systems (BESS) or pumped hydroelectric storage are frequently considered for storage.

14.2.2.4 Peak shaving and valley shifting

Peak shaving and valley shifting are when an energy storage system is employed to shift the load from peak to off-peak hours. The electrical system infrastructure is initially designed to satisfy the uppermost level of demand; however, this makes the system underutilized during off-peak periods [19]. In large industrial cities, the demand peak–valley gap can be as significant as 40–50% [20,21], and peak shaving and valley filling are necessary to diminish the cost of system infrastructure and develop its utilization. Figure 14.3 presents an illustration of peak shaving and load shifting concepts. This ancillary service may require few minutes to 6 h.

Peak-shaving strategies are preferred when it is feasible to apply high-demand electricity tariffs to commercial and industrial sites grounded on their high power consumption during the peak demands. In addition, when dealing with inflexible loads that cannot be shifted to different periods, peak-shaving techniques are

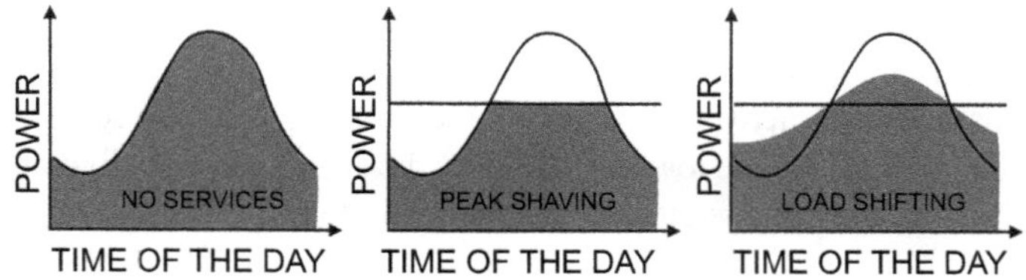

Figure 14.3 Peak shaving and load shifting concepts

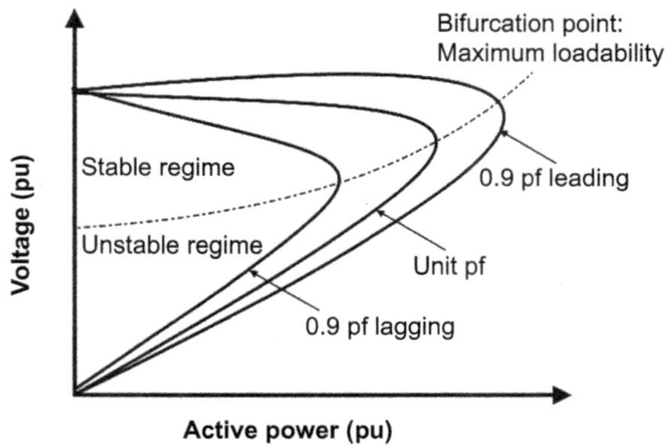

Figure 14.4 Nose curves showing critical active power values for different values of the reactive power [22]

recommended in which a battery system can be turned on to alleviate the burden on the utility. If a facility possesses its own generation capability, such as photovoltaic energy or some kind of thermal generators, these can be exploited to shave the peak during demand periods, reducing the quantity of electricity drawn from the electric utility. On-site generation is usually combined with storage, especially when using renewables like photovoltaic energy. Finally, demand response plans have been made to incentivize energy consumers to minimize electricity consumption throughout peak periods. They receive monetary incentives in the format of a discount on their electricity bill for their participation.

14.2.2.5 Voltage regulation and reactive power

While active power does the useful work like running an electric motor, illuminating a light bulb or heating water, reactive power is required to support system voltages and facilitate active power transfer. If the bus voltage is too low, real power cannot be delivered; reactive power is needed to provide the voltage levels necessary for real power transfer. Figure 14.4 shows the active power–voltage graph (referred to as the "nose" curve) for various values of power factors. For a growing load, the power–voltage curve moves from the top-left to the extreme of the "nose," which corresponds to the extreme power deliverable to the load and the limit of stability. Past this point, the load causes a descent in both voltage and power, and dives into the unstable regime.

For a characteristic network, the power factor is lagging, and reactive power is taken from the network by the loads. This limits the active power deliverable to loads without distressing the voltage stability, as shown in the figure as (a) on the extreme loadability graph connecting the voltage bifurcation points. This also suggests that voltage regulation requires injections of reactive power (capacitive or leading) into the network so that the active power transmitted can be augmented, and we can go from (a) to (b) or (c) indicated in Figure 14.3. Distributed energy sources fortified with inverters ready to inject reactive power into the grid, increase the capability to achieve voltage regulation, and facilitate the move from (a) to (b) to (c). This solution consists of primary control using reactive power–voltage (*Q*–*V*) droop curves and proportional–integral (PI) control, characteristically at the local device level. The adoption of distributed energy sources such as photovoltaic systems or BESS throughout the utility provides the opportunity to coordinate across multiple devices and the application of secondary control which focuses on optimizing grid voltages.

14.2.2.6 Ancillary services summary

Table 14.2 shows a brief summary about the AS that are needed to sustain the electric utility, and in which a large number of EVs may contribute.

14.3 EV integration to the smart grid

14.3.1 Uncontrolled recharge of EVs

Uncontrolled charging, also known to as dump charging, takes place when connecting the vehicle for a recharge without taking any concern of the utility status such as peak time or downtime, electricity's price, safety of power lines, and

Table 14.2 Ancillary services intervention period and grid-level intervention [22]

Ancillary service	**Intervention period**	**Grid-level intervention**	**Required power (Europe)**
Frequency regulation – primary control	30 s to 15 min	TSO (transmission system operator)	~1 MW
Frequency regulation – secondary control	30 s to 30 min	TSO	~2–5 MW
Frequency regulation – tertiary control	30 min to 6 h	TSO	~10 MW
Peak-shaving and valley-shifting	2–10 h	TSO and DSO	T: ~500 kWD: ~100 kW
Voltage regulation and reactive power	As long as it takes	DSO (distribution system operator)	~100 kVAR
Integration of renewables*	10 min to 2 h	TSO and DSO	~100 kW–2 MW

*Refer to the next section.

pment. The vehicle is plugged to the utility until its battery pack of charge (SOC), or until it is unplugged for utilization. The ing is the simplest charging method and suits the traditional ing of the infrastructure for the smart charging scheme. This ecomes impracticable as the quantity of EVs increases and the unbearable for the electric utility. Congestion of the power line ierous EVs are plugged simultaneously (usually around 7 pm n from work). The capacity factor of the electric utility is then esides to the demand variation, which will span from its mini-l) to the peak demand in a short period. Excessive load variation ses and premature aging of power equipment, especially power hough this type of charging is simple and doesn't necessitate nents, it increases the system losses and premature aging of equip- electricity's price would therefore increase and affect negatively the ners. The authors of [23–27] concluded that large EV penetrations cause an eased peak-to-valley difference, larger network losses, and overload and under-load effects. The following presents a brief description of these impacts.

14.3.1.1 Transformer increased losses

Transformers are viewed as the backbone of the transmission and distribution networks, and the costliest equipment. They are degraded by additional losses and hotspot temperature. The author of [28] found that EV uncontrolled charging potentially decreases transformer life by 93%. The author of [29] showed that even 10% penetration of EV recharge may induce additional transformer overloading beyond predicted overloading. Moreover, the investigators of [30] indicated that arbitrary charging of the EVs may cause bigger peak power, causing the transformer loss of life and its exponential aging behavior.

14.3.1.2 Bus voltage deviation

The authors of [31] demonstrated that the uncoordinated recharge of EVs is also influencing the voltage of the electric utility node where the vehicles are connected and charging. The massive connection of EVs to the distribution network for charging will hasten the aging infrastructure and aid voltage fluctuations, line loading, and power losses. Moreover, the unplanned charging process of EVs in several stations might overload the transformers and surge the energy losses in distribution network feeders.

14.3.1.3 Harmonic distortion

Modern EVs contain switching circuits acting as interface devices. These include chargers and filters that are highly nonlinear and thereby draw currents from the electric utility that includes harmonics. The authors of [32] studied the placement and optimum dispatch of passive filer banks for power quality upgrading in existence of EVs nonlinear loads. "Dirty electricity" affects the useful life of equipment and increases the maintenance costs. One major concern for power quality is the voltage distortion, which is usually evaluated for each odd harmonic component and then for the overall voltage distortion.

14.3.1.4 Peak power increment

Considering the culmination of a typical business day, many comı plug their EVs in the afternoon and at night, when peak power demanı affected by massive EV charging [33]. This would worsen an already cı demand that the utility operators must deal with before the use of EVs. reserves are usually used to lessen the peak request.

14.3.2 Controlled charging/discharging

Controlled charging, also known as smart charging, considers the status of utility grid and its components for proper charging. Moreover, the EV can of ancillary services to the utility grid when required. Thus smart charging ca potentially be of benefit to the utility and the EV owners. To achieve this objective, it is essential to implement a bidirectional charging scheme, in which electricity can flow from the electric utility to the EV batteries for recharge, and from the EV batteries to the utility, a building, or a home. This is referred to as V2G, V2B, vehicle-to-load (V2L), and eventually V2X. The EV communicates with the electric utility to sell the electricity stored in its batteries. Several studies have showed that an average city vehicle is parked over 90% of the time [34]. For instance, a small EV with 24-kWh battery capacity and 150-km autonomy, if charged with a level 2 charger of 3 kW, needs about 1.5–4 h to be fully charged. This makes the EV uselessly plugged, in utmost of the time when plugged-in to the utility. That's why the EVs fleets have high potential for utility service support in these periods. This services can be provided by taking into account the EV owner's preferences like departure time and the range to destination, the EV batteries SOC, the status, energy pricing, battery degradation, among other factors.

14.3.2.1 V2G

The vehicle-2-grid notion allows to temporary discharge of the EV batteries into the electric utility for regulation objectives. This is supported by financial incentives to the EV owner to promote these services to the electric utility and the massive purchasing of EVs. The available battery capacity to be sold to the electric utility is a key factor for the delivery of ancillary services. In addition, the deliverable power is more important and is limited by the charger. V2G technology provides services including frequency control, reactive power reparation, harmonic filtering, voltage regulation, balance of the local distribution system, spinning reserve, and peak power shaving.

Frequency regulation

Frequency regulation is often considered the most significant ancillary service since frequency is the measuring mechanism for the proper balance between demand and offer. To maintain stability, deviations in the frequency are corrected by adjusting power generation and/or consumption. Traditional power utility exploits spinning reserves that offer enough power to the electric utility.

Renewable energy sources (RES) provide fluctuating power since they are weather-dependent. This makes the smart grid more susceptible to frequency deviations [35]. By exploiting the V2G technics, EVs have the potential to compensate for the RES fluctuations while optimizing the exploitation of RES for EVs recharge and feeding the power utility. The authors of [36] have shown that the quantity of EVs plugged to the grid and engaging in V2G play a key role to support frequency control, and consequently the utility would exhibit reduced frequency deviations.

The authors of [37] proposed that EV owners could collaboratively bid in the regulation energy market, harmonized by a network of charging equipment, allowing EVs to provide frequency regulation services to the electric utility in V2G systems. Their research proved the capacity to propose time-varying rebates and install office and home chargers to significantly improve the V2G earnings.

Voltage control

Other than frequency control, EVs can contribute to a certain extent to ancillary services like voltage regulation, power factor correction, and reactive power compensation. Thanks to power electronic converters, the reactive power correction is attained by V2G by controlling the angle between current and voltage without any major consequence on the charging/discharging processes. The EV batteries are then used to deliver or consume reactive power from the electric utility by fine-tuning the charging/discharging energy flow in EVs as required to maintain a steady voltage and improve the power factor. EV bidirectional chargers can provide these services while operating within their remaining power capacity, especially when not transferring full active power during charging, with a minimal effect on the battery lifespan [38]. The authors of [39] proposed a voltage support system using EVs by implementing a droop control in an EV recharge station.

The authors of [40] applied a virtual synchronization control technique to the V2G solution. Simulation results demonstrated that this system possesses a favorable active and reactive following behavior and can efficiently react to utility's irregularities, control the charging/discharging power, and deliver reactive power back to the electric utility. Having the charging/discharging system designed allowing the virtual synchronization, factors like virtual inertia and damping in the controller are modified to allow V2G system to inherit the functioning characteristics of a traditional power utility.

Peak shaving and valley filling

Energy storage technologies, especially those using the electrochemical battery, are crucial for peak shaving. Throughout off-peak, energy consumers can store energy in these batteries thanks to a lesser energy price. Then, when the demand is high in peak hours, the stored energy can be dispatched to the electric utility, effectively shaving off peak demand and pressure on the electric utility.

EVs are supplied with battery packs that can be viewed as ESS to be exploited by the smart grid. The EVs are typically parked for many hours, and if plugged into the electric utility, they can offer the needed storage system for peak shaving. A study supporting this idea was proposed by the authors of [41]. Furthermore, the authors of [42] investigated the possibilities of peak-shaving facilities to the utility for EVs that remain idle in parking lots for an average period of 3–5 h only. They concluded that EV batteries can function as distributed energy storage that increases the reliability of smart grid.

The authors of [43] proposed the idea that the stored energy in EV might be exploited in valley filling and peak shaving to balance electricity supply and demand. This requires scheduling EVs to be charged during low-power demand periods subject to the charging flexibility of the EV, in other words the ratio between the minimum charging time and parking duration. In the case of higher demand, the stored energy in the EV flows back into the smart grid.

The EV recharge station is equipped with DC fast charging options having chargers of 360 kW and above. Battery storage systems can contribute in reducing the electricity drawn from the electric utility during peak demand periods to minimize the pressure.

Spinning reserve

Spinning reserve service is typically provided by online generators that instantly adapt their output power in response to major power shortages [44]. The spinning reserve requires a generation capacity with a quick response time. Electrochemical battery cells are very well suited for this mission [45,46].

In addition, the active power supplied must be synchronous with the electric utility. This is feasible for the EVs plugged to the utility with bidirectional chargers. However, for the effective implementation of spinning reserves using EV fleet, enough EVs should be plugged to the utility with sufficient electricity stored in their batteries. EV owners receive incentives for the services they provide and for their ability to supply power during contingencies [47].

Integration of renewable energy resources

The ever-changing weather conditions produce strong power imbalances. This is challenging for the utility operators because it disrupts conventional methods for planning the daily actions of the power grid. In grid-connected RES, if the power generated from the RES is too high, centralized power plants must decrease production to restore power balance, or the RES must be reduced.

In this situation, by adjusting their charging/discharging, EVs can aid in aligning consumption and generation. RES reduction balances the electric utility by resolving issues related to the utility congestion, and unmatched supply and demand. V2G technology assists in lowering RES curtailment, which is seen as unfavorable because curtailment increases operating costs and is an ineffective way of utilizing RES.

The authors of [48] employed V2G techniques in an AC microgrid having a V2G self-governing active/reactive control strategy using a virtual synchronous machine that simulates its transient behavior of a traditional generating system with the optimal design of virtual inertia and damping coefficients. The proposed solution automatically assures voltage and frequency regulation. This solution meets electricity utility demand and mends its stability, and rises the exploitation rate of renewable sources while decreasing the need of fossil fuels. EVs may be coupled with other renewables through V2G techniques to construct a novel energy management system.

The authors of [49] investigated EV battery sets to integrate wind energy into the grid, effectively enhancing the utility's capability to host wind power. As the wind has a stochastic nature, wind energy generation affects dramatically the stability of the electric utility. This phenomenon leads to wind power relinquishment in some cases.

Alleviating congestion and losses in distribution networks
The authors of [50,51] suggested that EVs can inject reactive power into the MG when not required to operate with a unity power factor. This reactive power dispatch from EVs lowers the distribution energy losses in the MG without affecting the EV's charging costs. By using information about the EVs like the departure time, energy prices, and batteries SOC, the EV charging periods may be improved to reduce peak congestion in the distribution network, minimize load variances, and thus minimize system losses [52,53].

Economic incentives
V2G services offer EV owners economic incentives by purchasing electricity at lower prices during off-peak periods and selling excess energy at higher prices during on-peak periods [54,55]. Current power generation units also benefit from V2G, reducing the requirement for expansion or new power generation plants, therefore reducing investment costs [56]. When there is a growing power demand, this increase might be provided for by EV battery storage. In addition, V2G backs a large penetration of RES into MGs, hence dropping the number of backup generators [57]. This is beneficial, as it decreases reliance on fossil fuels and allows the application of cost-effective RES. The authors of [58] have demonstrated reduction in costs associated with power systems by implementing V2G.

Scheduling and dispatch
The scheduling is very important for the grid operator, who can evaluate the dispatch needed to fulfill electricity demand, the EV charge and discharge schedule, and BESS exploitation. Smart grids use advanced metering infrastructure comprising smart sensors and meters capable of both measurements and wireless communication with other devices to efficiently exploit the integrated assets that are available online [59, 60]. Smart grids use smart sensors and smart meters to find when EVs are plugged, the SOC of the EVs. The data from the on-board sensors are available to the operator to decide day-ahead optimal scheduling, especially in combination with forecasting techniques for solar, wind, and the demand, EV charging patterns, etc. [61].

14.3.2.2 V2B and V2X

V2B (sometimes vehicle-2-home term is used) allows the owner of an EV to power home with electricity from the vehicle's battery pack. This feature is much more common among EVs than V2G. With the V2B concept, the EV can function as a source of backup power for a household for some days given the energy stored in the vehicle (from 10 s to 100 s kWh of electricity). For instance, the Nissan Leaf includes a V2B function with 7 kW of power available to an average home for about 2 days.

Finally, V2L concept feeds battery feed appliances plugged into the EV. A power limit, typically limited to 3.6 kW, is implemented in the vehicle during the V2L operation to protect the batteries of the vehicle from full depletion. V2L is currently accessible in vehicles such as Kia Niro EV, EV6, and EV9, and in the Hyundai Ioniq 5, Ioniq 6, and Kona. V2X describes the EV's capacity to deliver the electricity to other applications such as V2G, V2B, and V2L [62].

14.4 Challenges of V2G technology

While V2G technology potentially offers substantial advantages to the electric utility, its implementation is quite challenging. These challenges arise from the sociotechnical system and encompass economic, regulatory, and social aspects.

14.4.1 EV battery degradation

Battery degradation is the greatest challenge facing V2G systems, because of the repeated charge/discharge of the battery, effectively reducing its useful life. V2G in ancillary services may shorten battery's useful life, and battery degradation leads to capacity loss over time, thus affecting the range capability of EVs.

The cycle life of batteries varies significantly with the driving habits of the owner, the environmental influences, due to the chemical composition, and manufacturing techniques. Today, Li-ion is viewed to be an effective way for V2G because of their capability of providing long cycle life, acceptable deep cycling capacity, relatively high energy density, and high efficiency. A Li-ion battery tolerates between 2,000 and 4,000 deep cycles. In addition, in order to prevent degradation, circumventing deep discharging and overcharging and using smart charging devices are also highly recommended.

Engaging in V2G operations lowers the usable capacity of EV batteries by a frequent charge and discharge. EV owners find it problematic to take part in the V2G program because of the rapid ruin of batteries, which necessitates regular replacement at a cost [63–65].

14.4.2 Overall system efficiency degradation

A major issue with V2G systems is the overall efficiency of energy transmission to and from the electric utility to the batteries. Even though each converter is efficient (about 95%), when energy has to flow in several converters in cascading fashion, the overall efficiency loses up to 40%. The authors of [66] experimentally measured and examined the power losses for a V2G and found the measured one-way losses to vary from 12% to 36% as summarized in Table 14.3.

14.4.3 Smart grid infrastructure

The V2G implementation is accompanied by the enactment of smart grids. This requires high investment costs to upscale the power system. Enhancements in

Table 14.3 Percentage of energy loss per stage in the V2G system [66]

Component	AC current (A)	Percentage losses (%)	
		Charging	**Discharging**
EV battery	10	0.64	0.64
	40	1.69	1.91
EV PEU	10	6.28	16.67
	40	5.77	19.23
EVSE	10	0.10	1.42
	40	0.29	1.39
Breakers	10	0.00	2.80
	40	1.30	0.60
Transformer	10	10.20	14.60
	40	3.33	6.65
Total	10	17.22	36.13
	40	12.38	29.78

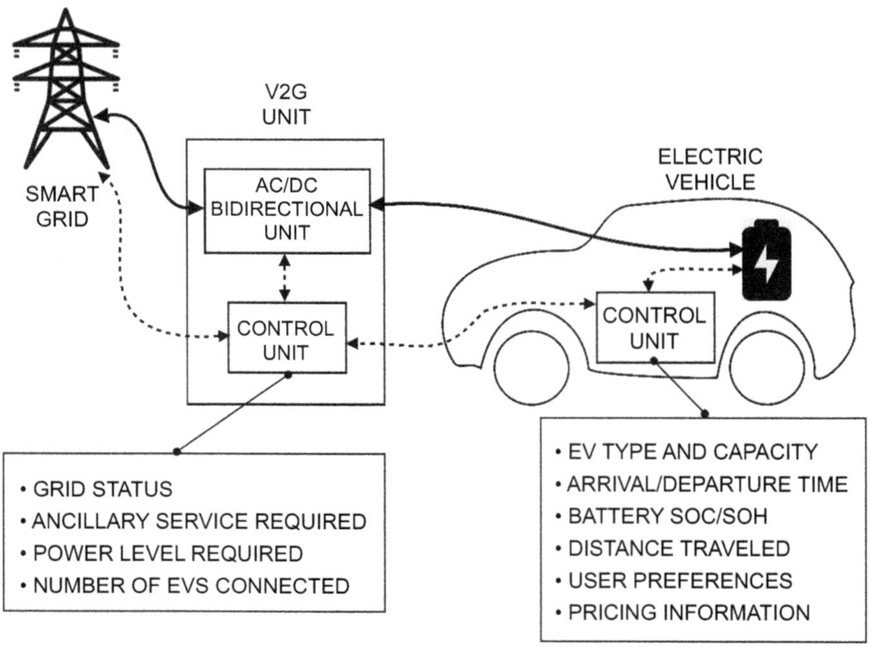

Figure 14.5 V2G infrastructure

hardware, software, and communication systems are required for the V2G adoption. From the electric utility side, it needs to know the SOC and state of health of the battery pack, the arrival and departure times, the distance to be traveled, etc. This is illustrated in Figure 14.5. Each EV that contributes in the V2G must be equipped by a bidirectional battery charger that communicates with the utility to know the utility's status, the needed ancillary service, pricing information, and more.

14.5 Conclusion

The massive adoption of EVs constitutes a critical challenge to the electric utility, especially when these vehicles are charged at random. This is referred to as uncontrolled recharge of EVs, and may cause transformers degradation and loss, bus voltage deviation, harmonic distortion, and a dramatic peak power demand. Smart grid technologies offer many solutions to the actual power utility in the form of bidirectional communications and two-way transfer of electricity. These two critical features allow the consumer of electricity to become a producer, adding the possibility of selling electricity from renewables to the utility. The "prosumer" can offer ancillary services to the electric utility and benefit from some type of cost reduction. Vehicle-to-grid technologies aim to facilitate the massive adoption of EVs and their integration to the electric utility. EVs are typically equipped with Li-ion batteries having high capacities and power contents. Most EVs are in parking position for 95% of their time. Hence, they offer the opportunity to explore their stored electricity for the needed ancillary services to the utility, like frequency regulation, voltage support, adoption of renewable sources, and others. However, this necessitates massive investments for the realization of a smart grid capable of executing such objectives. Moreover, the adoption of V2G technologies presents other challenges to the EVs, since a frequent charge/discharge of their batteries would affect their useful life. In addition, the power transfer from the EV to the electric grid undergoes several stages, which significantly increases the overall systems losses. These issues must be addressed and implemented for the massive adoption of EVs.

References

[1] V. Sultan, A. Aryal, H. Chang, and J. Kral, "Integration of EVs into the smart grid: A systematic literature review," *Energy Informatics*, vol. 5, no. 1, p. 65, 2022. doi:10.1186/s42162-022-00251-2.

[2] M. İnci, Ö. Çelik, A. Lashab, K. Ç. Bayındır, J. C. Vasquez, and J. M. Guerrero, "Power system integration of electric vehicles: A review on impacts and contributions to the smart grid," *Applied Sciences*, vol. 14, no. 6, p. 2246, 2024. doi:10.3390/app14062246.

[3] R. Asghar, M. H. Sulaiman, Z. Mustaffa, Z. Ali, and Z. Ullah, Integration of electric vehicles in smart grids: A review of the advantages and challenges of vehicle-to-grid technology. In: *2022 International Conference on IT and Industrial Technologies (ICIT)*. Piscataway, NJ: IEEE, 2022, pp. 1–7.

[4] *Energy ABCs*, "Smart grid vs. traditional grid: A comprehensive comparison," Energyabcs.com. Accessed: April 15, 2024. [Online]. Available: https://www.energyabcs.com/smart-grid-vs-traditional-grid/.

[5] M. I. Henderson, D. Novosel, and M. L. Crow, *Electric Power Grid Modernization Trends, Challenges, and Opportunities*. Accessed: April 15, 2024. [Online]. Available: https://www.ieee.org/content/dam/ieee-org/ieee/web/org/about/corporate/ieee-industry-advisory-board/electric-power-grid-modernization.pdf.

[6] C. Alaoui, and Z. M. Salameh, "On-board diagnostic and rejuvenation system for electric vehicles," In: *IEEE 58th Vehicular Technology Conference*, (IEEE Cat. No. 03CH37484). Piscataway, NJ: IEEE, 2003, pp. 3251–3256.

[7] C. Alaoui, and Z. M. Salameh, "Electric vehicle diagnostic and rejuvenation system (EVDRS)," *International Journal of Power and Energy Systems*, vol. 27, no. 2, p. 151, 2007.

[8] C. Alaoui, and Z. M. Salameh, "A novel system-on-chip system for diagnostic and rejuvenation for electric and hybrid vehicles," In: *International Symposium on Power Electronics, Electrical Drives, Automation and Motion*, SPEEDAM 2006. Piscataway, NJ: IEEE, 2006, pp. 920–925.

[9] W. Frye, "Smart grid transforming the electricity system to meet future demand and reduce greenhouse gas emissions." *Technical Report*. Cisco Internet Business Solutions Group, San Jose, CA, 2008.

[10] M. R. Hossain, M. T. Amanullah, and S. Ali, "Evolution of smart grid and some pertinent issues," In: *20th Australasian Universities Power Engineering Conference (AUPEC2010)*. http://www.ieeeexplore.com/stamp/stamp.jsp?tp=&arnumber=5710797&isnumber=5710678. Cited February 9, 2013.

[11] G. Graditi, M. G. Ippolito, E. Telaretti, and G. Zizzo,, "Technical and economical assessment of distributed electrochemical storages for load shifting applications: An Italian case study," *Renewable and Sustainable Energy Review* vol. 57, pp. 515–523, 2016.

[12] E. Alghsoon, A. Harb, and M. Hamdan, "Power quality and stability impacts of vehicle to grid (V2G) connection," In: *2017 8th International Renewable Energy Congress (IREC)*, Amman, 2017, pp. 1–6.

[13] C. Liu, K. T. Chau, D. Wu, and S. Gao, "Opportunities and challenges of vehicle-to-home, vehicle-to-vehicle, and vehicle-to-grid technologies," *Proceedings of the IEEE*, vol. 101, no. 11, pp. 2409–2427, 2013.

[14] V. Monteiro, J. G. Pinto, and J. L. Afonso, "Operation modes for the electric vehicle in smart grids and smart homes: Present and proposed modes," *IEEE Transactions on Vehicular Technology*, vol. 65, no. 3, pp. 1007–1020, 2016.

[15] M. Ansari, A. T. Al-Awami, E. Sortomme, and M. A. Abido, "Coordinated bidding of ancillary services for vehicle-to-grid using fuzzy optimization," *IEEE Transactions on Smart Grid*, vol. 6, no. 1, pp. 261–270, 2015.

[16] Y. Susowake, H. Yongyi, T. Senjyu, A. M. Howlader, and P. Mandal, "Optimum operation plan for multiple existing EV charging stations," In: *2018 IEEE PES Asia-Pacific Power and Energy Engineering Conference (APPEEC)*, Kota Kinabalu, 2018, pp. 611–615.

[17] E. Hirst, and B. Kirby, *Ancillary Services*. Oak Ridge National Lab (ORNL), Oak Ridge, TN (United States), 1996.

[18] B. Shively, Ancillary Services in a World of Renewable Energy. Accessed: April 17, 2024. [Online]. Available: https://www.enerdynamics.com/Energy-Currents_Blog/Ancillary-Services-in-a-World-of-Renewable-Energy.aspx.

[19] International Energy Agency, "Technology roadmap smart grids," 2011. Accessed: April 15, 2024. [Online]. Available: https://www.iea.org/reports/technology-roadmap-smart-grids.

[20] X. Bai, K. Clark, G. A. Jordan, N. W. Miller, and R. J. Piwko, "Intermittency analysis Project: Appendix B, impact of intermittent generation on operation of California Power Grid," *GE Energy Consulting, Report CEC-500-2007-081-APB*, 2007.

[21] International Energy Agency, "Prospects for large-scale energy storage in decarbonised power grids," 2009. [Online]. Available: https://www.iea.org/reports/prospects-for-large-scale-energy-storage-in-decarbonised-power-grids.

[22] P. Srivastava, R. Haider, V. J. Nair, *et al.*, "Voltage regulation in distribution grids: A survey," *Annual Reviews in Control*, vol. 55, pp. 165–181, 2023.

[23] K. Clement-Nyns, E. Haesen, and J. Driesen, "The impact of charging plug-in hybrid electric vehicles on a residential distribution grid," *IEEE Transactions on Power Systems*, vol. 25, pp. 371–380, 2010.

[24] M. Gonzalez Vaya, M. D. Galus, R. A. Waraich, and G. Andersson, "On the interdependence of intelligent charging approaches for plug-in electric vehicles in transmission and distribution networks," In: *Proceedings of the 2012 IEEE PES Innovative Smart Grid Technologies (ISGT Europe 2012)*, Berlin, Germany, October 14–17, 2012, pp. 1–8.

[25] S. Deilami, A. S. Masoum, P. S. Moses, and M. A. S. Masoum, "Real-time coordination of plug-in electric vehicle charging in smart grids to minimize power losses and improve voltage profile," In: *Proceedings of the 2011 IEEE PES Innovative Smart Grid Technologies (ISGT Europe 2011)*, Manchester, UK, December 5–7, 2011, vol. 2, pp. 456–467.

[26] M. Nour, H. Ramadan, A. Ali, *et al.*, "Impacts of plug-in electric vehicles charging on low voltage distribution network," In: *2018 International Conference on Innovative Trends in Computer Engineering (ITCE)*. Piscataway, NJ: IEEE, 2018, pp. 357–362.

[27] J. Yang, L. He, and S. Fu, "An improved PSO-based charging strategy of electric vehicles in electrical distribution grid," *Applied Energy*, 2014, vol. 128, pp. 82–92.

[28] C. Roe, E. Farantatos, J. Meisel, A. P. Meliopoulos, and T. Overbye, "Power system level impacts of PHEVs," In: *42nd Hawaii International Conference on System Sciences*, January 2009, pp. 1–10.

[29] S. Shao, M. Pipattanasomporn, and S. Rahman, "Challenges of PHEV penetration to the residential distribution network," In: *Power Energy Society General Meeting*, July 2009PES'09. Piscataway, NJ: IEEE, pp. 1–8.

[30] P. Grahn, J. Rosenlind, P. Hilber, K. Alvehag, and L. Soder, "A method for evaluating the impact of electric vehicle charging on transformer hotspot temperature," In: *2nd IEEE PES International Conference and Exhibition on Innovative Smart Grid Technologies (ISGT Europe)*, December 2011, pp. 1–8.

[31] F. Geth, N. Leemput, J. Van Roy, J. Buscher, R. Ponnette, and J. Driesen, "Voltage droop charging of electric vehicles in a residential distribution feeder," In: *3rd IEEE PES International Conference and Exhibition on Innovative Smart Grid Technologies (ISGT Europe)*, October 2012, pp. 1–8.

[32] M. A. S. Masoum, S. Deilami, and S. Islam, "Mitigation of harmonics in smart grids with high penetration of plug-in electric vehicles," In: *IEEE Power and Energy Society General Meeting*, 2010. Piscataway, NJ: IEEE, pp. 1–6.

[33] S. Sarabi, L. Kefsi, A. Merdassi, and B. Robyns, "Supervision of plug-in electric vehicles connected to the electric distribution grids," *International Journal of Electrical Energy*, vol. 1, no. 4, pp. 256–263, 2013.

[34] C. D. White, and K. M. Zhang, "Using vehicle-to-grid technology for frequency regulation and peak-load reduction," *Journal of Power Sources*, vol. 196, no. 8, pp. 3972–3980, 2011.

[35] D. Orihara, S. Kimura, and H. Saitoh, "Frequency regulation by decentralized V2G control with consensus-based SOC synchronization," *IFAC-PapersOnLine*, vol. 51, no. 28, pp. 604–609, 2018, doi:10.1016/j.ifacol.2018.11.770.

[36] F. Alfaverh, M. Denaï, and Y. Sun, "Optimal vehicle-to-grid control for supplementary frequency regulation using deep reinforcement learning," *Electric Power Systems Research*, vol. 214, p. 108949, 2023.

[37] M. Bayati, M. Abedi, G. B. Gharehpetian, and M. Farahmandrad, "Short-term interaction between electric vehicles and microgrid in decentralized vehicle-to-grid control methods," *Protection and Control of Modern Power Systems*, vol. 4, no. 1, p. 5, 2019, doi:10.1186/s41601-019-0118-4.

[38] V. Kuruvilla, P. V. Kumar, and A. I. Selvakumar, Challenges and impacts of V2g integration—A review. In: *Proceedings of the 8th International Conference on Advanced Computing & Communication Systems*, Coimbatore, India, March 2022, vol. 1, pp. 1938–1942, doi:10.1109/ICACCS54159.2022.9785095.

[39] S. Singh, S. Jagota, and M. Singh, "Energy management and voltage stabilization in an islanded microgrid through an electric vehicle charging station," *Sustainable Cities and Society*, vol. 41, pp. 679–694, 2018.

[40] Y. Ma, L. Chen, and Y. Zhao. "Research on virtual synchronous charge and discharge machine with power quality control for EVs," In: *8th Asia Conference on Power and Electrical Engineering (ACPEE)*. Piscataway, NJ: IEEE, 2023, pp. 1189–1195.

[41] D. M. Steward, "Critical elements of vehicle-to-grid (V2G) economics," *National Renewable Energy Laboratory*, USA, Tech. Rep. NREL/TP-5400-69017, 1390043, September 2017, doi:10.2172/1390043.

[42] S. Al-Rubaye, A. Al-Dulaimi, and Q. Ni, "Power interchange analysis for reliable vehicle-to-grid connectivity," *IEEE Communications Magazine*, vol. 57, no. 8, pp. 105–111, 2019, doi:10.1109/MCOM.2019.1800657.

[43] S. Chowdhury, H. Keshtkar, and F. D. Mohammadi, "Microgrid peak shaving implementation using vehicle-to-grid system," In: *Proceedings of the International Conference on Electrical, Computer and Energy Technologies (ICECET)*, Prague, Czech Republic, July 2022, pp. 1–5, doi:10.1109/ICECET55527.2022.9873000.

[44] S. S. Ravi and M. Aziz, "Utilization of electric vehicles for vehicle-to-grid services: Progress and perspectives," *Energies*, vol. 15, no. 2, p. 589, 2022, doi:10.3390/en15020589.

[45] J. K. Szinai, C. J. R. Sheppard, N. Abhyankar, and A. R. Gopal, "Reduced grid operating costs and renewable energy curtailment with electric vehicle charge management," *Energy Policy*, vol. 136, p. 111051, 2020, doi:10.1016/j.enpol.2019.111051.

[46] X. Tang, C. Sun, S. Bi, S. Wang, and A. Y. Zhang, "A holistic review on advanced bi-directional EV charging control algorithms," *ACM SIGENERGY Energy Informatics Review*, vol. 1, no. 1, pp. 78–88, 2022.

[47] S. Lotfi, M. Sedighizadeh, R. Abbasi, and S. H. Hosseinian, "Vehicle-to-grid bidding for regulation and spinning reserve markets: A robust optimal coordinated charging approach," *Energy Reports*, vol. 11, pp. 925–936, 2024, doi:10.1016/j.egyr.2023.12.044.

[48] C. Wu, H. Yu, Z. Yang, *et al.*, "Review of research on electric vehicle V2G schedule technology under multi perspective," In: *Proceedings of the Asia Conference on Electrical, Power and Computer Engineering*, 2022. pp. 1–5.

[49] B. Shang, N. Dai, L. Cai, C. Yang, J. Li, and Q. Xu. "V2G scheduling of electric vehicles considering wind power consumption," *World Electric Vehicle Journal*, vol. 14, no. 9, p. 236, 2023.

[50] J. Wang, G. R. Bharati, S. Paudyal, O. Ceylan, B. P. Bhattarai, and K. S. Myers, "Coordinated electric vehicle charging with reactive power support to distribution grids," *IEEE Transactions on Industrial Informatics*, vol. 15, no. 1, pp. 54–63, 2019, doi:10.1109/TII.2018.2829710.

[51] K. M. Tan, V. K. Ramachandaramurthy, J. Y. Yong, S. Padmanaban, L. Mihet-Popa, and F. Blaabjerg, "Minimization of load variance in power grids—Investigation on optimal vehicle-to-grid scheduling," *Energies*, vol. 10, no. 11, p. 1880, 2017, doi:10.3390/en10111880.

[52] S. Velamuri, S. H. C. Cherukuri, S. K. Sudabattula, N. Prabaharan, and E. Hossain, "Combined approach for power loss minimization in distribution networks in the presence of gridable electric vehicles and dispersed generation," *IEEE Systems Journal*, vol. 16, no. 2, pp. 3284–3295, 2022, doi:10.1109/JSYST.2021.3123436.

[53] S. Ray, K. Kasturi, S. Patnaik, and M. R. Nayak, "Review of electric vehicles integration impacts in distribution networks: Placement, charging/discharging strategies, objectives and optimisation models," *Journal of Energy Storage*, vol. 72, p. 108672, 2023, doi:10.1016/j.est.2023.108672.

[54] Y. Zhang, N. Rahbari-Asr, J. Duan, *et al.*, "Day-ahead smart grid cooperative distributed energy scheduling with renewable and storage integration," *IEEE Transactions on Sustainable Energy*, vol. 7, no. 4, p. 1739–1748, 2016.

[55] T.-C. Chiu, C.-W. Pai, Y.-Y. Shih, *et al.*, "Optimal day-ahead pricing with renewable energy for smart grid," In: *IEEE International Conference on Communications Workshops (ICC)*. Piscataway, NJ: IEEE, 2014, pp. 472–476.

[56] A. Raghavan, P. Maan, and A. K. B. Shenoy, "Optimization of day-ahead energy storage system scheduling in microgrid using genetic algorithm and particle swarm optimization," *IEEE Access*, vol. 8, pp. 173068–173078, 2020.

[57] U. Ur Rehman, M. Riaz, and M. Y. Wani, "A robust optimization method for optimizing day-ahead operation of the electric vehicles aggregator," *International Journal of Electrical Power & Energy Systems*, vol. 132, p. 107179, 2021.

[58] Y. Zheng, H. Yu, Z. Shao, *et al.*, "Day-ahead bidding strategy for electric vehicle aggregator enabling multiple agent modes in uncertain electricity markets," *Applied Energy*, vol. 280, p. 115977, 2020.

[59] B. K. Barman, S. N. Yadav, S. Kumar, *et al.*, "IOT based smart energy meter for efficient energy utilization in smart grid," In: *2nd International Conference on Power, Energy and Environment: Towards Smart Technology (ICEPE)*. Piscataway, NJ: IEEE, 2018, pp. 1–5.

[60] Y. Zhang, L. Hou, K. Liang, *et al.*, "Testing and evaluation of the impact sensor networks for power distribution and utilization in smart grid," In: *International Conference on Identification, Information and Knowledge in the Internet of Things (IIKI)*. Piscataway, NJ: IEEE, 2016, pp. 156–159.

[61] A. Nikoobakht, J. Aghaei, M. J. Mokarram, *et al.* "Adaptive robust co-optimization of wind energy generation, electric vehicle batteries and flexible AC transmission system devices," *Energy*, vol. 230, p. 120781, 2021.

[62] O. Elma, U. Cali, and M. Kuzlu, "An overview of bidirectional electric vehicles charging system as a vehicle to anything (V2X) under cyber–physical power system (CPPS)," *Energy Reports*, vol. 8, p. 25–32, 2022.

[63] B. Viswanath, D. K. Khatod, and N. P. Padhy, "Optimal vehicle-to-grid (V2G) scheduling of plug-in electric vehicles (PEVs) for grid peak demand management: A state-of-the-art review," In: *2022 IEEE International Conference on Power Electronics, Drives and Energy Systems (PEDES)*. Piscataway, NJ: IEEE, 2022. pp. 1–6.

[64] C. Alaoui, Z. M. Salameh, and W. A. Lynch, "Sealed lead acid electric vehicle batteries: A performance comparison study," *International Journal of Power and Energy Systems*, vol. 21, no. 3, pp. 174–179, 2001.

[65] C. Alaoui. "Electric vehicle energy management system," PhD thesis, Department of Electrical and Computer Engineering, University of Massachusetts Lowell, 2001.

[66] E. Apostolaki-Iosifidou, P. Codani, and W. Kempton, "Measurement of power loss during electric vehicle charging and discharging," *Energy*, vol. 127, pp. 730–742, 2017.

Chapter 15

Energy management for EV: state-of-charge estimation in Li-ion batteries with support vector regression hybrid approach

Elmahdi Fadlaoui[1] *and Noureddine Masaif*[1]

Abstract

Lithium-ion batteries serve as pivotal energy storage solutions across diverse do- mains, including smartphones, laptops, electric vehicles (EVs), and other applications. To ensure optimal performance and prevent sudden power loss, accurate estimation of the battery's state of charge (SOC) remains paramount. This chapter introduces a novel approach, using first the support vector regression (SVR) for SOC estimation, then exploits the gh-filter to refine this estimation, thus improving their accuracy. Notably, the training phase incorporates the US06 driving cycle, while the generalization ability of the SOC model is validated with LA92 and HWFET driving cycles. The main contributions of this study are first, adopting the Turnigy Graphene 5000 mAh 65C Li-ion battery data as a foundational dataset enhances the real-world applicability of the proposed methodology; also, unlike traditional methods, our SVR-gh approach reduces computational complexity, it bypasses the need for explicit battery model development, and finally, incorporating the gh-filter adds a layer of refinement, resulting in SOC estimates that are accurate and robust against noise. Through comprehensive testing on the LA92 and HWFET drive cycles, this study showcases the efficacy of the proposed approach in various operational scenarios. The achieved accuracy and adaptability substantiate its potential for broader adoption in diverse battery-powered systems. In summary, the SVR-gh method shows good SOC estimation accuracy and acceptable convergence ability under bad initial SOC.

Keywords: Battery management system; lithium-ion batteries; State of charge (SOC); electric vehicles; support vector machine (SVR); gh-filter

[1]Laboratory of Electronic Systems, Information Processing, Mechanics and Energetics, Ibn Tofail University, Morocco

15.1 Introduction

The global transition to sustainable energy systems has become increasingly urgent amid growing concerns about climate change and environmental challenges [1]. Rising greenhouse gas emissions and depletion of fossil fuel resources have accelerated the adoption of renewable energy technologies, particularly in the transportation sector. EVs have emerged as a promising solution to reduce carbon emissions and dependence on conventional fuel sources [2]. However, the effective integration of EVs into existing power systems requires sophisticated energy management strategies, particularly about battery performance and longevity [3]. Advances in battery technology, particularly lithium-ion batteries, have revolutionized the EV industry by offering higher energy density, longer cycle life, and improved reliability compared to traditional energy storage systems. These batteries are the cornerstone of modern EVs, making accurate monitoring and management of their SOC crucial for optimal performance and safety. SOC, which indicates the remaining capacity of a battery, plays a vital role in preventing unexpected power loss and extending battery life [4,5]. This shift towards electric transportation requires innovative approaches in energy management, particularly in developing robust battery state estimation methods and control strategies that can adapt to various driving conditions and usage patterns [5]. Despite the widespread adoption of lithium-ion batteries in EVs, accurate SOC estimation remains a significant challenge in battery management systems. The dynamic behavior of lithium-ion batteries, influenced by various factors such as temperature, aging, and driving conditions, makes accurate SOC estimation particularly complex. This complexity has a direct impact on vehicle range prediction, battery lifetime optimization, and overall system reliability [6,7].

Traditional SOC estimation approaches, especially those based on equivalent circuit models (ECMs), face several limitations. These methods often require parameter identification, which is a tedious procedure [8,9]. In addition, ECM-based approaches typically struggle to maintain accuracy when battery characteristics change due to degradation or varying environmental conditions. The need for efficient energy storage management systems in EVs is becoming increasingly critical as the automotive industry continues its electric transformation. Accurate SOC estimation not only ensures optimal battery utilization and avoids premature degradation but also addresses the range concerns of EV users. However, existing energy management systems often lack the adaptability and robustness required for real-world applications, especially to handle the dynamic nature of daily driving habits and varying environmental conditions [10,11]. Recent research efforts in the field of battery SOC estimation have mainly followed two distinct approaches: traditional model-based methods and machine learning methods. Conventional approaches, especially those using ECMs, have been widely studied by researchers. For example, Monirul *et al.* implemented an extended Kalman filter with ECM, achieving reasonable accuracy but requiring complex parameter identification procedures [12]. Similarly, Liu *et al.* developed an enhanced ECM model that, despite its improved accuracy, still suffered from computational complexity in parameter tuning [13]. The limitations of ECM-based methods have sparked increased interest in machine learning approaches.

Neural networks (NNs) and support vector machines (SVMs) have shown promising results in SOC estimation. Chandran *et al.* demonstrated the effectiveness of machine learning models [14], while Stighezza *et al.* explored SVM architecture with ant colony optimization algorithm [15]. However, these methods typically require large training data and computational resources, which limits their practical implementation in real-world applications. A significant research gap exists between traditional model-based methods and machine learning approaches for battery SOC estimation. Conventional approaches, especially those using ECMs, require complex parameter identification procedures, which can introduce errors. However, machine learning methods, such as NNs and SVMs, have shown promising results but typically require large training datasets that are not always available in practical real-world applications [16,17]. This research gap highlights the need for a method that can combine the interpretability of model-based approaches with the adaptability of machine learning techniques, while avoiding their respective limitations. To address these limitations, this chapter proposes a hybrid approach combining SVR with a gh-filter. This novel combination exploits the ability of SVR to capture nonlinear battery characteristics while using the gh-filter to smooth out estimation fluctuations, while requiring significantly less training data compared to conventional machine learning methods.

The structure of this chapter is organized to systematically present our research methodology and findings. Section 15.2 introduces our novel SVR-gh method for battery SOC estimation, beginning with the fundamentals of SVMs (section 15.2.1) and the gh-filter algorithm (section 15.2.2). The framework for online SOC estimation is detailed in section 15.2.3, encompassing both the development of the SVR-gh model (section 15.2.3.1) and the experimental setup including battery specifications and data collection methods (section 15.2.3.2). The experimental results and their comprehensive analysis are presented in section 15.3, while section 15.4 provides concluding remarks and future research directions.

15.2 SVR-gh method for battery SOC estimation

In this section, we present the theoretical foundations and methodology of our proposed hybrid approach. First, we introduce the SVM theory; then, we explain the gh-filter algorithm (Table 15.1) and its role in smoothing estimation fluctuations. Finally, we detail our novel SVM-gh framework, which combines the advantages of both techniques to achieve robust and accurate SOC estimation while minimizing data requirements and computational complexity.

Table 15.1 gh-filter algorithm

Initialization
Initialize the state of the filter
Initialize the gh-filter parameters
Predict
Predicted state estimate $X_{pred} = AX_{est} + BU$
Update
Compute residual between predicted state and measurement $Residual = (Z - X_{pred})$
Update state estimate $X_{est} = X_{pred} + g * Residual$

15.2.1 Support vector machines

An SVM is a supervised machine learning algorithm primarily used for classification tasks [18]. However, it can also be employed for regression, including tasks (SOC) estimation [19]. SVM aims to find a hyperplane in a multidimensional space that best separates data points into different classes or predicts a continuous output such as SOC (Figure 15.1).

The core concept behind SVM is to find the optimal hyperplane by maximizing the margin between data points and the hyperplane. In regression tasks, the goal is to find a hyperplane that best fits the data points [20]. The SVM regression equation can be written as follows:

$$f(x) = \sum_{i=1}^{n} \alpha_i K(x, x_i) + b \tag{15.1}$$

where

- $f(x)$ is the predicted SOC value,
- α_i are Lagrange multipliers,
- $K(x, x_i)$ is the kernel function, which can be linear, polynomial, or radial basis function (RBF).
- b is the bias term.

Once the SVM model is trained, it can be used to predict the class labels of new inputs by simply evaluating the decision function $f(x)$. The versatility of SVMs lies in their ability to handle nonlinear decision boundaries through the use of kernel functions, such as the RBF kernel, which was mentioned in the original methods section.

15.2.2 gh-Filter

The gh-filter, also known as the alpha-beta or alpha-beta-gamma filter, is a type of recursive linear filter commonly used in state estimation and tracking applications.

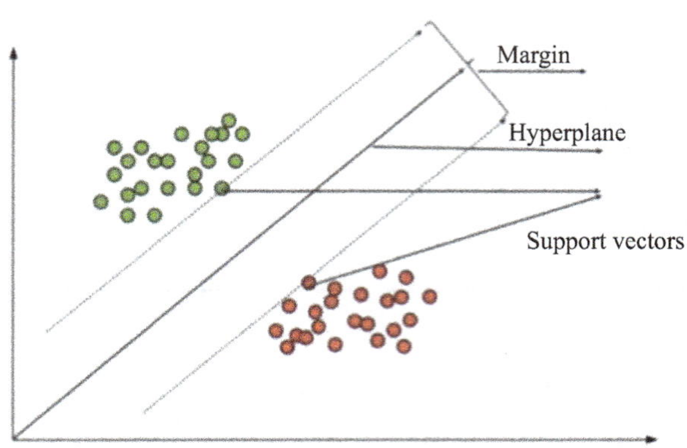

Figure 15.1 The SVM model

It is particularly useful for smoothing noisy measurements and providing reliable state estimates in real-time systems [21,22].

15.2.3 Online SOC estimation framework

The SVR-gh SOC estimation framework as shown in Figure 15.2, begins by predicting the SOC using the SVR model, which gives the $SOC_{measured}$. Using the Coulomb counting method, it then calculates the $SOC_{predicted}$. The $SOC_{estimated}$ is extracted using the gh-filter based on these two values. This helps to smooth the output of the SVR model (Figure 15.2).

$$SOC_k = SOC_{k-1} + \eta \frac{\Delta t}{Qn} I_k \tag{15.2}$$

15.2.3.1 SVR-gh model development

The SVR model was constructed using a RBF kernel, which was selected for its capability to capture nonlinear relationships in the battery's charging and discharging characteristics. The model architecture can be expressed as follows where, $K(x_i,x)$ represents the RBF kernel function:

$$K(x_i, x) = \exp(-\gamma||x_i - x||^2) \tag{15.3}$$

the state function used on the gh-filter is constructed as:

$$X_k = SOC_k, \quad A = 1, \; B = \eta \frac{\Delta t}{Q}, \quad U_k = I_k.$$

Also the gh-filter parameters used were $g = 1$ and $h = 0.5 \times 10^{-3}$.

15.2.3.2 Battery specifications and data acquisition

The experimental setup utilized a Turnigy Graphene Li-ion battery with specifications of 5,000 mAh capacity and 65 C discharge rate. The training dataset was derived from the US06 driving cycle, using: current (A), voltage (V), and temperature (°C) as input features, with the SOC serving as the target variable. To optimize the temporal resolution of the data, we implemented a resampling

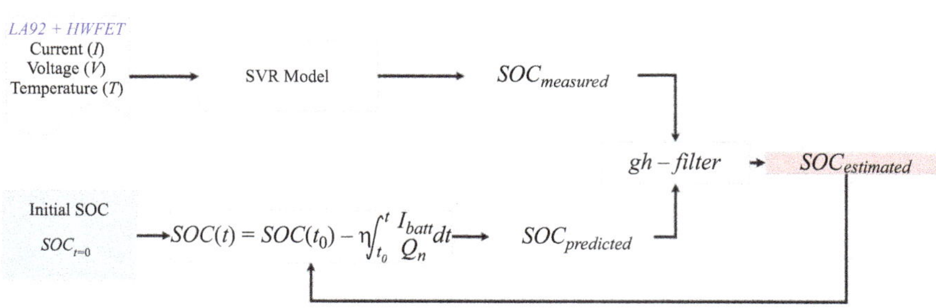

Figure 15.2 SVR-gh model framework

strategy, We selected one data point from every ten consecutive samples, resulting in a sampling rate of 1 sample per second. Data normalization was performed using the MinMaxScaler function, mapping all features to a standardized range of [0,1] to ensure consistent scaling across input variables.

15.3 Results and discussion

This section aims to validate the superior performance of the proposed hybrid method for battery SOC estimation compared to the solo SVR method. The evaluation of the proposed method's generalization ability will be evaluated using the following drive cycles: LA92 and HWFET at 0 °C, 10 °C, 25 °C, and 40 °C. Furthermore, this section examines the divergence ability of the method by adding a manual error to the initial SOC.

15.3.1 SOC estimation with SVR

The SVR is trained using the US06 drive cycle at different temperatures, and then the LA92 and HWFET datasets are used to test the generalization ability of the model at the temperatures (0 °C, 10 °C, 25 °C, and 40 °C). Figures 15.3 and 15.4 illustrate the SOC estimation results at different temperatures. The actual SOC

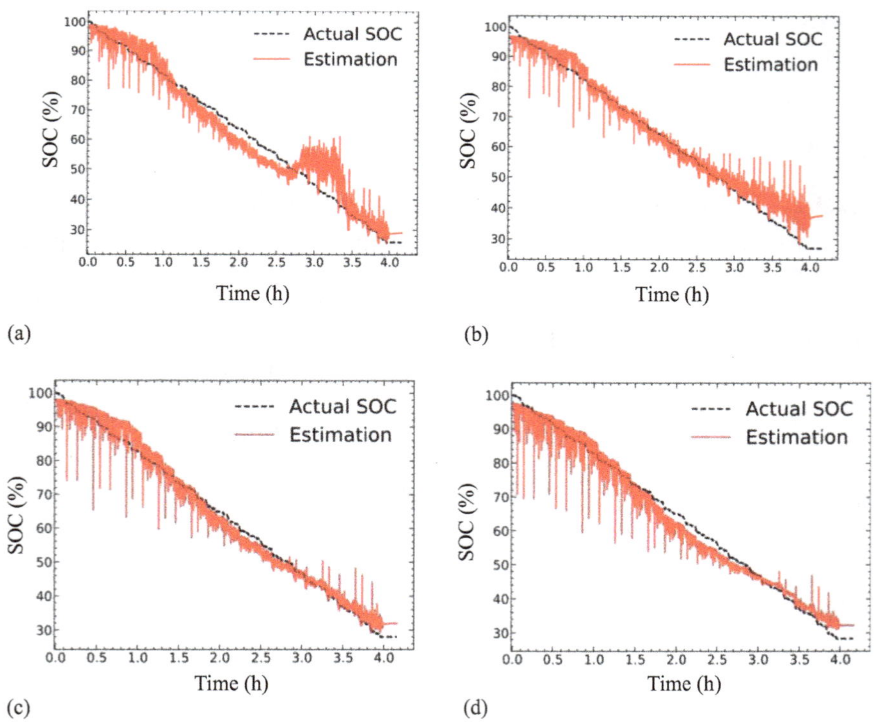

Figure 15.3 SOC estimation results of the SVR for LA92 at different temperatures: (a) 0 °C, (b) 10 °C, (c) 25 °C, and (d) 40 °C

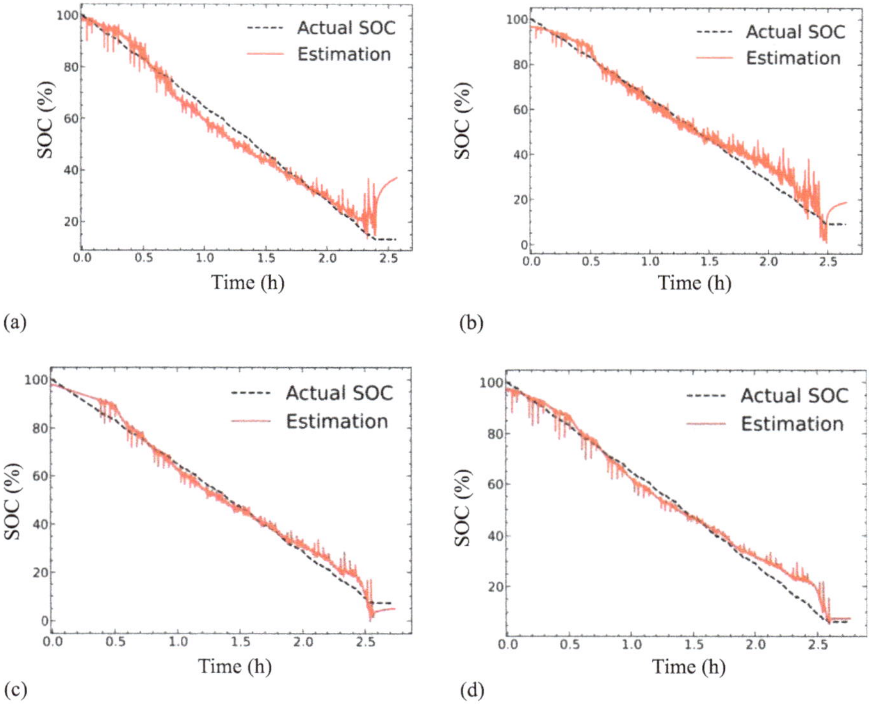

Figure 15.4 SOC estimation results of the SVR for HWFET at different temperatures: (a) 0 °C, (b) 10 °C, (c) 25 °C, and (d) 40 °C

Table 15.2 Comparison of SOC estimation for LA92 using solo SVR and the combined SVR-gh

Temperatures	**RMSE**		**MAE**	
	SVR (%)	**SVR-gh (%)**	**SVR (%)**	**SVR-gh (%)**
0	4.60	2.49	23.09	4.51
10	4.29	2.21	25.35	7.07
25	2.72	1.44	27.08	2.70
40	2.67	1.38	28.49	2.68

comes from the Coulomb counting method under specific situations. It can be seen from these figures that the SVR can capture the downward trend of the SOC with some deviations and high fluctuations. Tables 15.2 and 15.3 show that the RMSE of SOC estimation is within 4%, and MAE is within 27%. The instability of the results is due to many factors such as the complex battery behavior, limited training data, and noise in measurements. These complexities can lead to challenges in accurately modeling their dynamics, which may result in estimation errors. To solve these

Table 15.3 Comparison of SOC estimation for HWFET using solo SVR and the combined SVR-gh

Temperatures	RMSE		MAE	
	SVR (%)	SVR-gh (%)	SVR (%)	SVR-gh (%)
0	6.34	2.04	24.04	6.30
10	4.41	2.06	22.95	4.74
25	2.93	2.64	13.08	9.14
40	3.57	1.80	14.33	4.60

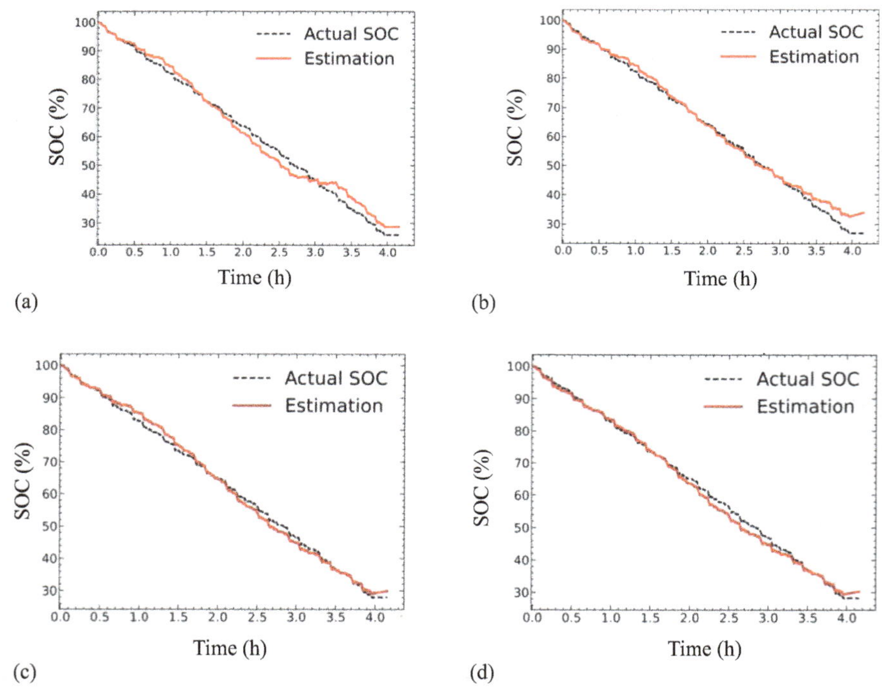

Figure 15.5 SOC estimation results of the SVR-gh-filter for LA92 at different tem peratures: (a) 0 °C, (b) 10 °C, (c) 25 °C, and (d) 40 °C

problems and improve the accuracy of SOC estimation, this chapter uses the gh-filter to reduce the RMSE and the fluctuations of the SVR.

15.3.2 SOC estimation after g-h filter

As seen in the previous section, the experimental results demonstrate high generalization ability for different drive cycles at various temperatures, but its output has obvious fluctuation and some deviations. The g-h filter is employed to improve the accuracy and stability of SOC estimation. Figures 15.5 and 15.6 show the SOC

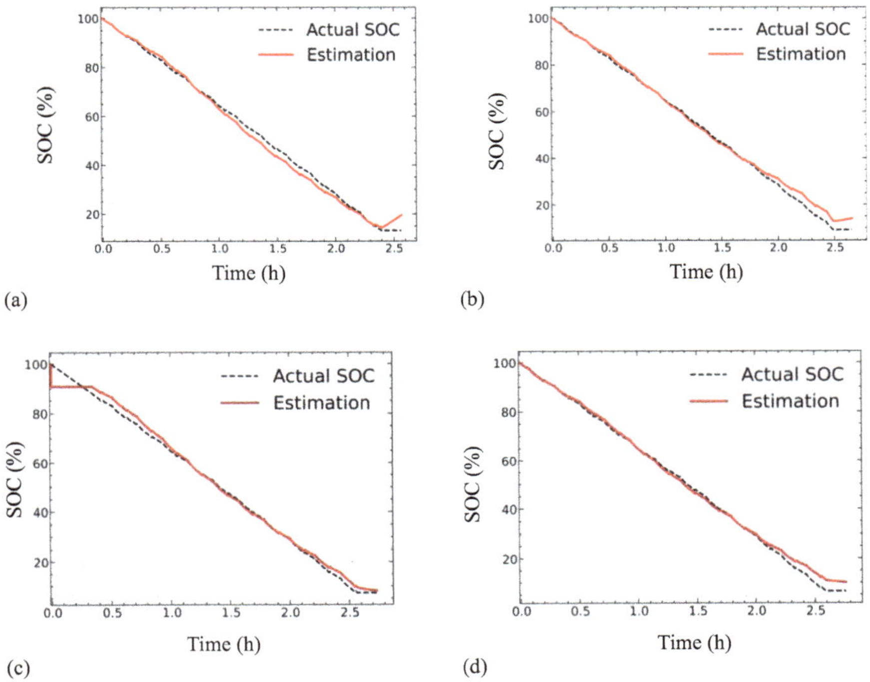

Figure 15.6 SOC estimation results of the SVR-gh-filter for HWFET at different temperatures: (a) 0 °C, (b) 10 °C, (c) 25 °C, and (d) 40 °C

estimation results after filtering the outputs at (0 °C, 10 °C, 25 °C, and 40 °C) under LA92 and HWFET drive cycles. By comparing Figures 15.5 and 15.6 with 15.3 and 15.4, we can observe clearly that the addition of the gh-filter can smooth the fluctuations from the original output. Tables 15.2 and 15.3 summarize the comparison of the solo SVR and the combined SVR-gh method.

Both the RMSE and MAE are reduced. For example, the RMSE and MAE of the LA92 at 40 °C reduced from 2.6% to 1.3% and 28.4% to 2.6%, respectively. However, the HWFET at 25 °C shows a low decrease in the error for both RMSE and MAE, the RMSE decreased from 2.9% to 2.6%, and MAE from 13% to 9%, and this is due to the bad measurements at the beginning of the experiments.

15.3.3 Validation of initial SOC convergence ability

To evaluate the convergence capability of the proposed method, we conducted experimental validation using the LA92 drive cycle at a temperature of 25 °C as shown in Figure 15.7. The results demonstrate the convergence characteristics of the gh-filter under significant initial SOC estimation error conditions. With an initial SOC estimation discrepancy of approximately 20% (estimated: 80%, actual: 100%), the filter demonstrates a moderate convergence rate, requiring approximately 1 h to achieve alignment with the actual SOC trajectory. This extended

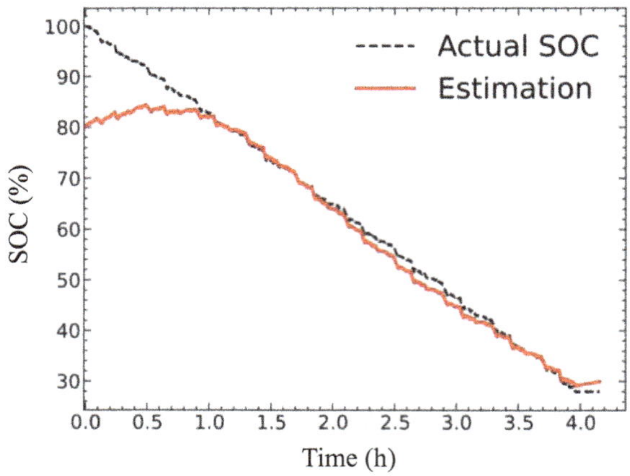

Figure 15.7 SOC estimation result with an initial SOC = 80% for the LA92 cycle

convergence period can be attributed to the fundamental simplicity of the gh-filter's algorithmic structure. Despite its computational efficiency, the filter's basic architecture, which relies solely on fixed gain parameters, lacks the sophisticated adaptive mechanisms inherent in more advanced estimation techniques such as Kalman filters family. This limitation manifests in the observed gradual convergence behavior when confronted with substantial initial estimation errors.

15.4 Conclusion

This chapter has presented a novel hybrid approach combining SVR and gh-filter for accurate SOC estimation in lithium-ion batteries. The proposed SVR-gh methodology successfully addresses several key challenges in battery management systems while offering significant advantages over traditional approaches. The experimental results, validated using the Turnigy Graphene 5,000 mAh 65 C Li-ion battery data across multiple driving cycles (US06, LA92, and HWFET), demonstrate the robust performance of our method. The SVR-gh approach has shown remarkable accuracy in SOC estimation, particularly in handling the nonlinear characteristics of lithium-ion batteries without requiring complex battery modeling. The integration of the gh-filter has proven effective in minimizing estimation fluctuations and enhancing the system's robustness against measurement noise.

References

[1] Child M, Koskinen O, Linnanen L, *et al*. Sustainability guardrails for energy scenarios of the global energy transition. *Renewable and Sustainable Energy Reviews*. 2018;91:321–334.

[2] Chen T, Jin Y, Lv H, *et al.* Applications of lithium-ion batteries in grid-scale energy storage systems. *Transactions of Tianjin University*. 2020;26(3):208–217.

[3] Hasan MK, Mahmud M, Habib AA, *et al.* Review of electric vehicle energy storage and management system: Standards, issues, and challenges. *Journal of Energy Storage*. 2021;41:102940.

[4] Li W, Rentemeister M, Badeda J, *et al.* Digital twin for battery systems: Cloud battery management system with online state-of-charge and state-of-health estimation. *Journal of Energy Storage*. 2020;30:101557.

[5] Wang Y, Tian J, Sun Z, *et al.* A comprehensive review of battery modeling and state estimation approaches for advanced battery management systems. *Renewable and Sustainable Energy Reviews*. 2020;131:110015.

[6] Manoharan A, Begam K, Aparow VR, *et al.* Artificial Neural Networks, Gradient Boosting and Support Vector Machines for electric vehicle battery state estimation: A review. *Journal of Energy Storage*. 2022;55:105384.

[7] Zhang M, and Fan X. Review on the state of charge estimation methods for electric vehicle battery. *World Electric Vehicle Journal*. 2020;11(1):23.

[8] Naseri F, Schaltz E, Stroe DI, *et al.* An enhanced equivalent circuit model with real-time parameter identification for battery state-of-charge estimation. *IEEE Transactions on Industrial Electronics*. 2021;69(4):3743–3751.

[9] Liu J, Yadav S, Salman M, *et al.* Review of thermal coupled battery models and parameter identification for lithium-ion battery heat generation in EV battery thermal management system. *International Journal of Heat and Mass Transfer*. 2024;218:124748.

[10] Sangswang A, and Konghirun M. Optimal strategies in home energy management system integrating solar power, energy storage, and vehicle-to-grid for grid support and energy efficiency. *IEEE Transactions on Industry Applications*. 2020;56(5):5716–5728.

[11] Zhou G, Li L, Lee SY, *et al.* Dual-strategy-encapsulated phase change materials with thermal immune functions for efficient energy storage and all-climate battery thermal management. *Composites Science and Technology*. 2023;243:110256.

[12] Monirul IM, Qiu L, and Ruby R. Accurate SOC estimation of ternary lithiumion batteries by HPPC test-based extended Kalman filter. *Journal of Energy Storage*. 2024;92:112304.

[13] Liu G, Lu L, Fu H, *et al.* A comparative study of equivalent circuit models and enhanced equivalent circuit models of lithium-ion batteries with different model structures. In: 2014 *IEEE Conference and Expo Transportation Electrification Asia-Pacific (ITEC Asia-Pacific)*. Piscataway, NJ: IEEE; 2014. pp. 1–6.

[14] Chandran V, Patil CK, Karthick A, *et al.* State of charge estimation of lithium-ion battery for electric vehicles using machine learning algorithms. *World Electric Vehicle Journal*. 2021;12(1):38.

[15] Stighezza M, Bianchi V, and De Munari I. FPGA implementation of an ant colony optimization based SVM algorithm for state of charge estimation in li-ion batteries. *Energies*. 2021;14(21):7064.
[16] Zhou W, Zheng Y, Pan Z, *et al.* Review on the battery model and SOC estimation method. *Processes*. 2021;9(9):1685.
[17] Tian J, Chen C, Shen W, Sun F, and Xiong R. Deep learning framework for lithium-ion battery state of charge estimation: Recent advances and future perspectives. *Energy Storage Materials*. 2023;61:102883.
[18] Noble WS. What is a support vector machine? *Nature Biotechnology*. 2006; 24(12):1565–1567.
[19] Hu J, Hu J, Lin H, *et al.* State-of-charge estimation for battery management system using optimized support vector machine for regression. *Journal of Power Sources*. 2014;269:682–693.
[20] Chen PH, Lin CJ, and Schölkopf B. A tutorial on ν-support vector machines. *Applied Stochastic Models in Business and Industry*. 2005;21(2):111–136.
[21] Jamil F, and Kim DH. Improving accuracy of the alpha–beta filter algorithm using an ANN-based learning mechanism in indoor navigation system. *Sensors*. 2019;19(18):3946.
[22] Khan J, Fayaz M, Hussain A, *et al.* An improved alpha beta filter using a deep extreme learning machine. *IEEE Access*. 2021;9:61548–61564.

Chapter 16
Conclusions and outlook

Badre Bossoufi[1]

This book provides a wide scope regarding the various emerging technologies, strategies, and methodologies for energy management systems (EMS) within microgrids that are increasingly being integrated with small-scale renewable energy sources. As energy demands globally rise and the effects of climate change intensify, it has never been more important to accelerate the transition to sustainable and resilient energy systems. This book is a substantial contribution to the continuous development of microgrid technologies with findings that show new insights into the optimization of energy flows, cost reduction, and increased reliability and efficiency of renewable energy integration.

16.1 Value of the work

The main value of the book consists of the complex approach to solving the key problems of modern energy systems. Each chapter explores one aspect of microgrid operation and energy management, bringing some new views into the knowledge base about decentralized energy generation. For example, the integration of renewable energy sources like wind, solar, and biomass into microgrids presents a route to power system decarbonization. The book emphasizes how these technologies can not only reduce greenhouse gas emissions but also enhance energy resilience in remote or underserved regions, a factor considered critical in the context of global energy insecurity.

Demonstrated in this research is that using advanced energy management strategies like those presented in Chapter 1 will contribute toward the optimum operation of a microgrid to realize better energy efficiency and reduce costs to support a policy for local energy self-sufficiency. The performances of performance improvement and cost reductions have also been well verified using metaheuristic optimization techniques that Chapter 3 discourses. This optimization capability is invaluable since fine-tuning of microgrid components ensures that distributed renewable generation is effectively integrated with storage systems and the grid.

[1]LIMAS Laboratory, Faculty of Sciences Dhar El Mahraz, Sidi Mohamed Ben Abdellah University, Morocco

The value of the work is further underlined by its detailed exploration of energy storage solutions, as underlined in Chapters 2, 5, and 9. Energy storage is one of the key factors in overcoming the intermittency of renewable sources. Through examining batteries, supercapacitors, and thermal storage systems, the book highlights how these solutions ensure that smart EMS maintains stability within the grid for reliable power whenever renewable generation fluctuates. The novel methodologies presented for energy storage management, such as applying battery management systems and integrating machine learning into predictive analytics, will continue to guide future research and development in this critical area.

As shown in Chapters 4 and 6, the application of machine learning and artificial intelligence (AI) to energy management shows in quite a crystal clear way the direction that smart grids will take. The predictability of demand, optimization of power flows, and autonomous management of energy storage units enabled by AI are huge improvements in energy efficiency. These are then validated with case studies and simulations to demonstrate how machine learning and optimization techniques can solve such complex challenges in real time. This would highly impact the development of highly responsive and adaptive microgrids with the possibility of optimizing performance under variable environmental and operational conditions.

16.2 Linking to future studies

While this book makes a vital contribution to the advancement of microgrid technologies, it simultaneously opens avenues for further research that will be necessary in surmounting the challenges future energy systems are bound to face. One of the major challenges put forth in this work is the intermittent nature of renewable energy sources, which is still limiting their widespread adoption. Chapter 15 offers an interesting approach by incorporating more advanced control strategies such as model predictive control and particle swarm optimization. The success of the application of these strategies to optimize renewable energy systems extends a platform for further studies that could apply similar techniques to even more complex hybrid energy systems or other forms of renewable energy integration.

Future research may extend this work on model predictive control and particle swarm optimization to more dynamic and larger-scale energy systems. In future, global energy systems that are rapidly moving toward decentralization and deployment of various renewable resources, the optimization of power flow, voltage regulation, and grid stability will necessitate advanced and scalable control systems. Fundamental research may investigate hybrid optimization techniques that would blend more than one control strategy in a way to be more robust and adaptive to extreme weather, sudden demands, or shifting grid dynamics.

Another critical area of investigation for the future will be in the integration of electric vehicles (EVs) into the grid. As indicated in Chapter 10, the prospect of vehicle to grid (V2G) provides an exciting, new solution to manage increased electricity demand from EVs. For V2G to work properly, major investments in infrastructure, regulation, and utility/owner coordination would be required. Future research should work on how to optimize V2G systems so that EVs become not only a demand resource but also a contributor to grid stability with ancillary services like frequency regulation and peak shaving.

Besides, further development and improvement of energy storage systems have remained one of the challenges for future energy systems. Although battery technologies have been developed and advanced, battery degradation, limited energy density, and high costs are still major issues that need to be resolved. Future research needs to be directed more toward efficiency, longevity, and cost-effectiveness of storage technologies. This may involve the invention of new materials, like solid-state batteries, which promise higher energy density and safety. In addition, advanced control algorithms working in concert with energy storage units can optimize charging/discharging cycles and help extend battery life, enhancing overall microgrid performance.

Another area that deserves more attention is the political and regulatory barriers to adopting microgrids, as discussed in Chapter 6. While technological development is important, the regulatory framework for energy markets, grid integration, and commercialization of microgrid technologies must evolve to accommodate the shift toward decentralized energy systems. Future research could investigate in detail the economic, policy, and regulatory frameworks that will lay the path for the wide-scale deployment of microgrids, which, in turn, will help all local utilities, governments, and communities take full advantage of these technologies. Overcoming the remaining barriers will depend heavily on the collaboration of industry leaders, academia, and policymakers in framing a favorable environment that better nurtures the growth of microgrids.

16.3 The road ahead

The insights from this book show that microgrids, powered by renewable energy sources and advanced storage solutions, form a key piece of the puzzle in the future of energy. However, further innovation in energy storage, optimization algorithms, and grid integration is necessary if microgrids are to reach their full potential. As the world moves toward a low-carbon, decentralized energy future, integration of renewables and optimization of energy systems will be key in emission reduction, enhancing energy security, and fostering local energy independence.

This work provides a broad overview of the current status of microgrid technologies and acts as a springboard for future research in driving the next generation of sustainable energy systems. The combination of state-of-the-art control strategies, renewable energy integration, and advanced energy storage provides an enabling roadmap for the creation of resilient, efficient, and sustainable energy systems to meet the demands of the future.

Index

www.ingramcontent.com/pod-product-compliance
Lightning Source LLC
Chambersburg PA
CBHW071802240525
27178CB00036B/60
9781837240906